Advanced Topics in Science and Technology in China

Volume 63

Zhejiang University is one of the leading universities in China. In Advanced Topics in Science and Technology in China, Zhejiang University Press and Springer jointly publish monographs by Chinese scholars and professors, as well as invited authors and editors from abroad who are outstanding experts and scholars in their fields. This series will be of interest to researchers, lecturers, and graduate students alike.

Advanced Topics in Science and Technology in China aims to present the latest and most cutting-edge theories, techniques, and methodologies in various research areas in China. It covers all disciplines in the fields of natural science and technology, including but not limited to, computer science, materials science, the life sciences, engineering, environmental sciences, mathematics, and physics.

This book series is indexed by the SCOPUS database.

If you are interested in publishing your book in the series, please contact Dr. Mengchu Huang (Email: mengchu.huang@springer.com).

More information about this series at http://www.springer.com/series/7887

Xiang Gao · Chenghang Zheng · Pen-Chi Chiang · Kefa Cen

Multi-Pollutant Control for Flue Gases

Principles and Applications

ZHEJIANG UNIVERSITY PRESS
浙江大学出版社

Springer

Xiang Gao
College of Energy Engineering
Zhejiang University
Hangzhou, Zhejiang, China

Chenghang Zheng
College of Energy Engineering
Zhejiang University
Hangzhou, Zhejiang, China

Pen-Chi Chiang
Graduate Institute of Environmental Engineering
Taiwan University
Taipei, Taiwan, China

Kefa Cen
College of Energy Engineering
Zhejiang University
Hangzhou, Zhejiang, China

ISSN 1995-6819 ISSN 1995-6827 (electronic)
Advanced Topics in Science and Technology in China
ISBN 978-981-16-1516-0 ISBN 978-981-16-1518-4 (eBook)
https://doi.org/10.1007/978-981-16-1518-4

Jointly published with Zhejiang University Press, Hangzhou, China
The print edition is not for sale in China Mainland. Customers from China Mainland please order the print book from: Zhejiang University Press.

This Springer imprint is published by the registered company Springer Nature Singapore Pte Ltd.
The registered company address is: 152 Beach Road, #21-01/04 Gateway East, Singapore 189721, Singapore

Preface

Since the Industrial Revolution, energy production from various fossil fuels (e.g., coal, crude oil, and natural gas) has rapidly increased worldwide. Although energy utilization has contributed significantly toward global economic and social developments, millions of tons of pollutants are emitted every year from electricity production, industries, transportation, and other sources. The resulting air pollution has massively affected the environment and human health, thus becoming a global concern. Many countries have adopted legislations to regulate the emission of various pollutants to alleviate their adverse effects. After years of development and unremitting governance, air quality all over the world has significantly improved. However, many parts of the world, particularly developing countries, are still facing severe air pollution.

Research efforts in China have led to the development of the best multi-pollutant control technologies in the thermal power industry that are now being employed in other industries. In particular, breakthroughs have been made in the development of an ultra-low emission technology for flue gas pollutants. This book thus provides a detailed overview of recent progresses in air pollutant control technologies ranging from single-pollutant treatment to multi-pollutant control. In the first chapter, types of air pollution and their control issues related to energy utilization are discussed in detail. The principles and applications of pollutant control technology, including catalytic, electric, and absorption methods, and multi-pollutant removal systems and a technology evaluation are then discussed in the subsequent four chapters.

This book provides scientists and engineers with fundamental and applied knowledge associated with multi-pollutant control of flue gases and can also be used as a reference for teachers and students in related fields at educational institutions. Furthermore, the emerging challenges, new methodologies, innovative technologies, and integrated strategies discussed herein could provide researchers, students, engineers, and policymakers with helpful insights for multi-pollutant control.

This book would not have been possible without many contributors whose names did not make it to the cover. We are greatly indebted to our colleagues for their valuable time to proofread this book and suggest changes. Special thanks to our colleagues and students in our research group, including Dr. Yang Yang, Dr. Zhengda Yang, Dr. Xuesen Du, Dr. Xiao Zhang, Dr. Qingyang Lin, Dr. Shaojun Liu, Dr. Yan

Lu, Dr. Longchao Yao, Dr. Weiguo Weng, Dr. Hao Song, Dr. Xinbo Zhu, Dr. Jun Zhang, Dr. Shuo Zhang, Dr. Yishan Guo, Dr. Ruiyang Qu, Dr. Wenshuo Hu, Dr. Qianyun Chang, Dr. Pei Yan, Dr. Haitao Zhao, Mr. Yifan Wang, Mr. Zhongyang Zhao, Mr. Qi Xin, Mr. Chang Liu, Mr. Lingyu Shao, Mr. Zhicheng Wu, Mr. Yi Dong, Mr. Mingchu Ran, Ms. Weijia Ren, Mr. Yujie Fu, Ms. Linhang Zhu, Ms. Ying Wang, Mr. Yuhao Shao, Mr. Shangbo Han, Mr. Hao Zhang, Mr. Wenjun Li, Mr. Cong Luo, Ms. Yu Zhang, Mr. Zhong Zheng, Mr. Zhesheng Hua, Mr. Jiaben Wang, Ms. Xiaofan Liu, Mr. Chunbo Pang, Mr. Yongxin Zhang, Mr. Weihong Wu, Ms. You Zhang, Mr. Can Zhou, and Ms. Dian Xu.

This work was supported by the National Science Fund for Distinguished Young Scholars of China (No. 51125025), National Natural Science Foundation of China (No. 50776079, No. U1609212, No. 51836006, No. 52076191), National Key Research and Development Program of China (No. 2016YFC0203701, No. 2017YFB0603201), National Key Technology R&D Program (2011BAA04B08, 2012BAB18B03), and Environmental Welfare Project of Ministry of Environmental Projection of China (No. 201509012).

Hangzhou, Zhejiang, China — Xiang Gao
Hangzhou, Zhejiang, China — Chenghang Zheng
Taipei, Taiwan, China — Pen-Chi Chiang
Hangzhou, Zhejiang, China — Kefa Cen

Contents

1 **Introduction** 1
1.1 Energy Utilization and Air Pollutant Emissions 1
1.2 Environmental and Health Influences of Air Pollutants 6
1.3 Global Actions for Clean Air 7
1.4 Development of Air Pollutant Control Technologies 14
References 17

2 **Pollutant Control by Catalytic Methods** 21
2.1 Introduction 21
2.2 Fundamentals of Selective Catalytic Reduction 24
2.2.1 Overview 24
2.2.2 Reaction Mechanism and Kinetics 25
2.3 NO_x Control Enhancement and Applications 34
2.3.1 Catalyst Poisoning and Resistance 34
2.3.2 Widening the Working Temperature Window 42
2.3.3 SO_2/SO_3 Conversion Inhibition 54
2.3.4 Applications 65
2.4 Simultaneous Control of Multiple Pollutants 84
2.4.1 Simultaneous Control of Mercury 84
2.4.2 Simultaneous Control of Volatile Organic Compounds 91
2.4.3 Applications 97
References 98

3 **Pollutant Control by Electric Methods** 105
3.1 Introduction 105
3.2 Fundamentals of Particulate Matter (PM) Control 107
3.2.1 Physicochemical Characterizations of PM 107
3.2.2 Mechanisms of PM Control in Electric Field 113
3.3 PM Control Enhancement and Applications 129
3.3.1 Particle Growth 130
3.3.2 Particle Conditioning 147
3.3.3 Particle Precharging 153
3.3.4 Electrode Optimization 159

3.4 Simultaneous Control of Multiple Pollutants ... 171
3.4.1 Simultaneous Control of SO_x ... 172
3.4.2 Simultaneous Control of NO_x ... 177
3.4.3 Simultaneous Control of Trace Elements ... 188
References ... 193

4 Pollutant Control by Absorption Methods ... 199
4.1 Introduction ... 199
4.2 Fundamentals of Pollutant Control by Absorption ... 200
4.2.1 Process Chemistry ... 200
4.2.2 Mass Transfer and Kinetic Reaction Modeling ... 226
4.3 SO_2 Control Enhancement and Applications ... 235
4.3.1 PH Adjustment ... 235
4.3.2 Desulfurization Additives ... 242
4.3.3 Scrubber Optimization ... 249
4.4 Simultaneous Control of Multiple Pollutants ... 262
4.4.1 Simultaneous Control of NO_x ... 262
4.4.2 Simultaneous Control of Hg ... 270
4.4.3 Applications ... 276
References ... 277

5 Multi-pollutant Removal System and Technology Evaluations ... 281
5.1 Introduction ... 281
5.2 Multi-pollutant Removal System ... 282
5.2.1 Ultra-Low Emission (ULE) Technologies ... 282
5.2.2 Technical Routes ... 286
5.3 Engineering Practice of ULE System ... 294
5.3.1 1000 MW Coal-Fired Power Unit ... 294
5.3.2 600 MW Coal-Fired Power Unit ... 296
5.3.3 300 MW Thermal Power Unit ... 301
5.3.4 50 MW Coal and Sludge Co-combustion Unit ... 304
5.4 Technology Evaluations ... 306
5.4.1 Economic Evaluation ... 307
5.4.2 Environmental Evaluation ... 318
References ... 322

Chapter 1
Introduction

1.1 Energy Utilization and Air Pollutant Emissions

Energy consumption from fossil fuels has been intensifying both in quantity and quality along with rapid economic and social development, causing severe air pollution. A lack of regulation and understanding of air pollution led to heavy smog in Europe and North America from the 1930s to 1950s, including the well-known Great Smog of London in 1952 and the Los Angeles smog episode in the 1940s–1970s. After years of development and increased government regulations, the air quality in developed countries and regions such as Europe and the United States (US) has significantly improved. The type and level of energy-related air pollution are often linked to a country's economic development stage. At low-income levels (e.g., many developing countries in Asia and Africa), the industrial and residential sectors rely heavily on burning solid biomass, leading to undesirable exposure to air pollutants. For example, the World Health Organization (WHO) and US Environmental Protection Agency (EPA) ranked New Delhi, the capital city of India, as the world's most polluted city due to industrial exhaust, vehicular emission, agricultural stubble burning, etc. in 2014 and 2016, respectively. Severe air pollution has also been observed in similar-sized cities such as Mexico City, Mexico and Beijing, China. As economies industrialize, the use of fossil fuels in power generation and the industrial sector generally increases, resulting in emissions of sulfur dioxide (SO_2), nitrogen oxides (NO_x), particulate matter (PM), and other pollutants. Further economic development leads to increased demand for electricity for appliances and oil for transport, leading to further increased emissions. Furthermore, urbanization often occurs in parallel with economic development, which can consequently aggravate the energy-related air pollution.

Pollutants such as SO_2, NO_x, PM, volatile organic compounds (VOCs), and ammonia (NH_3) are generated worldwide, as summarized in Fig. 1.1 [1]. Among these, SO_2, NO_x, and PM emissions are mostly attributable to energy production and utilization. In 2015, the energy sector was responsible for an estimated 80 million tons of SO_2 emissions globally, with over 45% generated from industrial processes

X. Gao et al., *Multi-Pollutant Control for Flue Gases*, Advanced Topics in Science and Technology in China 63, https://doi.org/10.1007/978-981-16-1518-4_1

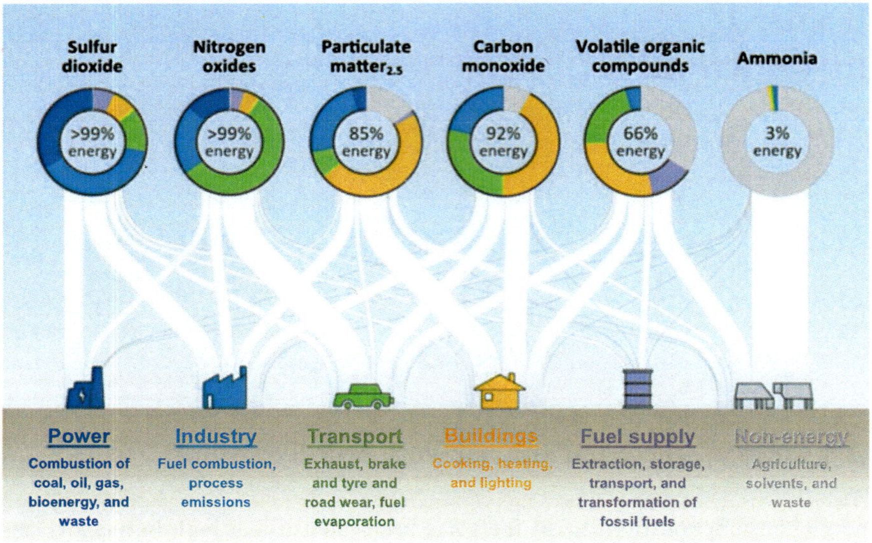

Fig. 1.1 Pollutants generated from global energy production and utilization [1]

and one-third from the power sector [2]. From the breakdown of energy-related SO_2 emissions by regions and sectors recorded in 2015 in Fig. 1.2a, over one-quarter of the total SO_2 emissions arose in China (22 Mt), where nearly two-thirds of the total SO_2 emissions are caused by industrial processes. India was the next highest source of SO_2 emissions (9 Mt). Conversely, the European Union (approximately 3 Mt) and Latin America (approximately 3 Mt) were accountable for only 5% of the world's SO_2 emissions. By sector, power generation is the primary source of SO_2 emissions in many countries, particularly India (approximately 53%), the United States (approximately 58%) and the Middle East (approximately 63%) [1].

Global energy-related NO_x emissions reached approximately 107 Mt in 2015 and have continued to increase; in 2015, the transportation sector accounted for the largest share (>50%), followed by the industrial (26%) and power (14%) sectors. The 2015 energy-related NO_x emissions presented region- and sector-wise in Fig. 1.2b. Increases in NO_x emissions in many developing countries have been rapid and have volumetrically outweighed the declines seen in several developed countries. China (23 Mt) and the US (13 Mt) account for one-third of global NO_x emissions. Transport is the largest source of such emissions in many regions; however, notable exceptions have been observed. In China, the industrial sector is the largest source of NO_x emissions. India's NO_x emissions have continually increased and reached a level similar to that of Europe in 2015, albeit with more than twice as large a population. Overall, stationary sources, such as power generation and industry, contribute to most of the pollutant emissions. Therefore, these two sources were selected for further analysis; their role in air pollution is discussed in the subsequent section.

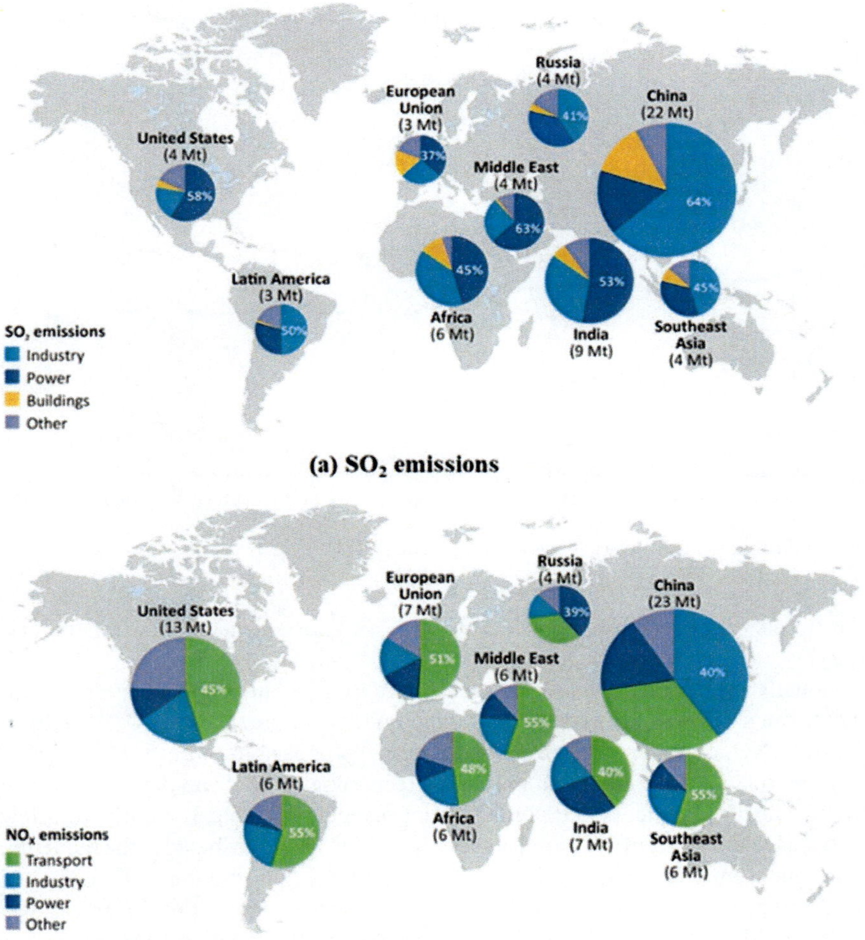

Fig. 1.2 Energy-related. **a** sulfur dioxide (SO_2) and **b** nitrogen oxides (NO_x) emissions by region and sector, 2015 [1]

a. *Power Generation*

The source-wise (based on the fuel) global electricity generation trend from 1986 to 2018 is shown in Fig. 1.3. Despite the recent rapid expansion in renewable sources (herein refers to solar, wind, biomass, geothermal and waste), most of the world's electricity continues to be generated from the combustion of fossil fuels. Coal remains the dominant fuel, although its share fell by 1.5% (36.4%) in 2019. Both natural gas and renewable sources were used at record levels in 2019 (23.3% and 10.4%,

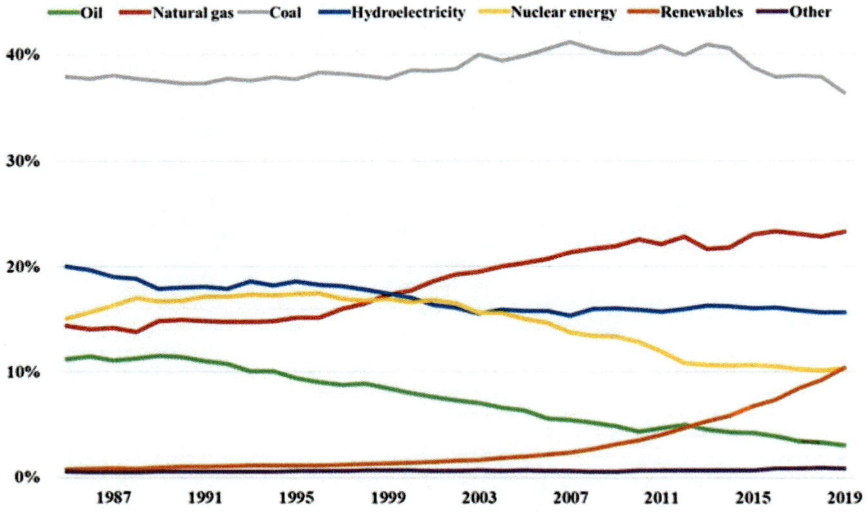

Fig. 1.3 Share of global electricity generation by source [3]

respectively), and renewable generation surpassed nuclear generation for the first time [3].

Globally, the burning of fossil fuels for electricity generation is the major source of air pollutants. Coal alone accounted for three-quarters of the sector's SO_2 emissions, 70% of its NO_x emissions, and over 90% of its $PM_{2.5}$ emissions in 2015 [1]. Emission levels are related to the pollution mitigation technologies used and the fuel type and quality. The development of air pollutant control technologies, particularly ultra-low emission technologies in coal-fired power generation, has contributed to the reduction of pollutant emissions [4]. Consistent with its minor contribution (4%) to global electricity generation, oil-fired power generation accounts for 7% of the global SO_2 emissions. Much of the oil-fired power generation comes from aging power plants—particularly in the Middle East—that have minimal pollution control technologies. Burning oil to generate electricity also produces a significant amount of NO_x and, in some cases, $PM_{2.5}$. In general, natural gas-fired plants emit fewer air pollutants than coal and oil-fired power plants; in 2015, natural gas-fired power generation accounted for nearly 20% of NO_x emissions from power generation but barely any SO_2 or $PM_{2.5}$ emissions [1]. Natural gas is typically pretreated to remove sulfur before commercialization, explaining the low sulfur emissions. Biomass currently plays a minor role in global power generation and thus contributes comparatively little to air pollution. However, in relative terms (emissions per unit of electricity output), direct burning of biomass has only slightly fewer NO_x emissions, moderately fewer $PM_{2.5}$ emissions, and much fewer SO_2 emissions than coal-fired power plants. With the increasing share of biomass-fired power plants, their emissions should also be addressed.

Over the last decade, global SO_2 emissions and, to a lesser extent, NO_x and $PM_{2.5}$ emissions, have been decoupled from the increase in coal-fired power generation. From 2005 to 2015, coal-fired power generation increased by 34%, while total power sector emissions of SO_2, NO_x, and $PM_{2.5}$ decreased by 55%, 34%, and 32%, respectively [5]. These reductions are attributable to either the usage of low-sulfur coal or advances in pollution control technologies to satisfy strict emission standards of coal-fired power plants.

b. *Industry*

In 2015, manufacturing and other transformation sectors, such as refining and mining, accounted for approximately half of the global energy-related emissions of SO_2 (37 Mt) and approximately 30% of both NO_x (28 Mt) and $PM_{2.5}$ (10 Mt) [1]. Similar to power generation, the source fuels and combustion operating conditions (e.g., pollution control technologies employed) in the related industry are crucial determinants of combustion-related pollutant emissions. Coal is the dominant fuel in manufacturing industries, accounting for 31% of the global industrial energy use and more than three-quarters of combustion-related SO_2 emissions. Meanwhile, NO_x emissions in the industry sector were split between coal and oil combustion (38 and 35%, respectively).

According to the World Energy Outlook Special Report [1], process-related pollutants from various industrial sources differed in amounts and concentrations. Worldwide, the highest specific industrial emitters of process-related SO_2 in 2015 were cement (13%), steel (9%), and refineries (9%), although the pulp and paper subsector has by far the most considerable specific emissions intensity (on a global basis 3.7 kt SO_2/Mt of paper compared with other industries that range from 0.6 kt SO_2/Mt of cement to 0.2 kt SO_2/Mt of crude steel). Conversely, process-related NO_x emissions were mostly released in cement making (1.5 kt NO_x/Mt of cement, accounting for over 50% of global process-related NO_x emissions), followed by pulp and paper production (1.2 kt NO_x/Mt of paper). Aluminum refining stands out as the largest emitter of $PM_{2.5}$ per unit of output (2.1 kt $PM_{2.5}$/Mt of aluminum), whereas other industrial processes activities were below 0.7 kt $PM_{2.5}$/Mt of product. The most considerable portion of global process-related $PM_{2.5}$ emissions arose in the cement, iron, and steel subsectors (combined, approximately 70%).

From a regional perspective (considering both combustion and process emissions), China, India, and the United States were the top emitters of SO_2, NO_x, and $PM_{2.5}$ in 2015 in the manufacturing sector owing to their significant share of manufacturing. Particularly, SO_2, NO_x, and $PM_{2.5}$ emissions from manufacturing in China in 2015 accounted for 44% of SO_2 global emissions, almost 40% of NO_x emissions, and 35% of $PM_{2.5}$ emissions. China was followed by India in SO_2 (9%) and $PM_{2.5}$ (14%) emissions and by the US in NO_x emissions (11%). When considering emissions per unit of industrial output, South Africa topped the ranking of combustion-related SO_2 intensity (0.5 kt SO_2 per petajoule), followed by the Caspian region. Chile, followed by Northern African countries, the Caspian region, and the United States, had the highest combustion-related NO_x emission intensity (0.4 kt NO_x per petajoule). In general, the global reduction of pollutants requires emission reductions from industrial sources.

1.2 Environmental and Health Influences of Air Pollutants

Air pollution has a detrimental effect on the environment, and, as a result, people's lives, and has thus become an essential obstacle to global development. Although the atmospheric environment may be affected by natural factors, anthropogenic activities account for the vast majority of negatively influencing factors, such as industrial emissions, automobile exhaust, and garbage incineration. The resulting pollution has long-lasting detrimental effects on people's living environments and can severely harm and significantly shorten human lifespans.

a. *Environmental Influences*

The impacts of air pollution on the environment manifest primarily as acid rain, smog, ozone layer destruction, and global warming [6]. Acid rain, one of the earliest environmental problems faced by humankind, contains harmful acids formed by sulfur oxides and NO_x primarily from energy utilization and causes soil acidification, substantial crop production reduction, and corrosion of buildings and industrial equipment [7]. In 1952, a severe acid rain event occurred in London, wherein the pH of rainwater was as low as 1.4–1.9 and caused the death of many animals. As governments realized the gravity of pollution-caused acid rains, they imposed emission regulations. Still, the presence of haze in many regions worldwide continued to demonstrate the destructive nature of air pollution. Haze is traditionally an atmospheric phenomenon wherein dust, smoke, and other dry particulates obscure visibility and the sky's clarity, color, and texture. Haze impacts roads, railways, aviation, shipping, power supply systems, and crop growth. The presence of haze has gradually increased owing to the increase in primary particle emissions and secondary particle generation. Nowadays, most parts of the world are still affected by haze, particularly East Asia, Southeast Asia, and South Asia [8].

Globally, air pollution has led to ozone (O_3) generation and global warming [9]. High O_3 concentrations cause photochemical smog; furthermore, there is a close relationship between O_3 and the emission of NO_x and VOCs, which can be understood in terms of a fundamental split into NO_x-sensitive and VOC-sensitive (or NO_x-saturated) chemical regimes [10]. Global warming occurs when carbon dioxide and other air pollutants and greenhouse gases collect in the atmosphere and absorb sunlight and solar radiation that have bounced off the surface of the Earth, leading to a gradual increase in the Earth's average temperature and causing rising sea levels and increased extreme weather events [11]. Emissions of air pollutants and carbon dioxide from energy utilization significantly contribute to O_3 generation and global warming.

b. *Health Influences*

Decades of research have shown that air pollutants such as $PM_{2.5}$ and O_3 increase the prevalence and severity of lung and heart disease and other health problems. According to the WHO, more than 80% of the world's urban-dwelling population live where the concentration of air pollutants, including chemicals, soot, or other damaging substances introduced into the atmosphere, exceeds the WHO's Air Quality Guidelines, leading to millions of premature deaths annually and considerable costs to the global economy [12]. While the precise mix of pollutants and the severity of the resulting health influences vary by location, the air quality in each case reduces life expectancy. Air pollution is the fourth-largest overall risk to human health, after high blood pressure, dietary risks, and smoking [13], and is an ongoing health crisis for which many causes and cures are found in the energy sector. $PM_{2.5}$, even at low concentrations, is associated with a range of severe illnesses and has the most significant influence on human health owing to its propensity to penetrate deep into the lungs. O_3, nitrogen dioxide, and SO_2 also have negative health influences.

The countries with the largest number of premature deaths caused by air pollution are mostly in Asia and Africa. However, when adjusted to a country's population, the highest mortality rates from air pollution span across other parts of the world, including Eastern Europe. The global influence of air pollution has increased over time, although this hides large regional disparities. Premature deaths caused by air pollution have declined in Europe, the US, and many other countries. However, they have increased in many countries, particularly in Asia and Africa [8].

1.3 Global Actions for Clean Air

Air pollution threatens human health globally. In 1972, the first major international conference on the environment and its relationship with humans was held in Sweden [Stockholm] under the auspice of the United Nations. The resulting influential Stockholm Declaration, designed to inspire and guide the world in the preservation and enhancement of the human environment, was adopted. Since then, as summarized in Table 1.1, a series of international conferences have been held and various environmental protection conventions have been signed, leading to international cooperation toward preventing or alleviating the effects of acid rain, smog, ozone layer destruction, and global warming.

Individual countries or regions have also taken measures and formulated policies to reduce air pollutant emissions and improve air quality; the relevant emission regulations or standards in the United States, the European Union, Japan, and China are summarized in Table 1.2, as they currently have the most stringent emission standards globally. Here, italics is used to represent the standards related to power generation. Clearly, the critical standards have been amended several times. Early on, prevention and control methods were still in their infancy, and urban areas dominated

Table 1.1 Typical conferences and conventions regarding global environmental protection

Year	Location	Conferences	Keynotes
1972	Stockholm, Sweden	The United National Conference on the Human Environment	A stern warning "Only one Earth" was first issued by this international conference
1979	Geneva, Switzerland	Convention on Long-range Transboundary Air Pollution	International cooperation should be strengthened in controlling air pollution, particularly SO_2 emissions to prevent and control acid rains
1985	Vienna, Austria	Vienna Convention for the Protection of the Ozone Layer	The knowledge of the effects on the ozone layer and new policies aimed at limiting chlorofluorocarbon emissions should be promoted through sharing international research
1987	Montreal, Canada	Montreal Protocol	A preliminary consensus on the specific implementation of ozone layer protection and the establishment of the compliance mechanism were formed
1988	Toronto, Canada	The World Conference on the Changing Atmosphere	Global warming was treated as a political issue for the first time
1989	Noordwijk, the Netherlands	Noordwijk Climate Conference	A binding agreement on CO_2 emissions was issued
1990	Geneva, Switzerland	Second World Climate Conference	The Global Climate Observing System was established
1997	Kyoto, Japan	Kyoto Protocol	It is the first legally binding treaty to reduce greenhouse gas emissions and prevent global warming
2009	Copenhagen, Denmark	Copenhagen Protocol	The meeting was described as "the last chance to save humanity"
2015	Paris, France	Paris Agreement	The parties reached an agreement to limit on the rise in global average temperatures less than 2 °C above preindustrial times
2019	New York, United States (US)	The 2019 UN Climate Action Summit	The Paris Agreement was further improved, and the parties tried to limit the global average temperatures within 1.5 °C

Table 1.2 Typical regulations/standards in different countries or regions

Country/region	Actions	Year	Keynotes
US	The Air Pollution Control Act	1955	–
	Clean Air Act	1970, 1977, 1990	The most influential environmental law in the US, providing a law foundation for the treatment of acid rain, ozone depletion, and toxic air pollution
	The National Ambient Air Quality Standards	1971–2012	Specified emission limit on the supreme permissible effluent concentration of PM, SO_2, and NO_x in coal-fired power plants to 20, 184, and 135 (mg/m^3), respectively
	New Source Performance Standards	1971, 1978, 1997, 2005, 2007, 2011	Established emission standards for many industries, such as boilers, landfills, petroleum refineries, and stationary gas turbines
	National Emission Standards for Hazardous Air Pollutants	2008	Regulated the emission of thermal power plants > 73 MW
	Climate Action Plan	2013	Cut carbon emissions from power plants, the largest emitters of greenhouse gases, and boost renewable energy
	Clean Power Plan	2015	Established a 2030 target to cut carbon emissions from power plants from 2005 levels by 32%
	Affordable Clean Energy Initiative	2019	Repealed the Clean Power Plan to alleviate investment of pollutant control for existing coal-fired power plants
European Union	Convention on the Long-range Transboundary Air Pollution	1979	Controlled air pollution, particularly SO_2 emissions to prevent and acid rain

(continued)

Table 1.2 (continued)

Country/region	Actions	Year	Keynotes
	Oslo Agreement	1985	Set limitation on sulfur emissions
	Gothenburg Agreement	1990	Set limitation on emissions of SO_2, NO_x, NH_3, and VOCs by the year 2010
	Limitation of Emissions of Certain Pollutants into the Air from Large Combustion Plants (88/609/EEC)	1987	Limited the emission of SO_2, NO_x, and PM according to the fuel properties and thermal power, and specify the overall emission limit for each country in the EU
	Large Combustion Plants Directive (2001/80/EC)	2002	Defined large combustion plants as those with a rated thermal input $\geq$ 50 MW, regardless of the type of fuel used
	Directive 2004/35/EC Relating to the Environmental Responsibility for Preventing and Remedying Environmental Damage	2004	The EU's first Paying for Pollution Directive defined "damage to the environment" and set baseline conditions for assessing environmental damage
	The Energy Efficiency Directive	2012	Set targets for greenhouse gas emissions, the share of renewable energy, and energy efficiency standards by 2020 and 2030
	National Energy and Climate Program (NECPs)	2018	Built a common energy union for the EU
	Clean Vehicles Directive	2019	Accelerated the transition to low- and zero-emission vehicles

(continued)

Table 1.2 (continued)

Country/region	Actions	Year	Keynotes
	2030 Climate Target Plan	2020	Established targets on reduction of greenhouse gas emissions, the proportion of renewable energy, and energy efficiency are given for the EU in 2030
Japan	Dust Emission Control Act	1962	The first control act for industrial dust
	Air Pollution Prevention and Control Act	1968, 1970, 1974	Limited the emission of SO_x, NO_x and soot from boilers, waste incinerators, and diesel equipment below the required values
	Special measures concerning the reduction of the total emission of nitrogen oxides and particulate matter emitted by motor vehicles in specific areas	2001	Low-emission vehicles popularized in specific regions based on total emission and the vehicle exhaust emission standards strengthened further to control NO_x, photochemical oxidants, and dust pollution
	Electric Power Environmental Protection Act	2004	Limited the maximum pollutant emission standards for coal-fired power plants in Japan
	Basic Global Warming Countermeasures Act	2010	Reduced greenhouse gas emissions by 25% from 1990 levels by 2020
	Countermeasures to Promote Global Warming	2021	Aimed to achieve a carbon-free society by 2050
	Law of the People's Republic of China on the Prevention and Control of Atmospheric Pollution	1987, 1995, 2000, 2015	Established the law foundation for all emission regulations

(continued)

Table 1.2 (continued)

Country/region	Actions	Year	Keynotes
China	Tryout Standards of "3-Waste" Discharge in Industry	1973	Imposed emission regulations on thermal power started in China sources
	Ambient Air Quality Standard	1982, 1996, 2000, 2012	Set the concentration limits of common pollutants (SO_2, NO_x, $PM_{2.5}$, and O_3) over an average period
	Emission Standards of Air Pollutants for Coal-fired Power Plants/Emission Standard of Air Pollutants for Thermal Power Plants	1991, 1996, 2003, 2011, 2014	Set emission limits according to the type of thermal power plants (newly built and existing) and the plant's site, particularly regarding SO_2, NO_x, and PM
	Action Plan on Coal-fired Power Emission Reduction and Upgrading (2014–2020)	2014	Required the newly built coal-fired power plants in Eastern China to realize ultra-low emissions (PM < 10 mg/m^3, SO_2 < 35 mg/m^3, and NO_x < 50 mg/m^3)
	Plan for Full-implement of Ultra-Low Emission Upgrading in Coal-Fired Power Plants	2015	Required all conditioned coal-fired power plants to realize ultra-low emissions by 2020
	Regulations Management of Pollutant Discharge Permits	2016	The kind of pollutants permitted to be discharged, their concentration, the total amount, and their discharge pattern and period stipulated by pollutant discharge permits
	Three-year Action Plan to Fight Air Pollution	2018	Aimed to significantly reduce the emissions of PM, SO_2, NO_x and the corresponding air pollution days in China

Italics represent standards related to power generation

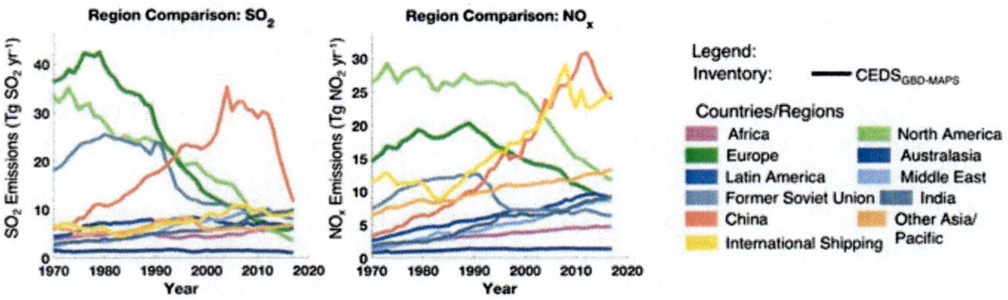

Fig. 1.4 Time series of global annual Community Emissions Data System (CEDS) emissions of SO_2 and NO_x by all sectors and fuel types in 11 main regions and countries, 1970–2017 [14]

the scope of air pollution. The primary pollutant of concern was suspended particulate matter (TSP). Prevention and control methods mainly focused on reforming boilers, eliminating smoke and dust, and controlling industrial point-source pollution. Continued rapid economic development and sharp increases in energy consumption led to severe SO_2 pollution in urban ambient air. Consequently, regional acid rain appeared. Accelerated urbanization and the high-speed development of urban traffic and the automobile industry have resulted in increasingly severe motor vehicle emissions. As a result, combined air pollution problems in economically developed countries and regions became increasingly prominent, and NO_x emission control was also addressed.

Historically, the development of air pollutant control technologies has been closely related to the imposed emission standards. Regulations of $PM_{2.5}$, SO_2, and NO_x emissions have been introduced in many countries; the US, Europe, and Japan have been regulating air pollution from coal-fired power plants since the 1970s. Since the turn of the century, many emerging economies have also introduced regulations. Initially, regulation was centered around reducing PM, followed by SO_2 (owing to increasing concerns about the effects of acid rain), and then NO_x. A time series of global annual emissions of SO_2, NO_x in 11 main regions and countries from 1970 to 2017 is shown in Fig. 1.4 [14]. The US and the European Union were pioneers in SO_2 control and thus showed an early significant pollutant reduction; total SO_2 emissions decreased by 89.1% from 1973 to 2017 in North America and by 86.1% from 1979 to 2017 in Europe. Conversely, SO_2 emissions increased by 83.1% from 1970 to 2004 in China. In 2011, strict emission policies were implemented to strengthen environmental protection, leading to a rapid decrease of SO_2 emissions by 61.6% by 2017. A small fluctuation in NO_x emissions was seen from 1970 to 1998 in North America, after which they began to decline, falling by 55.1% by 2017. In Europe, NO_x emissions rose by 28.4% from 1970 to 1989 and then fell by 65% by 2017. China's NO_x emissions peaked in 2012, an increase of approximately 89.5% over 1970. Owing to the implementation of strict policies in 2011, NO_x emissions were then reduced by approximately 22.6% by 2017.

1.4 Development of Air Pollutant Control Technologies

The relationship between fossil fuels, their use, and the resulting emissions is not always straightforward. With appropriate countermeasures, emissions of many of the discussed air pollutants per unit of a specific fuel depend on variables that could be reduced or removed at various stages, including before combustion (e.g., coal washing and oil refining), the combustion process (e.g., staged combustion or low-NO_x burning), and the postcombustion process. Among them, air pollutant control technologies in the postcombustion process most effectively reduce pollutants, often allowing poor-quality fuel to satisfy strict emission standards. Worldwide, many research institutes (to name a few, Zhejiang University, Tsinghua University, Huazhong University of Science and Technology, Southeast University, North China Electric Power University, Chinese Academy of Sciences, University of Cincinnati, Georgia Institute of Technology, Kyoto University, US Electric Power Research Institute, German Karlsruhe Institute of Technology, European Commission Joint Research Centre, etc.) have worked to develop air pollutant control technologies. International companies (e.g., General Electric, Babcock & Wilcox, Alstom, Cormetech, Siemens, Mitsubishi, Hitachi, Zhejiang Feida Environmental Science & Technology, Fujian Longking, Zhejiang Tiandi Environmental Protection Technology, and Beijing Guodian Longyuan Environmental Engineering) have promoted the application of air pollutant control technologies in various industries. In general, the developed post-combustion pollution control technologies have many similarities. Flue gases from power generation or industries are processed through physical separation or chemical reactions; these technologies can be sorted by their working principles into catalytic, electric, and absorption methods.

Catalytic methods. Catalytic methods usually employ catalysis to convert and eliminate complex atmospheric pollutants, such as NO_x, VOCs, and Hg [15–17]. For example, selective catalytic reduction (SCR), or NH_3-SCR, uses NH_3 to reduce NO_x emission and has been widely applied to lower NO_x emissions from stationary sources [18]. SCR is regarded as state-of-the-art technology and can now provide a NO_x reduction efficiency of 90%.

SCR technology has been improved via NH_3 injection mixing, flow field optimization, and catalyst designing. Novel catalysts have been developed to improve the adaptability, stability, and efficiency of this technology [19, 20]. During usage, the catalyst may be poisoned and deactivated by other elements found in the flue gas, such as alkali/alkaline earth metals, As, and Pb [15, 21, 22]. These components reduce the acidity of the catalyst and inhibit the reactivity of the catalyst's redox sites, leading to catalyst deactivation. Moreover, V_2O_5–WO_3/TiO_2-based commercial catalysts have a narrow working temperature window of approximately 300–400 °C [23], whereas the typical temperature of flue gases from industrial processes such as cement, glass, and steel manufacturing can be as low as 150–200 °C. The sulfur oxides (e.g., SO_2) in the flue gas will react with water vapor and the reducing agent (NH_3) to form ammonium bisulfate and deposit on the surface of the catalyst, covering the active sites, deactivating the catalyst, and ultimately reducing the denitrification efficiency of the system and increasing NO_x emissions [24]. Industrial waste gas often contains VOCs, heavy metals, and many other pollutants. Thus, to maintain a high removal efficiency of NO_x for long-term stability, efforts should be made to improve the low-temperature activity of the catalyst, widen the temperature window of the catalyst, improve the catalyst resistance to poisoning elements and SO_2, and optimize the entire system for collaborative treatment of multiple pollutants.

Electric methods. Here, an external power source is introduced through a discharge electrode to charge and separate particle-phase pollutants (i.e., solid particles or liquid droplets) from gaseous streams and oxidize gaseous pollutants (e.g., SO_2, NO_x, and Hg).

Researchers from many scientific research institutes have developed advanced electric methods for pollutant control. Electric and chemical agglomeration technologies can promote the growth of fine particles and reduce particle re-entrainment [25–27]. Using a high-frequency and three-phase power supply has been demonstrated to increase the particle migration velocity and reduce the system energy consumption [28, 29]. High-efficiency particle removal technologies such as low–low temperature electrostatic precipitators (ESPs) and rotating-electrode ESPs [30–33] have been widely used in power plants worldwide, such as in China, the US, Japan, and Australia. Recently, wet ESPs have been demonstrated to reduce fine particles and sulfuric acid mists [34–36]. High-temperature ESPs have also been used in large coal-fired power plants to protect the catalysts used in SCR from the mechanical wear and chemical toxicity of fly ash [37–39]. Although these technologies have helped industries meet the ultra-low emission level of coal-fired power plants, there are ongoing efforts to reduce operating costs and improve coal quality, load adaptability, and system stability. Chemical conversion of gaseous pollutant has also been achieved by increasing the electric field energy. Nonthermal plasma (NTP) technology allows for multi-pollutant removal, high reduction efficiencies, and rapid chemical reactions and can thus provide a feasible electric method for treating complex and toxic flue gases [40, 41].

Absorption methods. Here, one or more pollutants are selectively absorbed onto an absorbent; absorption methods are usually applied to remove gaseous pollutants from flue gases, particularly SO_2. Desulfurization technologies can be primarily classified into wet flue gas desulfurization (WFGD) and semidry FGD. WFGD is commonly used in large-scale power plants, whereas semidry FGD is more widely used in industrial boilers, steel sintering, and sludge incineration.

Efforts to improve WFGD have involved enhancing the mass transfer and reaction between the two phases. Absorbent dissolution, SO_2 absorption, oxidation, and product solidification are vital factors affecting the absorption efficiency [42–46]. In recent years, the influence of absorbent type, characteristic parameters, active additives, and operating parameters (pH value, temperature, and liquid–gas ratio) on absorbents' solubility characteristics and pollutant removal reaction activity have been studied [46–48]. SO_2 removal enhancement technologies, including pH dual-loop technology and desulfurization additives, have been applied in coal-fired units ranging from 50 to 1000 MW and have achieved desulfurization efficiencies of >98%. Furthermore, significant progress has been made on the scrubber design, which is vital to improve SO_2 absorption, by employing trays, liquid redistributors, venturi layers, swirling coupling devices, and deflectors [49–51]. These internal components can improve the uniformity and effective contact area between the gaseous and liquid phases in the scrubber. Simultaneous removal of multiple pollutants, including Hg and NO_x, during SO_2 absorption has become promising [52–55]; simultaneous removal of Hg and NO_x during SO_2 absorption can significantly reduce the land footprint and cost required.

Multi-pollutant removal systems. Air pollutant control technologies require continuous development to meet new requirements, including (i) high-efficiency and low-cost technologies to further decrease the emissions of coal-fired power plants; (ii) high-reliability technologies to handle complex flue gases generated from high-ash, high-sulfur, and high-alkali coals; (iii) simultaneous removal of multiple pollutants, including conventional pollutants (NO_x, PM, and SO_2) and other pollutants (SO_3, $PM_{2.5}$, heavy metals, and VOCs), through a single system; and (iv) higher-applicability technologies adaptable to multiple industrial uses, including power plants, sintering machines, cement kiln, coking plants, industrial boilers, and iron smelting.

Combining air pollutant control technologies can address these requirements while meeting emission standards. Advancements made to separate technologies can promote advances in multi-pollutant control systems. The systematic combination and optimization of catalytic, electric, and absorption methods can deliver an efficient, reliable, and low-cost multi-pollutant control system capable of meeting ultra-low emission standards (i.e., nearing emission levels of natural gas combustion). These multi-pollutant control systems, or ultra-low emission systems, have been employed in coal-fired power plants in China since 2014 to remove both pollutants from flue gases and satisfy emerging emission requirements.

Scope of the book. Continuous development and application of air pollutant control technologies have helped reduce pollutant emissions, thus improving air quality worldwide. Research and development into ultra-low emission technologies provide perspectives on air pollutant control. This book therefore aims to introduce readers to advancements in air pollutant control technologies from single pollutant treatment to multi-pollutant control, from simple to complex working conditions, and from research to application. Such a summary and case analyses of key technologies can provide a reference for researchers and decision-makers worldwide to reduce environmental pollution caused by the burning of fossil fuels, especially coal. This book is organized by pollutant control method, as summarized in Fig. 1.5. In Chap. 2, topics surrounding catalytic methods are discussed in detail, including the fundamentals of SCR, practical perspectives of SCR catalysts and systems, and the control of other pollutants by catalytic methods. Electric methods and related issues, including the fundamentals of $PM_{2.5}$ control, enhanced $PM_{2.5}$ control, and simultaneous multi-pollutant control are then discussed in Chap. 3. In Chap. 4, the fundamentals of pollutant removal by absorption, SO_2 absorption enhancement methods, and multi-pollutant absorption methods are discussed. Finally, multi-pollutant removal systems and a performance evaluation of such a system are introduced in Chap. 5. Each

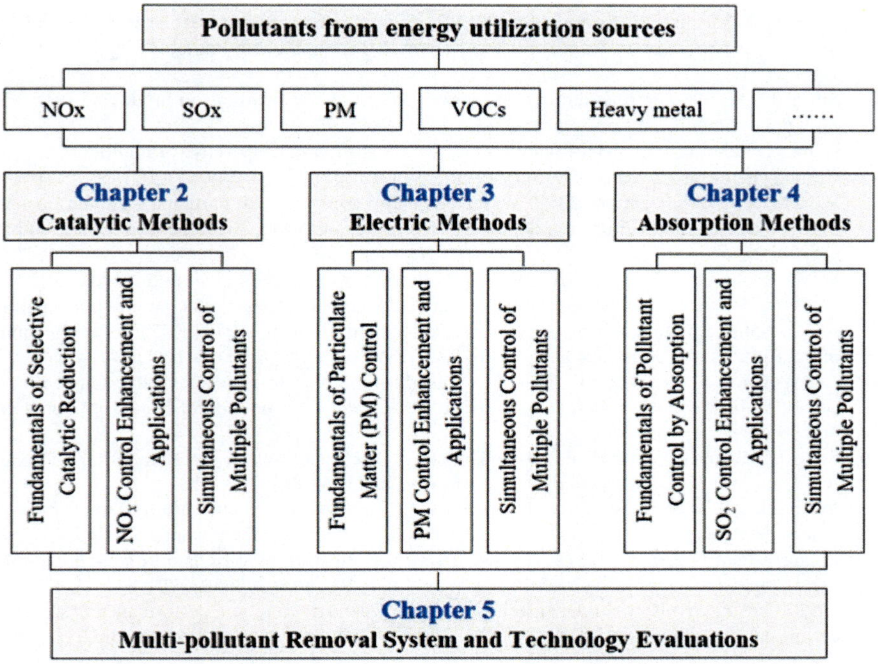

Fig. 1.5 Organizational framework of the book

chapter covers the content of theory basis, key technologies, and typical applications, which provides engineers and scientists with fundamental and applied knowledge as-associated with multi-pollutant control of flue gases.

References

1. OECD I (2016) Energy and air pollution: world energy outlook special report 2016
2. Rafaj P, Kiesewetter G, Gül T et al (2018) Outlook for clean air in the context of sustainable development goals. Glob Environ Chang 53:1–11
3. Looney B (2020) Full report–BP statistical review of world energy 2020. BP PLC, London
4. Tang L, Qu J, Mi Z et al (2019) Substantial emission reductions from Chinese power plants after the introduction of ultra-low emissions standards. Nat Energy 4(11):929–938
5. Bhati P, Singh M, Issar S (2017) Indonesia's coal power emission norms: lessons from India and China. Centre for Science and Environment, New Delhi
6. Manisalidis I, Stavropoulou E, Stavropoulos A et al (2020) Environmental and health impacts of air pollution: a review. Front Public Health 8:13
7. Grennfelt P, Engleryd A, Forsius M et al (2020) Acid rain and air pollution: 50 years of progress in environmental science and policy. Ambio 49(4):849–864
8. Institute HE (2019) State of global air: a special report on global exposure to air pollution and its disease burden

9. Andrady A, Aucamp PJ, Austin AT et al (2017) Environmental effects of ozone depletion and its interactions with climate change: progress report 2016. Photochem Photobiol Sci 16(2):107–145
10. Sillman S (1999) The relation between ozone, NO_x and hydrocarbons in urban and polluted rural environments. Atmos Environ 33(12):1821–1845
11. Hoegh-Guldberg O, Jacob D, Taylor M et al. (2019) The human imperative of stabilizing global climate change at 1.5 degrees C. Science 365(6459):1263
12. World Health Organization (2019) WHO air quality guidelines for particulate matter, ozone, nitrogen dioxide and sulfur dioxide. World Health Organization, 2006, Geneva. Designed by soapbox co uk
13. World Health Organization (2017) Burden of disease from household air pollution for 2012
14. McDuffie EE, Smith SJ, O'Rourke P et al (2020) A global anthropogenic emission inventory of atmospheric pollutants from sector- and fuel-specific sources (1970–2017): an application of the Community Emissions Data System (CEDS). Earth Syst Sci Data 12(4):3413–3442
15. Zhang S, Liu S, Hu W et al (2019) New insight into alkali resistance and low temperature activation on vanadia-titania catalysts for selective catalytic reduction of NO. Appl Surf Sci 466:99–109
16. Song H, Zhang M, Yu J et al (2018) The effect of Cr addition on HgO oxidation and NO reduction over V_2O_5/TiO_2 catalyst. Aerosol Air Qual Res 18(3):803–810
17. Yang Y, Liu S, Zhao H et al (2019) Promotional effect of doping Cu into cerium-titanium binary oxides catalyst for deep oxidation of gaseous dichloromethane. Chemosphere 214:553–562
18. Zhang R, Liu N, Lei Z et al (2016) Selective transformation of various nitrogen-containing exhaust gases toward N_2 over zeolite catalysts. Chem Rev 116(6):3658–3721
19. Qu R, Gao X, Cen K et al (2013) Relationship between structure and performance of a novel cerium-niobium binary oxide catalyst for selective catalytic reduction of NO with NH_3. Appl Catal B 142–143:290–297
20. Meng D, Xu Q, Jiao Y et al (2018) Spinel structured $Co_aMn_bO_x$ mixed oxide catalyst for the selective catalytic reduction of NO_x with NH_3. Appl Catal B 221:652–663
21. Hu W, Zhang Y, Liu S et al (2017) Improvement in activity and alkali resistance of a novel V-$Ce(SO_4)_2/Ti$ catalyst for selective catalytic reduction of NO with NH_3. Appl Catal B 206:449–460
22. Zhou L, Li C, Zhao L et al (2016) The poisoning effect of PbO on Mn-Ce/TiO_2 catalyst for selective catalytic reduction of NO with NH_3 at low temperature. Appl Surf Sci 389:532–539
23. Han L, Cai S, Gao M et al (2019) Selective catalytic reduction of NO_x with NH_3 by using novel catalysts: state of the art and future prospects. Chem Rev 119(19):10916–10976
24. Ye D, Qu R, Liu S et al (2019) New insights into the decomposition behavior of NH_4HSO_4 on the SiO_2-decorated SCR catalyst and its enhanced SO_2-resistant ability. ACS Omega 4(3):4927–4935
25. Yan J, Chen L, Yang L (2016) Combined effect of acoustic agglomeration and vapor condensation on fine particles removal. Chem Eng J 290:319–327
26. Chang Q, Zheng C, Gao X et al (2015) Systematic approach to optimization of submicron particle agglomeration using ionic-wind-assisted pre-charger. Aerosol Air Qual Res 15(7):2709–2719
27. Chang Q, Zheng C, Yang Z et al (2017) Electric agglomeration modes of coal-fired fly-ash particles with water droplet humidification. Fuel 200:134–145
28. Zereini F, Wiseman CLS (2011) Urban airborne particulate matter. Springer, Berlin
29. Parker KR (1996) Applied electrostatic precipitation. Springer, Netherlands
30. Gao S, Xie B (2020) The guarantee strategy for reliable operation of the precipitation with mobile electrode electrostatic technology. Modern Econ Manage Forum 1(1)
31. Li X, Zhou C, Li J et al (2019) Distribution and emission characteristics of filterable and condensable particulate matter before and after a low-low temperature electrostatic precipitator. Environ Sci Pollut Res 26(13):12798–12806
32. Qi Z, Li J, Wu D et al (2017) Particulate matter emission characteristics and removal efficiencies of a low-low temperature electrostatic precipitator. Energy Fuels 31(2):1741–1746

33. Wang G, Ma Z, Deng J et al (2019) Characteristics of particulate matter from four coal–fired power plants with low–low temperature electrostatic precipitator in China. Sci Total Environ 662:455–461
34. Bologa A, Seifert H, Paur HR et al (2008) Collection of fine particles by novel wet electrostatic precipitator. IEEE Trans Ind Appl 45(6):2170–2177
35. Huang J, Zhang F, Shi Y et al (2016) Investigation of a pilot-scale wet electrostatic precipitator for the control of sulfuric acid mist from a simulated WFGD system. J Aerosol Sci 100:38–52
36. Huang J, Wang H, Shi Y et al (2016) Performance of a pilot-scale wet electrostatic precipitator for the control of sulfuric acid mist. Environ Sci Pollut Res 23(19):19219–19228
37. Forestier LL, Libourel G (2008) High temperature behavior of electrostatic precipitator ash from municipal solid waste combustors. J Hazard Mater 154(1–3):373–380
38. Xu X, Gao X, Yan P et al (2015) Particle migration and collection in a high-temperature electrostatic precipitator. Sep Purif Technol 143:184–191
39. Xu X, Zheng C, Yan P et al (2016) Effect of electrode configuration on particle collection in a high-temperature electrostatic precipitator. Sep Purif Technol 166:157–163
40. Li S, Huang Y, Wang F et al (2014) Fundamentals and environmental applications of non-thermal plasmas: multi-pollutants emission control from coal-fired flue gas. Plasma Chem Plasma Process 34(3):579–603
41. Ma S, Zhao Y, Yang J et al (2017) Research progress of pollutants removal from coal-fired flue gas using non-thermal plasma. Renew Sustain Energy Rev 67:791–810
42. Wang H (2013) Studies on key technologies of enhancement for calcium-gypsum WFGD. Zhejiang University, China (in Chinese)
43. Chen Y (2012) Experimental study on the promoting dissolution of limestone and strengthening SO_2 absorption by WFGD additives. Zhejiang University, China (in Chinese)
44. Miu M (2011) Modeling and experimental study on limestone dissolution in wet desulfurization. Chin J Environ Eng 5(11):179–183
45. Wang H (2012) Effect of impurities on the structure and properties of flue gas desulfurization gypsum. China Building Materials Academy, China (in Chinese)
46. Wu C, Zheng C, Zhang J et al (2014) Molecular dynamics simulations of components diffusion process and SO_2 dissolving mechanism in WFGD slurry. Acta Science Circumstance 34(11):2904–2910
47. Wu C (2014) Enhancement of mass transfer and absorption in WFGD by molecular dynamics simulation. Zhejiang University, China (in Chinese)
48. Fu H (2013) Phase transition and crystallization of calcium sulfite and calcium sulfate in $CaCl_2$ solution. Zhejiang University, China (in Chinese)
49. Sui J, Song J, Fan J et al (2012) Modelling and experimental study of mass transfer characteristics of SO_2 in sieve tray WFGD absorber. J Energy Inst 3:176–181
50. Gao H, Li C, Zheng G et al (2011) Flue gas desulphurization based on limestone-gypsum with a novel wet-type PCF device. Sep Purif Technol 76(3):253–260
51. Wang Z (2013) Experimental investigation on wet flue gas desulfurization with electrostatically-assisted twin-fluid atomization. Environ Eng Manag J 12(9):1861–1867
52. Zheng C, Xu C, Gao X et al (2014) Simultaneous absorption of NO_x and SO_2 in oxidant-enhanced limestone slurry. Environ Prog Sustainable Energy 33(4):1171–1179
53. Zheng C, Xu C, Zhang Y et al (2014) Nitrogen oxide absorption and nitrite/nitrate formation in limestone slurry for WFGD system. Appl Energy 129:187–194
54. Zhao Y, Guo T, Chen Z (2011) Experimental study on simultaneous desulfurization and denitrification from flue gas with composite absorbent. Environ Prog Sustain Energy 30(2):216–220
55. Xu C (2014) Experimental study on the integrated and intensive absorption of NO_x/SO_2 from coal-fired flue gas. Zhejiang University, China (in Chinese)

Chapter 2
Pollutant Control by Catalytic Methods

2.1 Introduction

Catalysis has been demonstrated to be crucial for removing flue gas pollutants; NH_3-SCR is the most commonly used catalytic technology applied to remove flue gas pollutants from stationary sources. An SCR system generally comprises an injection of gaseous NH_3 and a two- or three-layered catalyst bed comprising oxides of metals (e.g., tungsten, vanadium, or titanium) formed into a honeycomb shape through which the flue gas passes. The SCR system was firstly developed and patented by Engelhard (now BASF Catalysts LLC) in 1957; by the 1970s, it was applied to industrial usage in Japan. Subsequently, SCR was widely applied in Europe, particularly in Germany, and later (late 1980s) in the US to control the emission of flue gas pollutants produced by coal combustion. In the twenty-first century, SCR has been further developed to meet strict local pollutant emission restrictions and improve the air quality in China. SCR is currently used widely in power plants and is considered the most promising technology for future control of flue gas pollutants emitted from various industrial boilers.

In recent decades, the removal of NO_x has become an environmental issue with increasing concern. NO_x is a generic term that represents mainly NO and NO_2, which are primarily produced in the combustion of fossil fuels in power stations, industries, and automobiles. NO_x produced during coal combustion can be classified as thermal, prompt, and fuel NO_x. Fuel NO_x originates from the thermal decomposition of organic heterocyclic nitrogen oxides of coal in the high-temperature flame and further oxidation. Thermal NO_x is a result of the combination reaction between nitrogen and oxygen at high temperatures. The relationship between NO_x generation and boiler temperature is shown in Fig. 2.1; generated NO_x was mainly thermal and fuel based.

NO_x deteriorates aquatic ecosystems and threatens the lives of terrestrial animals and plants. NO_x released into the atmosphere dissolves into water vapor present in the air and forms a weak nitric acid solution (i.e., acid rain). Once NO_x is mixed with VOCs and is exposed to sunlight, photochemical smog is formed, leading to breathing

X. Gao et al., *Multi-Pollutant Control for Flue Gases*, Advanced Topics in Science and Technology in China 63, https://doi.org/10.1007/978-981-16-1518-4_2

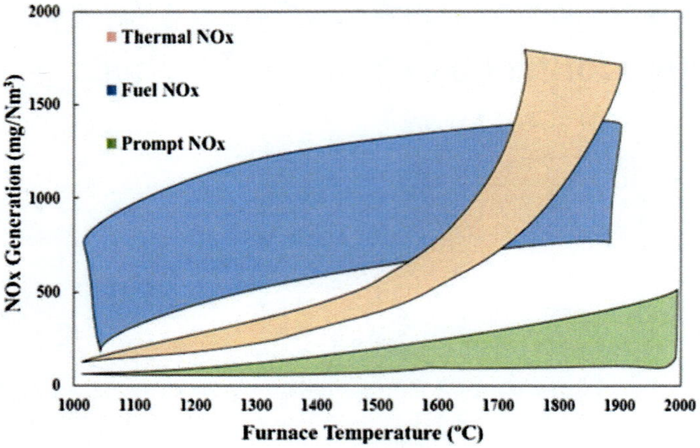

Fig. 2.1 Types of NO_x generated in the boiler

problems, headaches, reduced lung function, and eye irritation. SCR converts NO_x into N_2 and H_2O using NH_3 as a reducing agent in the presence of a catalyst with a denitrification (deNO_x) efficiency of 90–97% to decrease NO_x emissions. Still, this technology faces many challenges in industrial applications; some recent research efforts are detailed below.

High resistance to poisoning: The presence of alkali/alkaline earth and heavy metals in the flue gas can poison and deactivate the catalyst. Coal used in many facilities in China contains high levels of alkali/alkaline earth metals; further, exhaust gas from power plants, particularly waste incineration plants, usually contains arsenic (As) and lead (Pb). These elements impede the catalytic performance of commercial catalysts by reducing the acidity and inhibiting the reactivity of the catalyst's redox sites.

Wide temperature window: The catalyst most commonly used for SCR, V_2O_5-$WO_3(MoO_3)/TiO_2$, has a narrow working temperature window. Under 300 °C, the catalytic activity is insufficient; above 400 °C, NH_3 is oxidized to N_2O and NO, decreasing the NO_x removal efficiency. In many industrial processes, however, such as cement, glass, and steel manufacturing, the temperature of the flue gas can be lower than 200 °C. Reheating the flue gases above 300 °C would lead to unnecessary energy consumption. Furthermore, the fluctuation in a power plant generator's heavy load can cause the inlet temperature of the flue gas to fluctuate between 260 and 450 °C (Fig. 2.2).

Low sulfur conversion: Sulfur oxides in the flue gas react with water vapor and NH_3 to form ammonium bisulfate, which deposits on the surface of the catalyst and covers the active sites on the surface, resulting in catalyst deactivation and corrosion of downstream equipment by the generated ammonium bisulfate. Aside from the

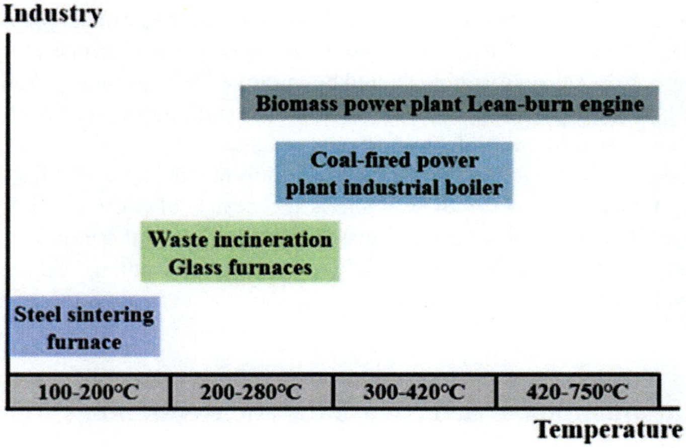

Fig. 2.2 Temperature range of flue gas by source

increased maintenance and equipment costs, this also leads to a reduction in the system's deNO_x efficiency and an increase in emitted NO_x.

Catalyst regeneration: The SCR system is generally placed between the economizer and air preheater. The high ash content of the flue gas at the inlet of the SCR system can cause chemical poisoning, physical blocking, abrasion, sintering, and the volatilization of active components of the catalyst. As a result, the lifespan of SCR catalysts is generally only three to four years. Catalyst regeneration could thus prolong the chemical life of the catalyst and reduce the waste associated with deactivated catalysts.

Ammonia mixture reinforcement: To maintain a high deNO_x efficiency in the SCR system, NO_x must be dispersed and mixed with the flue gas evenly. As insufficient mixing leads to low deNO_x efficiency, optimizing the integrated technology of the SCR system can strengthen the mixing between NO_x and NH_3 and improve the efficiency of deNO_x.

Synergistic removal of multiple pollutants: In addition to NO_x, industrial gaseous wastes contain various pollutants, including VOCs. If the concentration of VOCs in the flue gas is high, they can be removed economically by adsorption or catalytic combustion techniques. If the concentration of VOCs in the flue gas is low, simultaneous removal of NO_x and VOCs via a catalytic converter is more technically and financially viable than building additional facilities. Furthermore, the coal-fired power industry is the largest anthropogenic source of Hg emissions; however, construction costs of Hg emission control facilities are considerably high. Thus, a coordinated removal of multiple pollutants without the large-scale modification of existing equipment would be of great value.

Overall, SCR can effectively reduce NO_x emissions in industrial applications. For high deNO_x efficiency and long-term system stability, the low-temperature activity and resistance to catalyst poisoning should be improved and the temperature window of the catalyst should be widened. Furthermore, optimizing the entire system and simultaneously treating multiple pollutants is crucial.

In this chapter, the basic principles and reaction mechanisms of SCR are introduced first. Critical issues the SCR, such as the design of poison-resistant, wide-temperature window and low-sulfur conversion catalysts and the industrial preparation and regeneration of deactivated catalysts, are then presented. Finally, research on and applications of multi-pollutant synergistic catalytic treatments are discussed.

2.2 Fundamentals of Selective Catalytic Reduction

During SCR, NO_x is selectively converted into inert N_2 and H_2O through a catalyst using NH_3, CO, or hydrocarbons as a reducing agent. In this section, an overview of the reaction process is presented, followed by a discussion of representative SCR catalysts. Recent developments on the SCR reaction mechanism are presented at the end of this section.

2.2.1 *Overview*

NH_3 is the most commonly used reducing agent in contemporary SCR processes (i.e., NH_3-SCR) to control the NO_x emission levels from stationary sources. As NO is the dominant NO_x species (approximately 95%) in flue gas, the NH_3-SCR reaction generally proceeds according to the following reaction (standard SCR reaction) [1].

$$4NH_3 + 4NO + O_2 \rightarrow 4N_2 + 6H_2O \tag{2.1}$$

The presence of NO_2 could considerably improve the NO_x removal rate via the so-called fast SCR reaction, which has a reaction rate an order of magnitude higher than that of the standard SCR reaction above 200 °C [2]:

$$2NO + 2NO_2 + 4NH_3 \rightarrow 4N_2 + 6H_2O. \tag{2.2}$$

Several side reactions also occur during SCR, including the formation of N_2O and the oxidation of NH_3 [3]:

$$8NH_3 + 12NO + 5O_2 \rightarrow 10N_2O + 12H_2O, \tag{2.3}$$

$$4NH_3 + 3O_2 \rightarrow 2N_2 + 6H_2O, \tag{2.4}$$

$$4NH_3 + 3O_2 \rightarrow 2N_2 + 6H_2O, \tag{2.5}$$

$$2NH_3 + 2O_2 \rightarrow N_2O + 3H_2O. \tag{2.6}$$

However, N_2O is the primarily causes ozone depletion in the twenty-first century and has a greenhouse effect 300 times stronger than that of CO_2 [4]; N_2O formation should thus be avoided. Moreover, the oxidation of NH_3 can decrease the deNO_x efficiency. Therefore, selectivity is another critical factor when selecting a catalyst for SCR.

Significant efforts have been devoted to developing a variety of catalysts for SCR that have been prepared and characterized under various operating conditions; these catalysts can be generally divided into metal oxide, zeolite, and noble metal catalysts. Among metal oxide catalysts, vanadium-based catalysts exhibit excellent deNO_x activity and N_2 selectivity at high temperatures (320–380°C). Conversely, iron-, cerium-, and manganese-based catalysts demonstrate suitable catalytic activities at lower temperatures. Copper- or iron-doped zeolite catalysts are mainly used in vehicle exhaust treatment; however, their complicated synthesis process and low sulfur resistance have limited their application to stationary sources. Noble metal catalysts exhibit good activity at low temperatures but demonstrate low N_2 selectivity and have high costs. As a result, V_2O_5-WO_3 (MoO_3)/TiO_2 catalysts have been the most commonly used SCR catalysts in power plants for several decades. Still, the insufficient low-temperature activity of vanadium-based catalysts and secondary pollution caused by vanadium volatilization has led researchers to investigate vanadium-free alternatives.

2.2.2 *Reaction Mechanism and Kinetics*

An understanding of the reaction mechanisms and kinetics is essential when designing and optimizing SCR catalysts. The kinetics of the commercially available V_2O_5/TiO_2 catalysts doped with WO_3 or MoO_3 has been extensively studied. The reaction orders of the reactants NH_3 and NO were measured to be approximately 0 and 1, at the normal SCR operation temperature range (between 320 and 380 °C), respectively [5, 6], indicating that the SCR reaction follows the Eley–Rideal (E-R) reaction mechanism. However, in the low-temperature range, this reaction occurs via the Langmuir–Hinshelwood (L-H) reaction mechanism [7].

A six-step mechanism based on the general Mars-van Krevenlen (MvK) has also been proposed [8]:

$$NH_3 + V^{5+}{-}OH \leftrightarrow V{-}ONH_4, \tag{2.7}$$

$$V{-}ONH_4^* + V = O \leftrightarrow V{-}ONH_3 - V^{4+}{-}OH, \tag{2.8}$$

$$V{-}ONH_3{-}V^{4+} - OH + NO \rightarrow N_2 + HO + V^{5+}{-}OH, \tag{2.9}$$

$$2V^{4+}{-}OH \leftrightarrow H_2O + V^{3+} + V = O, \tag{2.10}$$

$$2V^{3+} + O_2 \rightarrow 2V = O, \tag{2.11}$$

$$H_2O + V^{5+}{-}OH \leftrightarrow V^{5+}{-}OH_3O. \tag{2.12}$$

According to the E-R reaction mechanism, NH_3 adsorbs on acid sites to form NH_3/NH_4^+, which can then be oxidized by redox sites with a high valence to form NH_2. The adsorbed NH_2 or NH_3 reacts with gaseous NO to form active NH_2NO/NH_3NO intermediates, which then decompose into N_2 and H_2O. According to the L-H reaction mechanism, NH_3 adsorbs on acid sites to form NH_4^+, whereas NO is oxidized by redox sites with high valence to form active bidentate nitrate, bridged nitrate, and monodentate nitrite intermediates, which then react with NH_4^+ to form NH_4NO_2/NH_4NO_3 that decomposes into N_2 and H_2O. Both reaction mechanisms comprise the reduction and reoxidation (by O_2) of redox sites with a high valence to finish the redox cycle. The overall reaction scheme encompassing both the E-R and L-H mechanisms is summarized in Fig. 2.3 [7].

Although significant efforts have been devoted to understanding the reaction mechanism of SCR, specific details about the nature of the catalytic active sites, reaction pathways, and rate-determining steps are still unclear. Fortunately, with advancements in theoretical chemistry and characterization techniques, particularly in situ/operando spectroscopies and transient response methods (TRMs), new insights have helped elucidate the detailed mechanism [9].

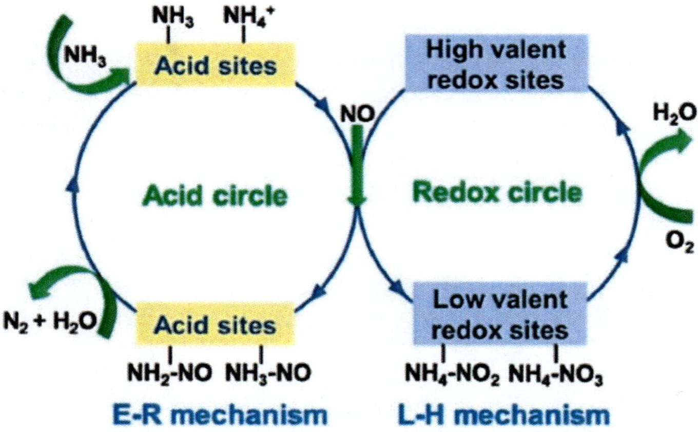

Fig. 2.3 Schematic of the SCR reaction pathway encompassing both the Eley–Rideal (E-R) and Langmuir–Hinshelwood (L-H) reaction mechanisms

2.2.2.1 Theoretical Prediction of SCR Reactivity

With the rapid development of quantum chemistry and insights into the NH_3-SCR process, a theoretical method based on density-functional theory (DFT) was proposed to predict the reactivity of metal oxides in NH_3-SCR [10]. As shown in Fig. 2.4, NH_3 adsorption, oxidation of intermediates, and reoxidation of the reduced catalysts are the three important steps of the NH_3-SCR process [8, 11], indicating that the acidity and redox capability of catalysts could have significant influences on the reactivity of metal oxide catalysts.

In this method, the lowest unoccupied molecular orbital (LUMO) energy and hydrogenation energy are used to measure the acidity and oxidizing ability, respectively. A lower LUMO energy facilitates the attraction of electrons from NH_3, whereas a higher LUMO energy implies poorer acidity and weakens the NH_3 adsorption ability. High hydrogenation energy represents the strong ability of the catalyst to oxidize the intermediates. A positive correlation between the LUMO and hydrogenation energies was found for the selected metal oxides, as shown in Fig. 2.5. These metal oxides were thus divided into three categories: active components (Zone A, including Cr^{5+}, Mn^{4+}, Fe^{3+}, Co^{2+}, Cu^{2+}, Ru^{4+}, Rh^{4+}, Ce^{4+}, and V^{5+} oxides), promoters (Zone B, including W^{6+}, Zn^{2+}, Nb^{5+}, Mo^{6+}, and Sb^{5+} oxides), and inactive components/support (Zone C, including Ti^{4+}, Al^{3+}, Si^{4+}, Ga^{3+}, Ge^{4+}, Zr^{4+}, In^{3+}, and Sn^{4+} oxides). The oxides in Zone A possess adequate acidity and oxidizing ability to be the main active components promoting the SCR reaction. In Zone B, though they are more acidic than those in Zone C, the oxides possess an insufficient oxidizing ability to act as main active components although they can be used as promoters. The gray

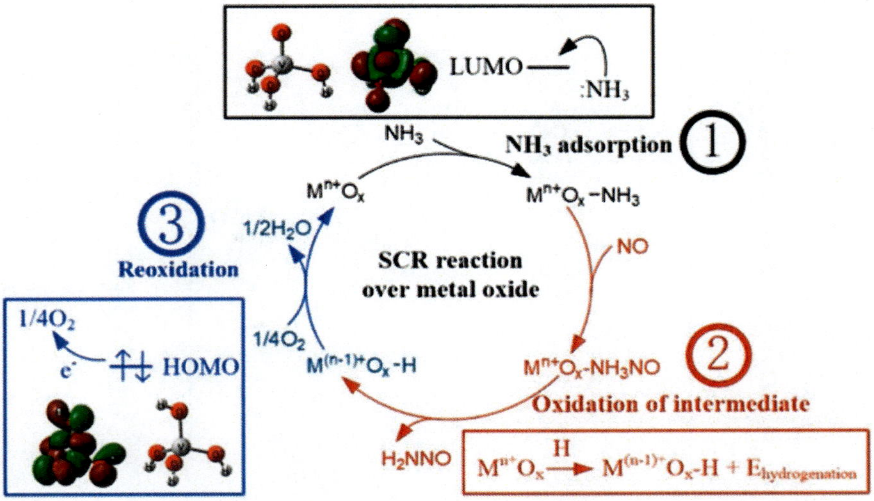

Fig. 2.4 Relationship between catalyst properties and theoretical parameters of the three important steps of the NH_3-SCR process [8, 11]

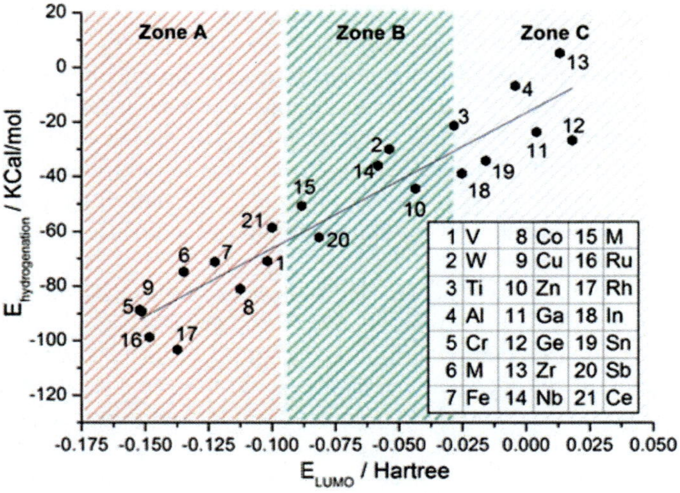

Fig. 2.5 Relationship between the hydrogenation and LUMO energies of various metal oxides [10]

Zone C includes the oxides with insufficient acidity and oxidizing ability but those that can act as porous supports of the SCR catalyst.

The HOMO energies of these metal oxides were then calculated to characterize the reoxidation ability of the reduced oxide. As shown in Fig. 2.6, the HOMO energies of the reduced Cu, Mn, Ru, and Rh oxides are close to that of reduced $V^{5+}O_x$, indicating a similar reoxidation behavior of these oxides. However, owing to their more negative

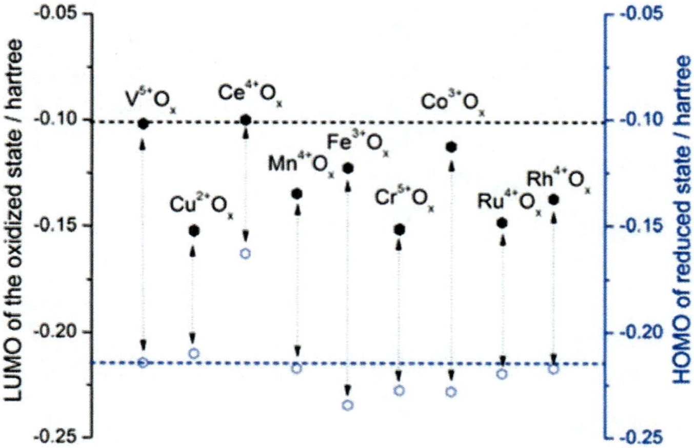

Fig. 2.6 LUMO energies of the oxidized state and HOMO energies of the reduced state of the oxides in Zone A. Black solid circles represent the LUMO energy of the oxidized state; blue hollow circles represent the HOMO energy of the reduced state [10]

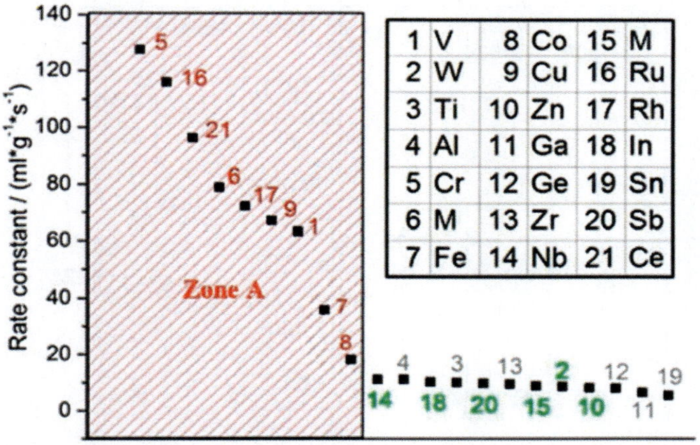

Fig. 2.7 Rate constants of SCR of NO with NH_3 over titania-supported metal oxides at 350 °C. Oxide number colors correspond to their theoretical category (red: Zone A, green: Zone B, and gray: Zone C) [10]

HOMO energies, the reduced forms of $Fe^{3+}O_x$ and $Co^{3+}O_x$ do not reoxidize easily and thus may have poorer reactivities than that of $V^{5+}O_x$ in SCR processes.

The SCR activities of these metal oxides were measured by supporting them on titania carriers. As shown in Fig. 2.7, the metal oxides in Zone A demonstrated higher SCR activities than the metal oxides in Zones B and C. The titania-supported Cr- and Ru-oxide catalysts exhibited the highest reaction rates among the candidates studied. These experimental results thus verified the theoretical catalyst activity model's efficiency in predicting the reactivity of metal oxides to aid in the screening of high-performance NH_3-SCR catalysts.

2.2.2.2 Quantification of Contributions from the E-R and L-H Models

As the SCR reaction mechanisms are closely related to the type of active sites, the relative contributions of the E-R and L-H mechanisms under different reaction conditions must be quantified to improve the catalyst. Using a Ce–Nb oxide catalyst to explore the reaction pathways and investigate the contributions of the E-R and L-H mechanisms [12, 13], both the E-R and L-H mechanisms were demonstrated to occur, as shown in Fig. 2.8.

In situ diffuse reflectance infrared Fourier transform spectroscopy (DRIFTS) of the transient reactions at 200 °C demonstrated that the surface of the catalyst was first saturated with NH_3, followed by NO and O_2, as shown in Fig. 2.8a. Here, the bands at 1188 and 1598 cm^{-1} and 1436 and 1661 cm^{-1} are attributable to the symmetric and asymmetric bending of coordinated NH_3 species on Lewis acid sites and adsorbed NH_4^+ species on Brønsted acid sites, respectively. The bands at 1370 and 1561 cm^{-1} are attributable to the oxidized products of NH_3. NO and O_2 were then introduced to

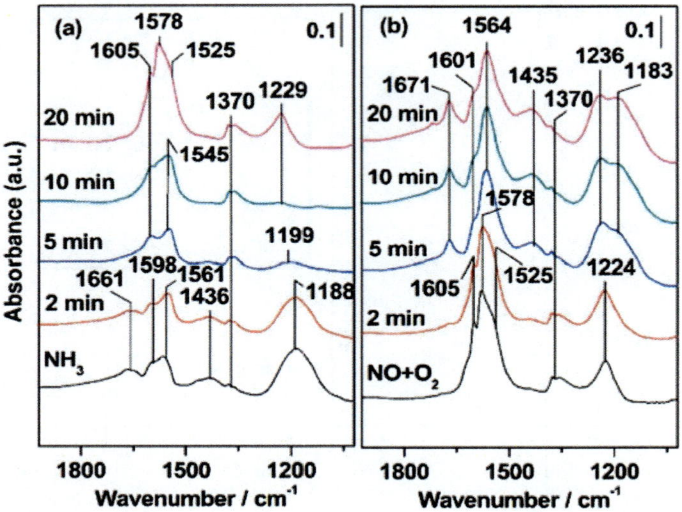

Fig. 2.8 In situ diffuse reflectance infrared Fourier transform spectroscopy (DRIFTS) of **a** adsorbed NH_3 reacting with gaseous NO and O_2 and **b** adsorbed NO and O_2 reacting with gaseous NH_3 [13]

the infrared (IR) cell; the NH_3 species absorbed on both the Lewis and Brønsted acid sites were completely consumed after 5 min. Meanwhile, some bands attributable to monodentate, bidentate, and bridging nitrate species appeared. The catalyst's surface was nearly entirely covered by these nitrate species after approximately 20 min, resulting in a spectrum similar to that of NO and O_2 adsorption on the catalyst surface. The IR cell was then purged with N_2 for 15 min before NH_3 was reintroduced to examine the reactions between the adsorbed nitrate species and NH_3, as shown in Fig. 2.8b. After 5 min, the peak for gaseous NO_2 at 1605 cm^{-1} and monodentate nitrate at 1525 cm^{-1} diminished, whereas bands attributable to adsorbed NH_3 species at 1183, 1433, 1601, and 1671 cm^{-1} appeared. Furthermore, the bidentate nitrate peak at 1578 cm^{-1} shifted to 1564 cm^{-1} and the bridging nitrate peak at 1224 cm^{-1} shifted to 1236 cm^{-1}. A mass spectrometer (MS) was used to detect the produced N_2. As N_2 ($m/z = 28$) was detected in both cases, the disappearance of these peaks was due to the SCR reaction.

By calculating the contributions of the E-R and L-H reaction mechanisms to SCR at different reaction temperatures, the contributions from the L-H mechanism were demonstrated to gradually decrease with an increase in reaction temperature, as shown in Fig. 2.9. The L-H mechanism thus has a more critical effect on the activity of the catalyst at low temperatures. In contrast, the E-R mechanism has a more obvious effect on the activity of the catalyst at high temperatures.

From Fig. 2.8b, the monodentate nitrate peak diminishes and the bidentate and bridging nitrate peaks shift after 5 min of NH_3 passing through the NO_x-saturated catalyst, suggesting that bidentate and bridging nitrates can capture NH_3 species, causing this shift of the characteristic bands. Unlike bidentate and bridging nitrates,

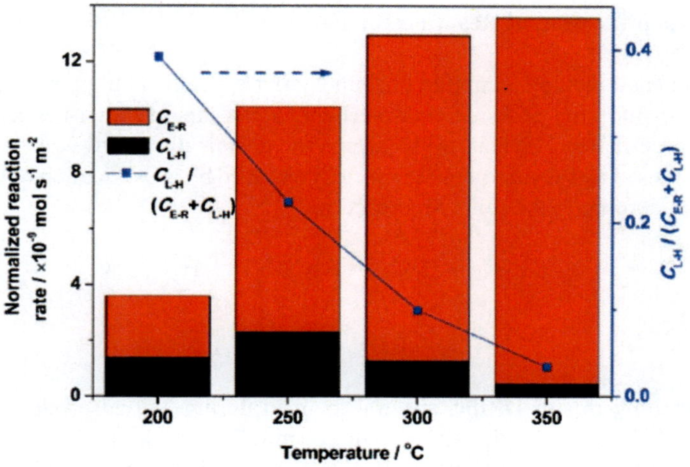

Fig. 2.9 Contribution of the E-R and L-H mechanisms to SCR over a SCR catalyst. Reaction conditions: $[NH_3]$=[NO] = 500 ppm, $[O_2]$ = 5%, total flow rate = 200 mL min^{-1}, catalyst mass = 0.02 g, gas hourly space velocity (GHSV) = 600,000 mL g^{-1} h^{-1}

monodentate nitrate may be involved in the reaction; this has also been noted by prior researchers [14]. The nitrate route [15, 16] has been found to drastically enhance the low-temperature catalytic activity. Therefore, after ammonia attacks the monodentate nitrate species, the newly formed ammonium nitrate can participate in the reaction owing to the low stability of monodentate nitrate. The reaction pathway can thus be described as follows:

$$NH_3(g) + Nb^{5+}_{surf}{-}OH \rightarrow Nb^{5+}_{surf}{-}O \cdots H{-}NH_3, \tag{2.13}$$

$$NO(g) + Ce^{4+}_{surf} = O + \frac{1}{2}O_2 \rightarrow Ce^{3+}_{surf}{-}O{-}NO_2, \tag{2.14}$$

$$Ce^{3+}_{surf}{-}O{-}NO_2 + Nb^{5+}_{surf}{-}O \cdots H\text{-}NH_3 \rightarrow Nb^{5+}_{surf}{-}O \cdots NH_4NO_3{-}Ce^{3+}_{surf}, \tag{2.15}$$

$$Nb^{5+}_{surf}{-}O \cdots NH_4NO_3{-}Ce^{3+}_{surf} + 2NO + 2NH_3 \rightarrow 3N_2{+}4H_2O + Nb^{5+}_{surf}{-}OH + Ce^{3+}_{surf}{-}OH, \tag{2.16}$$

$$Ce^{3+}_{surf}{-}OH + \frac{1}{4}O_2{+} \rightarrow Ce^{4+}_{surf} = O + \frac{1}{2}H_2O. \tag{2.17}$$

2.2.2.3 Identification of Reaction Intermediates

An experimental method combining in situ DRIFTS, chemical trapping, and TRMs was used to study the SCR reaction mechanism and identify intermediate species over a VO_x/ceria (CeO_2) catalyst [17]. Here, the chemical trapping agents BaO and Al_2O_3 were used to trap intermediate species generated by NO oxidation (e.g., HNO_2 or N_2O_3) according to the following reactions:

$$2HNO_2 + BaO \leftrightarrow Ba(NO_2)_2 + H_2O, \tag{2.18}$$

$$N_2O_3 + BaO \leftrightarrow Ba(NO_2)_2. \tag{2.19}$$

The resulting in situ DRIFT spectra obtained during NO adsorption at 473 K on a VO_x/CeO_2 catalyst without and with chemical trapping agents are shown in Figs. 2.10 and 2.11, respectively. Without the chemical trapping agents, prominent bands are observed at 1157 cm^{-1} and 1601–1612 cm^{-1}. The former band can be assigned to nitrites. However, the assignment of the latter band is not clear: both bridging nitrates and adsorbed NO_2 have been proposed. When VO_x/CeO_2 was mixed with chemical trapping agents BaO/Al_2O_3, the spectra differed substantially. Nitrites (1160–1221 cm^{-1}), particularly the newly formed broad mode at 1221 cm^{-1}, are the dominant species, suggesting that a mobile nitrite precursor may be the product of oxidative NO activation.

Surface reaction experiments using the TRM were then conducted to clarify the SCR reaction mechanism over VO_x/CeO_2. Briefly, NH_3 was first fully adsorbed on VO_x/CeO_2, H-SSZ-13, and a combination of VO_x/CeO_2 and H-SSZ-13, before a

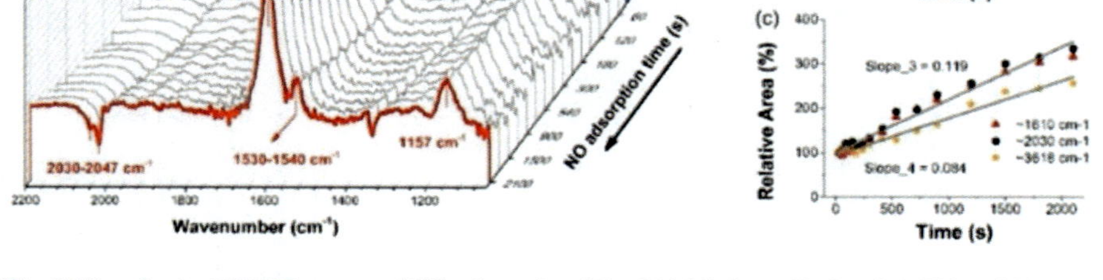

Fig. 2.10 **a** In situ DRIFT spectra of NO adsorption ($T = 200$ °C, $O_2 = H_2O = 0\%$, NO = 350 ppm) on VO_x/CeO_2, and temporal evolution of normalized area of **b** NO_x ad-species and **c** surface V sites. Inset: temperature-programmed desorption (TPD) spectra of the 1530 and 1610 cm^{-1} bands at 350 and 400 °C. Linear regression (gray lines) and fitted slopes are also shown

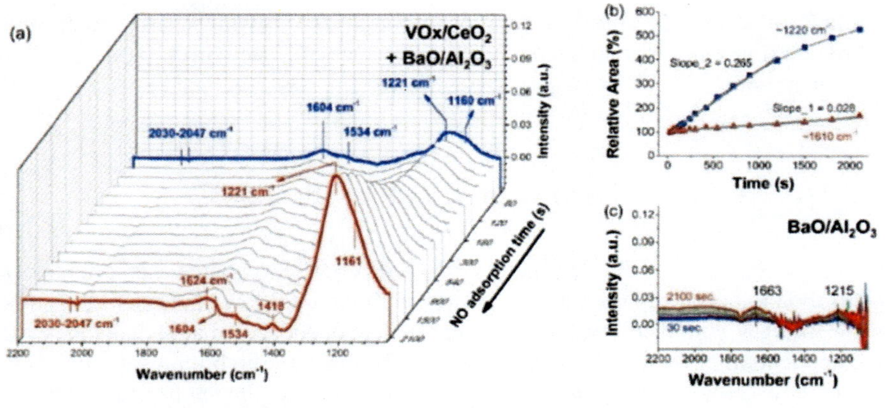

Fig. 2.11 **a** In situ DRIFT spectra of NO adsorption on the mechanical mixture of VO_x/CeO_2 and BaO/Al_2O_3, **b** temporal evolution of the normalized area of the prominent NO_x adsorption species, and **c** in situ DRIFT spectra of NO adsorption on BaO/Al_2O_3 alone. Linear regression (gray lines) and fitted slopes are also shown

NO pulse was introduced. Here, VO_x/CeO_2 effectively oxidized NO but could not adsorb NH_3, as shown in Fig. 2.12, whereas H-SSZ-13 was able to adsorb NH_3 but could not oxidize NO. Thus, neither VO_x/CeO_2 nor H-SSZ-13 could generate N_2 in a surface reaction. However, N_2 was generated on a VO_x/CeO_2 + H-SSZ-13 complex, suggesting that SCR over VO_x/CeO_2 proceeds in an NO oxidation route that may

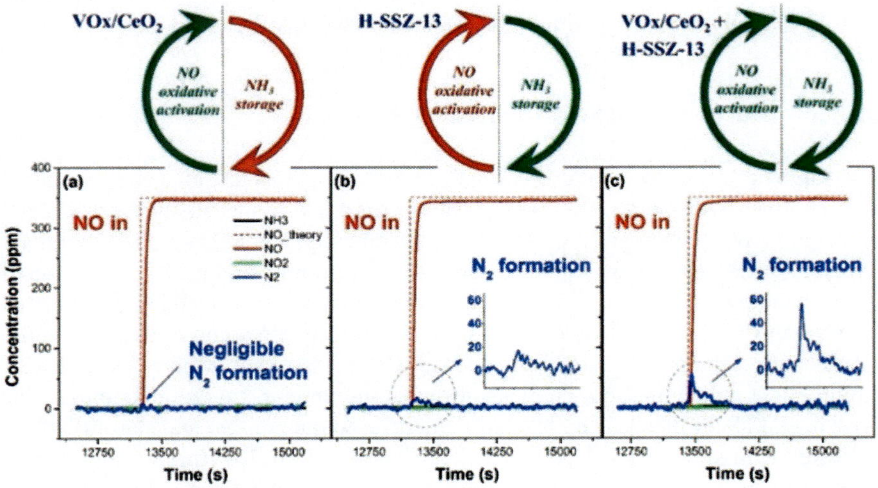

Fig. 2.12 Schematization of surface acidity, redox properties, and transient response method (TRM) results: NH_3 adsorption (not shown) + transient NO pulse over pretreated **a** VO_x/CeO_2 (0.3 g), **b** H-SSZ-13 (0.03 g), and **c** mechanical mixture of VO_x/CeO_2 + H-SSZ-13 (0.3 g + 0.03 g). T = 473 K, $O_2 = H_2O = 0\%$, NO = 350 ppm, flow rate = 3.33 cm^3s^{-1} NTP

dominate at low temperatures. In addition, mixing catalysts with separate acidic and redox sites may be a promising strategy to enhance SCR activities, particularly at low temperatures.

2.3 NO_x Control Enhancement and Applications

Although SCR using the commercial V_2O_5-WO_3/TiO_2 catalyst can effectively control NO_x emissions, such that they are within the ultra-low emissions standards from flue gas within the temperature range of 300–400 °C, many challenges for the application of SCR technology still exist. To maintain high removal efficiency, ensure long-term stability, and widen the application field, the temperature window of the catalyst should be widened and the catalyst's resistance to poisoning and SO_2 must be improved. Moreover, industrial preparation and regeneration techniques and system-wide optimization have also drawn significant research attention.

2.3.1 Catalyst Poisoning and Resistance

Alkali/alkaline earth and heavy metals (particularly As and Pb) present in coal are released into the flue gas during combustion and deposited on the catalyst when passing through the SCR system, leading to catalyst deactivation. Researchers have thus aimed to clarify the mechanism behind catalyst deactivation by poisoning during SCR and develop catalysts with high poisoning resistance.

2.3.1.1 Deactivation of SCR Catalyst by Poisoning

As a large portion of the coal used in coal-fired power plants in China contains alkali and alkaline earth metals, the mechanisms behind alkali and alkaline earth metal poisoning have been aggressively studied. Specifically, K and Na are of significant interest as they have shown the most substantial poisoning effects. A comparison of the poisoning of alkali (Na and K) with alkaline earth metal (Mg and Ca) conducted over a nanoV_2O_5-WO_3/TiO_2 catalyst [18] indicated that Na and K exhibit a more significant poisoning effect than Mg and Ca whereas Mg shows the weakest deactivation effect among the four elements tested at the same metal oxide loadings. The presence of alkali metals (i.e., Na and K) decreased the amount and stability of the Brønsted acid sites of the V_2O_5-WO_3/TiO_2 catalyst [19, 20]; further, the amount of chemisorbed oxygen was reduced by their presence (chemisorbed oxygen in the presence of Mg > Ca > Na > K) [18]. The Na- and K-doped catalysts have also revealed weakened reducibility of vanadium species when compared with the Ca- and Mg-doped catalysts [21]. Furthermore, Na and K ions could affect the reduction degree of tungsten species, whereas no obvious change was observed for the tungsten species over the Ca- and Mg-doped V_2O_5-WO_3/TiO_2 catalysts.

By experimentally and theoretically studying the reactions between the alkali metal oxides and catalysts and the afterward influence on the catalysts [18], both NH_3 adsorption and alkali metal poisoning occur on the acid sites. The V–OH Brønsted acid produced by hydration is highly active for NH_3 adsorption and activation on the vanadia (V_2O_5)-based catalyst. According to frontier molecular orbital theory, the neutralization of the V_2O_5 Brønsted acid by sodium and potassium oxides is more exothermic than NH_3 adsorption is (see Fig. 2.13). Thus, the active V–OH Brønsted acid sites are consumed by the doped alkali oxides, as verified by the loss of acidity in NH_3 sorption experiments and DFT calculations (Fig. 2.14). In addition, the Mulliken charges of the K/Na ions in V–O–K/Na are higher than those of the H ion in V–OH, which causes the reduction of the adjacent vanadium atom. This reduction makes the hydrogenation of the catalyst more difficult, weakening the reducibility of the catalyst by doping of alkali ions. Performing H_2-TPR confirmed the drop of reducibility of the catalyst. Moreover, the results of the NO reduction experiment

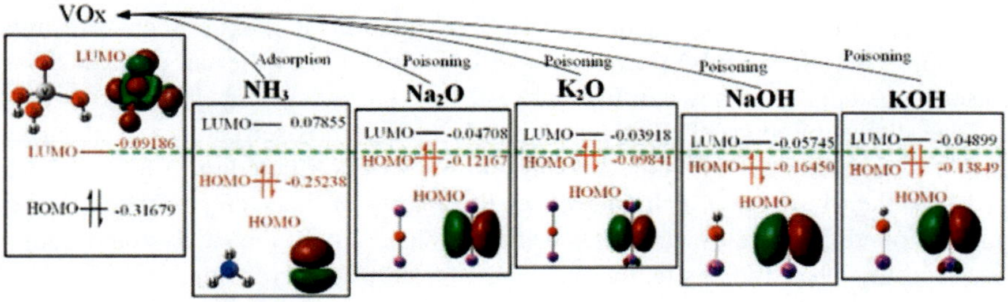

Fig. 2.13 Frontier molecular orbitals of VO_x, NH_3, Na_2O, K_2O, NaOH, and KOH [18]

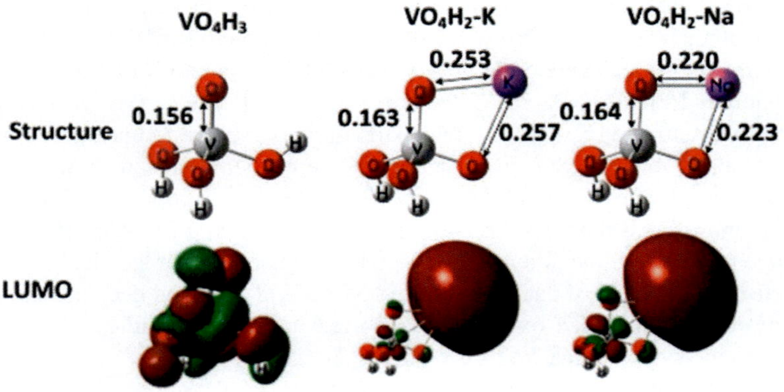

Fig. 2.14 Structures and LUMO snapshots of the fresh and alkali-doped catalyst [18]

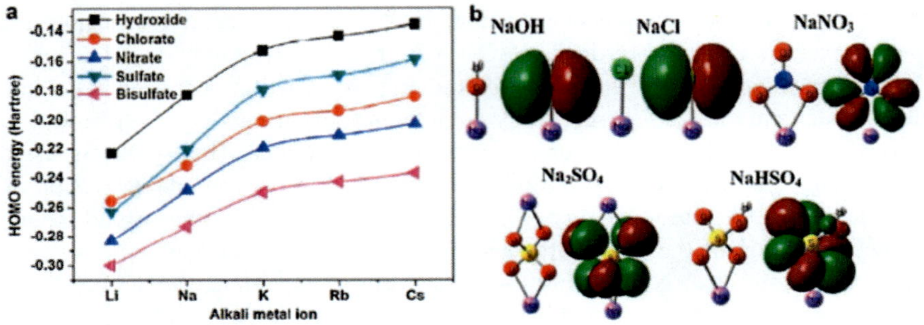

Fig. 2.15 **a** HOMO energies of various alkali species and **b** HOMO snapshots of Na-containing molecules [19]

demonstrated that the loss of Brønsted acid sites and the drop of reducibility caused by alkali oxide reduces the SCR reactivity.

The presence of other alkali metals (e.g., Rb, Cs, and Li) also causes the deactivation of SCR catalysts. The poisoning strength of alkali cations, Cs > Rb > K > Na > Li, is in accordance with theoretical DFT calculations and experimental results, confirming that the HOMO energy of alkali species follows the same order as the poisoning strength, as shown in Fig. 2.15a [19].

Overall, the influence of alkali metals on the vanadium-based catalyst used during SCR has been extensively studied. Many authors have attributed the effect of alkali metals to their basicity, which neutralizes the acidity of the catalyst surface. Other authors found that alkali metals both decrease the acidity of the catalyst and inhibit the reactivity of the redox site, thereby causing catalyst deactivation.

Compared with commercial V_2O_5-based catalysts, transition-metal-based MnO_x catalysts show superior low-temperature activity [20] but still suffer from alkali metal poisoning owing to the lowered reducibility of the MnO_x species, decreased specific surface areas, and damaged acid sites. For example, a Mn/TiO_2 catalyst prepared via a sol–gel method was deactivated by K and Na [21]. Although Mn/TiO_2 catalysts have demonstrated a NO conversion of 90% at 140–200 °C, Na and K doping caused the conversion to decrease to 78 and 27%, respectively [22]. Mn–Ce/TiO_2 catalysts can be strongly deactivated by Na_2SO_4, which researchers used to simulate the combined effects of alkali ions and SO_2 in flue gas [23]. Here, a simultaneous pore occlusion and sulfation effect were found. Further, NH_3-temperature-programmed desorption (TPD) demonstrated that NH_3 adsorption decreased when a Mn–Ce/ZrO_2 catalyst was doped with alkali metal ions (K, Na, and C), indicating that the alkali metals on the surface of the catalyst can destroy the surface acid sites and decrease the redox property and chemisorbed oxygen. As reported in previous studies, K was more harmful to the catalyst than Na or Ca [24, 25].

Other common poisoning elements released during coal combustion or municipal solid waste incineration include As and Pb.

As arsenic can significantly damage commercial catalysts, researchers have performed theoretical calculations using the Gibbs free energy correction to understand As transformations [26]. According to DFT, the As species present in the flue gas varies by temperature: trivalent As molecules are thermodynamically stable at high temperatures, whereas pentavalent species are stable at low temperatures. As summarized in Fig. 2.16, different colors represent different levels of stability, decreasing in the order of black, red, blue, green triangles. Thus, dehydrated compounds are the primary species at high temperatures; these compounds are hydrated and oxidized by O_2 as the temperature declines. At 25 °C, H_3AsO_4 is the most thermodynamically stable species.

Fresh and deactivated catalyst samples were collected from a power plant in southwest China and tested in a quartz reactor to further understand As's poisoning effect [27]; results are summarized in Fig. 2.17, where the black, red, and blue points indicate the fresh, windward (close to the inlet), and leeward (close to the outlet) catalyst samples, respectively. As does not remarkably change textural properties such as surface area and TiO_2 phase but dramatically decreases both the Lewis and Brønsted acid sites, as is clear from the sharp decline in catalyst activity. Furthermore, V–OH sites are replaced with less-reactive As-OH sites, which causes catalyst deactivation.

DFT calculations conducted to clarify the influence of Pb on V_2O_5-based catalysts have also demonstrated that Pb doping causes a significant change in the surface electronic property. Each Pb atom can influence two active sites, as shown in Fig. 2.18, thereby weakening the V=O double bond and increasing the LUMO energy [28]. The calculated Mulliken charge and spectra obtained from X-ray photoelectron spectroscopy (XPS) clarified that the Pb in the Pb-doped catalyst is more positively charged than it is in PbO. This electronic interaction between Pb and the surface of the

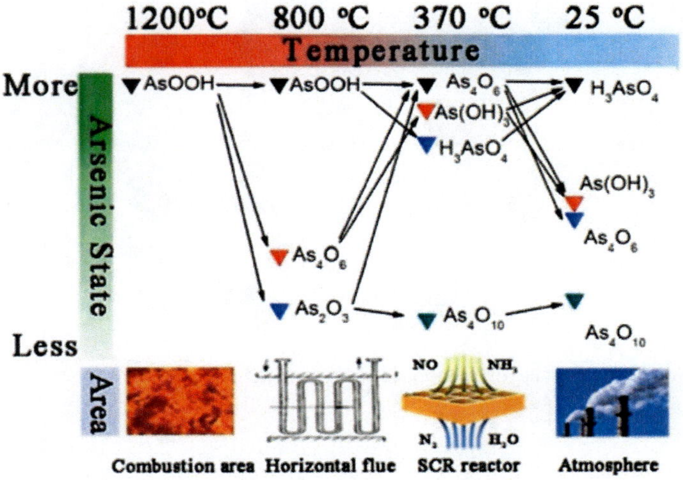

Fig. 2.16 Thermodynamically stable arsenic (As) species and their transformations in different areas of the gas flue [26]

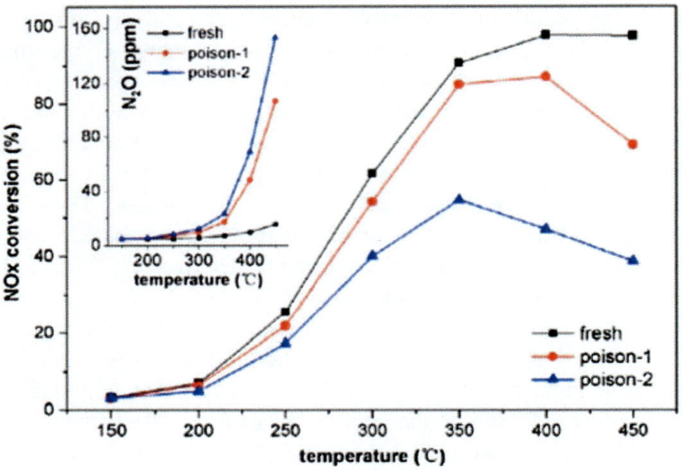

Fig. 2.17 Experimentally obtained NO_x conversion of fresh and poisoned catalysts during SCR [27]

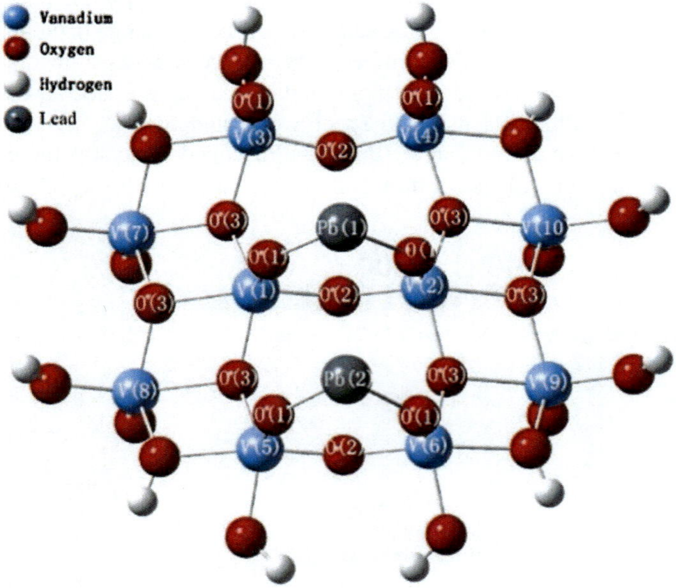

Fig. 2.18 Stationary structure of a Pb-doped catalyst model [28]

catalyst causes surface reducibility. Redshifts in the V=O stretching frequency were also observed during Pb doping. The theoretical difference in electrostatic potential between the undoped and doped models shows a decreased Brønsted acid formability caused by the Pb doping. Experimental NH_3-TPD and H_2-TPR have demonstrated a decrease in acid sites and surface reducibility, which is in accordance with the theoretical results.

Experimental investigations to clarify the catalyst activity and characterize fresh and $PbCl_2$-poisoned V_2O_5-based catalysts were also performed using BET surface area and BJH pore size distribution measurements, XRD, SEM, XPS, NH_3-TPD, NH_3-chemisorption, and FTIR spectroscopy techniques [29]. As a result, $PbCl_2$ caused severe deactivation of the catalysts used for the SCR of NO with NH_3. The BET surface area and total pore volume decreased with increasing $PbCl_2$ loadings, although no obvious change in the crystal structure of the different samples was visible. The presence of $PbCl_2$ reduced some V^{5+} to V^{4+} on the catalyst's surface, decreasing the reducibility of the V_2O_5 species. Similarly, XPS, NH_3-TPD, chemisorption, and FTIR results demonstrated a substantial decrease in surface acidity, particularly at Brønsted acid sites, leading to a decrease in adsorbed NH_3 and reduced catalytic activity.

2.3.1.2 Poison-Resistant Catalysts

There have been extensive studies regarding the deactivation caused by alkali metals on commercial V_2O_5-WO_3/TiO_2 catalysts, Mn-containing oxide catalysts, and iron or copper-exchange zeolite catalysts that have indicated that the reduced deNO_x efficiency is attributable to fewer acid sites, fewer active species in the exchanged site, and a decreased reducibility of the catalysts. As poisoning caused by alkali metals shortens a catalyst's service life, improving the resistance of SCR catalysts to alkali metals is of great practical value.

As such, researchers have aimed to improve the alkali resistance of the most commonly used catalysts for SCR, i.e., TiO_2-supported V_2O_5 promoted by WO_3 or MoO_3. In one such experiment, V_2O_5/TiO_2 catalysts with different V_2O_5 loadings were prepared to understand how the loading of active sites affects the catalyst's alkali resistance [30]. By increasing the V_2O_5 loading from 1 to 10%, the NO removal efficiency of the catalysts increased by 70% at 200 °C, as shown in Fig. 2.19a. However, when poisoned by K, as shown in Fig. 2.19b, the activity of catalysts with high and low V_2O_5 loadings alike dropped at all temperatures studied. The V_4/Ti catalyst ($V_2O_5 = 4$ wt%, represented by the green line) displayed the best performance, suggesting that optimizing the V_2O_5 content is key to increasing the alkali resistance of V_2O_5/TiO_2 catalysts.

Some have aimed to optimize the catalyst's alkali resistance by adding other components; the addition of some transition metals (e.g., Ce, Sb, and Nb) to catalysts was demonstrated to enhance their resistance to alkali metal poisoning during SCR reaction [31, 32]. Others have loaded Cu and Fe sulfates onto the Ce–Ti mixed

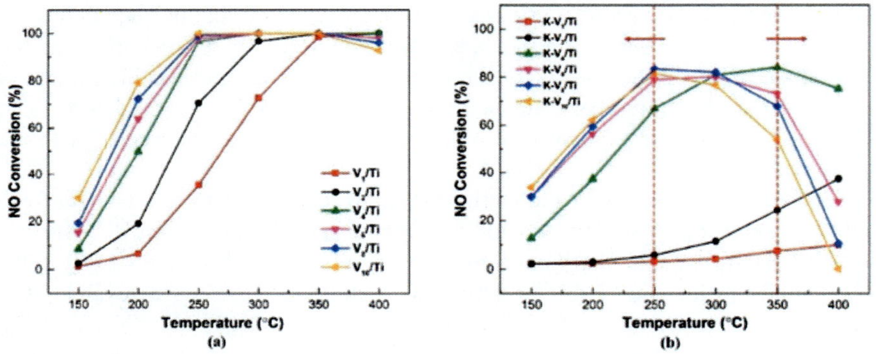

Fig. 2.19 SCR over fresh and K-doped V_x/Ti catalysts [30]

oxide [33]. Here, doping with 1.2 wt.% K_2O significantly decreased the NO reduction, whereas the sulfate-loaded Ce–Ti oxide maintained a high NO conversion rate. Valence analysis revealed that the metal sulfate species on the Cu/Ce-Ti and Fe/CeTi surface could react with K^+ to produce K_2SO_4 and a metal oxide, as summarized in Fig. 2.20, thereby allowing the V_2O_5 to maintain reducibility when interacting with K_2O.

$Ce(SO_4)_2$ may also increase the alkali resistance [34] because a V-0.5$Ce(SO_4)_2$/Ti catalyst synthesized by impregnation demonstrated an excellent low-temperature activity and alkali resistance during SCR reaction when compared with commercial V_2O_5-WO_3/TiO_2 catalysts, as shown in Fig. 2.21. This synthesized catalyst also showed excellent durability to SO_2 and H_2O, indicating its prospects for industrial application. Characterization results suggested that after sodium poisoning,

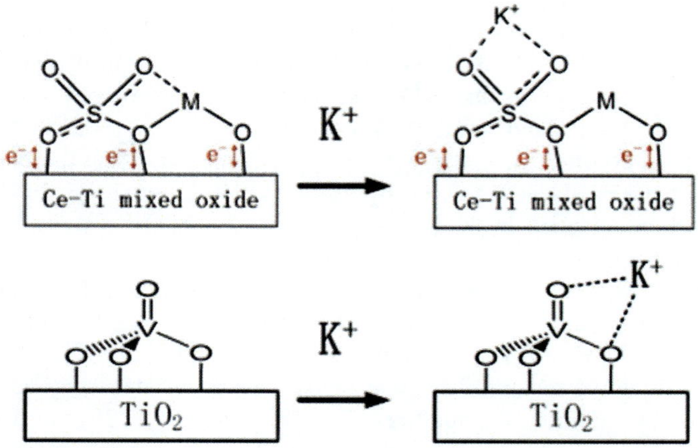

Fig. 2.20 Interactions between K^+ and the oxygen atoms present in V_2O_5 and metal sulfate [33]

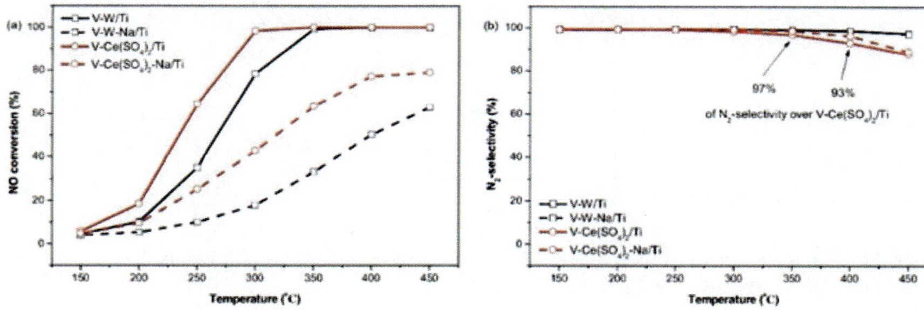

Fig. 2.21 **a** Activity comparison of fresh and Na-poisoned catalysts. **b** N_2 selectivity of relevant samples [34]

the improvement of redox properties and surface acidity improved by $Ce(SO_4)_2$ component didn't disappear. Further, more V–OH sites were preserved owing to the interaction between Na and O=S=O that protects the active sites.

The catalyst support can also be modified to improve the resistance to alkali metal poisoning. For example, a high amount of V_2O_5 can be dispersed on a ZrO_2 and sulfated-ZrO_2 support [35]. Here, a ZrO_2 support was prepared by calcination of hydrous zirconia at 500 °C for 4 h in air. Sulfation was performed by impregnating ZrO_2 with a 1-M solution of H_2SO_4 at room temperature (acid volume/powder weight ratio was 30 ml/g). These supports were loaded with vanadium oxalate ($V_2(C_2O_4)_3$) by incipient wetness impregnation. The formed vanadium oxide species were located preferentially on the strong Lewis acid sites of the support. Results indicated that V_2O_5/sulfated-ZrO_2 shows a higher resistance toward alkali poisoning than the non-sulfated systems; the presence of SO_2 in the reaction mixture was found to enhance the stability and activity of the V_2O_5/sulfated-ZrO_2, likely owing to the regeneration of surface sulfated groups. However, the introduction of potassium oxide (K/V ratio < 0.5) drastically decreased the acidity of the samples and weakened the remaining acid sites, particularly in the non-sulfated systems.

Efforts have also been made to improve resistance to arsenic and lead poisoning. The aforementioned metal sulfates can also deal with arsenic poisoning [27] because SO_4^{2-} can provide remarkable reactive surface acidities, thereby supplying Lewis and Brønsted acid sites destroyed by arsenic. NH_3 can adsorb on the S^{6+}, which are newly formed acid sites, as the blue "ads" shown in Fig. 2.22. Including SO_4^{2-} can thus increase the resistance of commercial SCR catalysts to arsenic. A novel CeO_2-WO_3-Al_2O_3 catalyst has also been reported to provide excellent arsenic resistance; at the same arsenic loading (2.1 wt%), this catalyst demonstrated an activity loss of 10%, compared with an activity loss of nearly 40% when using a CeO_2-WO_3 catalyst [36]. Overall, three aspects are required to obtain higher arsenic resistance, including the i) protection of Lewis acid sites, ii) release of oxygen vacancies from arsenic restriction, and iii) confinement of As^{5+} oxidizing capacity. Addressing both As and Pb poisoning, an Fe–Ti-MMT catalyst synthesized via an interlayer pillaring

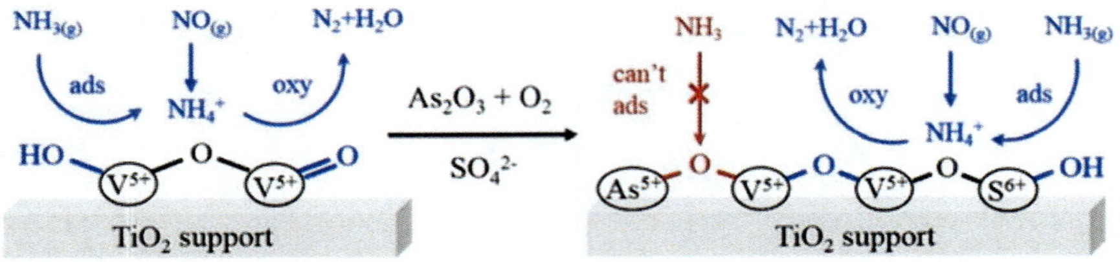

Fig. 2.22 The mechanism of increasing the resistance of commercial SCR catalysts to arsenic by including SO_4^{2-} (ads: adsorption, oxy: oxidized) [27]

method exhibited superior activity during SCR and boosted resistance to alkali and heavy metal within the range of 200−450 °C [37].

2.3.2 *Widening the Working Temperature Window*

Flue gas temperatures in industrial processes are often in the range of 100–200 °C; heating these flue gases to > 300 °C for SCR using a V_2O_5-WO_3/TiO_2-based catalyst wastes a lot of energy. Furthermore, many industries and applications of SCR experience large fluctuations in flue gas temperatures. As a result, NO_x emissions can often exceed those observed under optimal conditions and thus ultra-low emission standards. Researchers have thus aimed to develop wide-temperature window catalysts for deNO_x. In addition, SO_2 in the flue gas reacts with NH_3 to produce NH_4HSO_4, which adversely affects the SCR reaction and is considered the main barrier to the commercialization of low-temperature SCR systems.

2.3.2.1 Developing an SCR Catalyst with a Wide Temperature Window

The primary commercial catalyst used for SCR is V_2O_5-$WO_3(MoO_3)/TiO_2$, which is available in honeycomb, plate, and corrugated forms; it has been widely used for flue deNO_x in power plants and has an active temperature window of 300–400 °C [38]. However, a wider temperature range is required for catalysts used for SCR to meet the increasingly stringent environmental standards and the broad application of SCR technology to many industrial processes.

Metal doping of the traditional V_2O_5-WO_3/TiO_2 catalyst may help fill this requirement. CeO_2, for example, traditionally used for its oxygen storage and redox properties in the three-way catalyst (TWC) used for gasoline engine emission control, is known to store and release oxygen via the redox shift between Ce^{4+} and Ce^{3+} under oxidizing and reducing conditions [39]. Ce doping was thus applied to enhance the catalytic activity of the V_2O_5-WO_3/TiO_2 catalyst during NH_3-SCR; a catalyst with 10 wt.% Ce showed an increasing NO_x conversion from 150 to 250 °C compared to V_2O_5-WO_3/TiO_2 [40].

Using CeO_2 as the active component, a CeO_2-WO_3 catalyst demonstrated excellent catalytic performance during NH_3-SCR of NO_x in the broad temperature range of 175–450 °C, with a conversion rate of nearly 100% between 200 and 450°C [41]. Furthermore, a CeO_2/TiO_2 catalyst prepared using a single-step sol–gel method demonstrated excellent SCR activity under varying NO concentrations between 250 and 450 °C, as shown in Fig. 2.23 [42], thereby expanding the temperature window. High surface area and good redox ability are essential for the catalytic activity. The intense interaction between Ce and Ti as well as the high concentration of amorphous or highly dispersed nanocrystalline ceria thus likely caused this excellent performance. As the Ce loading (i.e., the mass ratio of CeO_2/TiO_2) was increased from 0 to 0.6, the NO conversion increased markedly but then decreased at higher loadings; the most active catalyst was obtained with a Ce loading of 0.6 [43]. Overall, the increased activity may be attributable to a high Ce loading, the strong interaction between Ce and Ti, the high concentration of amorphous Ce on the catalyst's surface, or the increase of chemisorbed oxygen or/and weakly bonded oxygen species, resulting from the presence of Ce^{3+}. Furthermore, when compared with the Ce/Ti oxide catalyst prepared by the coprecipitation method, the catalyst doped with Cu demonstrated a higher activity at temperatures lower than 350 °C and was more resistant to SO_2, as shown in Fig. 2.24 [44].

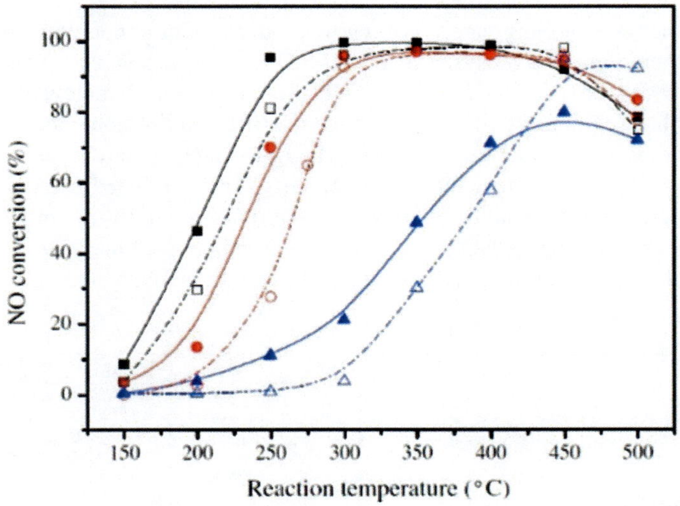

Fig. 2.23 NO conversion as a function of reaction temperature for different catalysts (filled square, open square) Ce/Ti (sol–gel), (filled circle, open circle) Ce/Ti (impregnation), (filled triangle, open triangle) Ce/Ti (coprecipitation). Reaction conditions: [O] = 3%, balance N_2, total flow rate = 500 ml/min, GHSV = 50,000 h^{-1}, (filled square, filled circle, filled triangle) [NO] = [NH_3] = 500 ppm, (open square, open circle, open triangle) [NO] = [NH_3] = 1000 ppm [42]

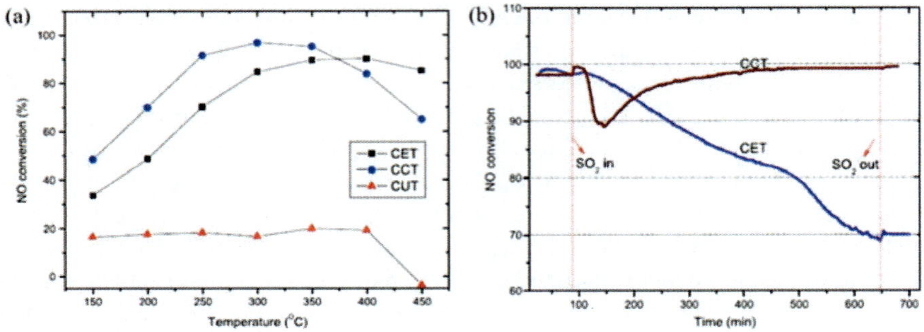

Fig. 2.24 **a** NO conversion as function of temperature; **b** Influence of SO_2 on NO conversion as function of time for CeTi and CuCeTi at 350 °C. Reaction conditions: [NO] = [NH_3] = 1000 ppm, [O_2] = 5%, and [SO_2 in] = 1000 ppm [44]

To investigate the effect of Sb-doping, a commercial honeycomb catalyst doped with Sb underwent long-term SCR under low-load operating conditions [45]. Characterizations of the original and modified catalysts indicated that the activity of the Sb-doped catalyst was improved in the temperature range of 275–325 °C, reaching a removal efficiency of 64.5% during the 30 h stability test at 275 °C, as shown in Fig. 2.25.

Researchers developing catalysts for low- and medium-temperature SCR have recently focused on catalyst systems using MnO_2 or CeO_2 as an active component; MnO_x–CeO_2 mixed oxides and MnO_x/TiO_2 catalysts have excellent low-temperature activity but have demonstrated limited applications in the high-temperature range [46, 47]. Zhang et al. [48] proposed a combined MnO_x–CeO_2/TiO_2 and V_2O_5–WO_3/TiO_2 catalyst for a highly efficient NH_3–SCR process. The combined catalyst exhibits a wider operating temperature window (150–400 °C) than either single MNO_x–CeO_2/TiO_2 or V_2O_5–WO_3/TiO_2 catalysts. A series of Ce–Nb binary oxide catalysts synthesized via coprecipitation showed good catalytic activity and selectivity, as

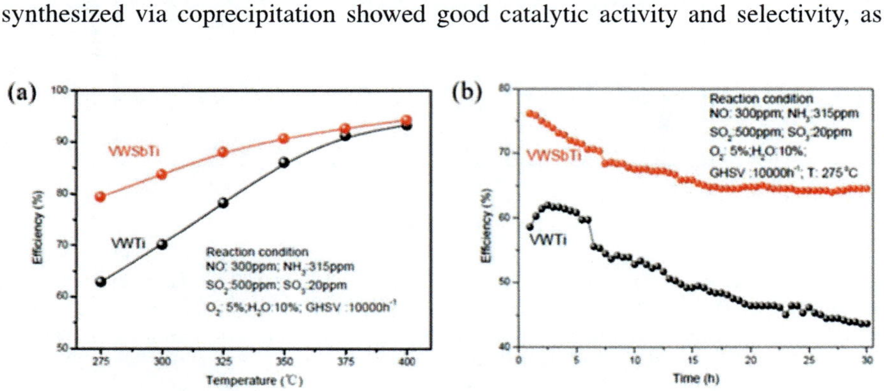

Fig. 2.25 **a** DeNO_x performance at different temperatures and **b** stability tests of commercial and Sb-doped catalysts [45]

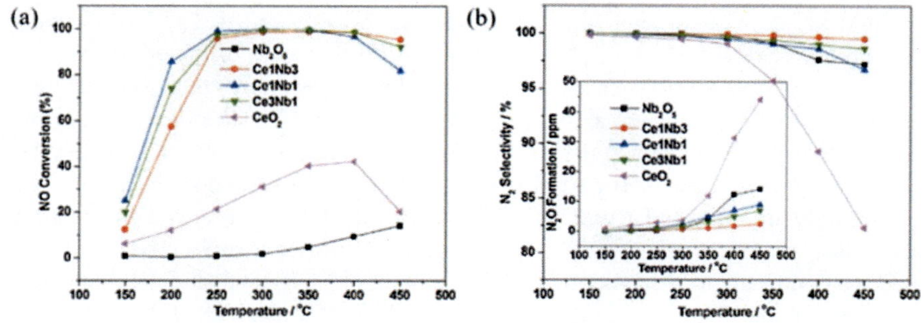

Fig. 2.26 **a** SCR performances and **b** N_2 selectivity and N_2O formation over Ce_xNb_y catalysts. Reaction condition: $[NH_3]=[NO] = 500$ ppm, $[O_2] = 5\%$, total flow rate = 200 mL min^{-1}, catalyst mass = 0.1 g, and GHSV = 120,000 mL g^{-1} h^{-1} [49]

shown in Fig. 2.26 [49]. Here, NO conversion of 80% was achieved using the CeNb catalyst across a wide temperature range (200–450 °C) at a flue gas hourly space velocity (GHSV) of 120,000 mL g^{-1} h^{-1}. The surface acid sites were mainly provided by Nb oxide species; the Nb–OH and Nb=O bonds were related to the Brønsted and Lewis acid sites, respectively. The abundance of surface adsorbed oxygen, the key factors for the catalytic behaviors, might arise from the short-range activation effect of Nb oxide species to Ce oxide species. In addition, water vapor only slightly and reversibly inhibited the catalytic performance of the CeNb catalyst, making it a promising catalyst for practical SCR applications.

It has been widely accepted that the surface acidity and redox properties of catalysts determine their overall SCR performance [50–53]. To demonstrate the NH_3 desorption properties and examine the surface acidity of V-W/Ti and Mn/Ce, NH_3 adsorption and TPD experiments were conducted; their results are shown in Fig. 2.27. In the right panel, a typical NH_3 adsorption and TPD protocol is shown;

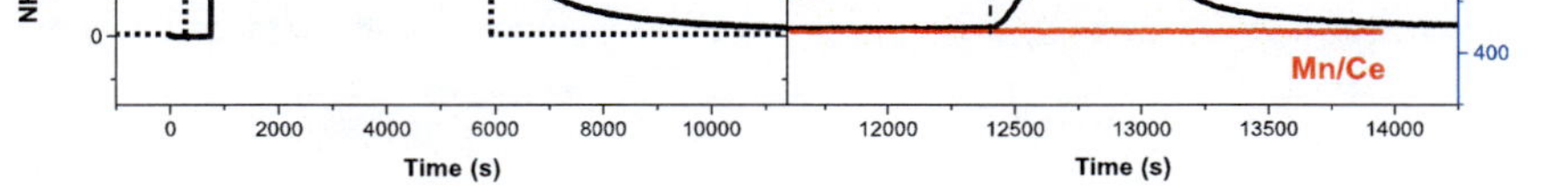

Fig. 2.27 NH_3 adsorption and TPD experiments over V-W/Ti and Mn/Ce. Left: isothermal NH_3 adsorption at 473 K and subsequent He purge over V-W/Ti and right: temperature-programmed NH_3 desorption over V-W/Ti and Mn/Ce [54]

during desorption, a completely different scenario was observed. The V-W/Ti catalyst exhibits typical NH_3 desorption as expected for this class of materials; Mn/Ce, however, shows negligible desorption during the temperature ramp, suggesting a rather weak NH_3 storage capacity on the surface of Mn/Ce. These features further indicate that the LT-SCR activity of Mn/Ce is restricted by a lack of surface NH_3 adsorbates [54] because NH_3 participates in SCR via an adsorbed state [38, 55, 56]. Accordingly, mixing Mn/Ce with V-W/Ti, because the latter saturates with adsorbed NH_3 at low temperatures [38, 55, 57], may provide sufficient surface NH_3 for SCR and thus contribute to the synergistic enhancement of SCR catalyst development.

To exploit the NO_x storage capacity of BaO/Al_2O_3, it was mixed with V-W/Ti and Mn/Ce catalysts to track the evolution of surface NO_x species during isothermal NO adsorption using time-resolved in situ FTIR spectroscopy [58]. The results suggest an oxidative activation of NO to HONO on the catalytic component, as detailed in Eq. (2.20), where M represents metal, which then moves to adjacent BaO/Al_2O_3 species and gets stabilized to $Ba(NO_3)_2$.

Further, intensities of nitrites on Mn/Ce (adsorbed HONO: 1198–1324 cm^{-1} [17]) are relatively higher than those on VW/Ti (HONO formation: about 1219 cm^{-1} [17]). In the presence of NH_3, nitrites/nitrite-precursors can easily react with adsorbed NH_3 to generate NH_4NO_2 (via Eq. (2.21), expressed in terms of HONO), an unstable species that readily decomposes to N_2 and H_2O [via Eq. (2.22)] [52, 59–61], which thus rationalizes the higher NO conversions on Mn/Ce in LT-SCR.

$$NO + M^{\delta+} - OH \leftrightarrow HONO + M^{(\delta-1)+} \tag{2.20}$$

$$HONO + NH_3{*} \leftrightarrow NH_4NO_2 \tag{2.21}$$

$$NH_4NO_2 \rightarrow N_2 + 2H_2O \tag{2.22}$$

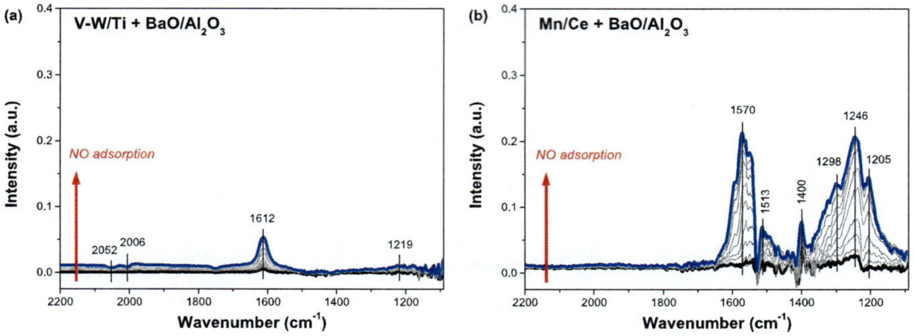

Fig. 2.28 Time-resolved in situ FTIR spectra of NO isothermal adsorption (T = 473 K) for 900 s on the mechanical mixture of **a** V-W/Ti and BaO/Al_2O_3 and **b** Mn/Ce and BaO/Al_2O_3. Spectra of NO adsorption for 30 s and 900 s are displayed in black and blue, respectively [58]

To explore whether this enhanced NO oxidative activation and nitrites/nitrite-precursors formation on Mn/Ce contribute to the observed synergistic enhancement, the mechanism behind LT-SCR was explored via a combination of chemical-trapping methods using BaO/Al_2O_3 as a NO_x-intermediate trap, TPD, and in situ FTIR spectroscopy over a series of metal-oxide catalysts. The results show a linear correlation between LT-SCR rates and nitrite generation from NO oxidative activation, as shown in Fig. 2.29, which is then stored on BaO/Al_2O_3, thereby implying the key role of nitrites and nitrite-precursors in LT-SCR.

Thus, as the acid and redox properties vitally affect the performance of catalysts used during SCR, understanding such effects provides insights into the correlations between the physicochemical properties and catalyst performance to provide a theoretical basis for designing new catalysts.

According to the SCR reaction mechanisms, acid properties are crucial under high temperatures. The first step of SCR reaction is the adsorption of NH_3 on the acid sites on the surface of the catalyst, which is important for SCR reaction. The relationship between the physicochemical properties (i.e., acid and redox properties) and performance of $H_4SiW_{12}O_{40}$-modified Fe_2O_3 catalysts was established [62]. Here, in situ DRIFTS experiments of NH_3 adsorption were conducted at 100 °C to investigate the detailed acid properties of the series catalysts. When a pure Fe support was used, only Lewis acid sites were available for NH_3 adsorption. Conversely, Brønsted acid sites begin to develop as the HSiW content was increased, as shown in Fig. 2.30. Calculating the ratio of the areas of the peaks indexed to NH_4^+ on the Brønsted acid sites and NH_3 on the Lewis acid sites demonstrated variations

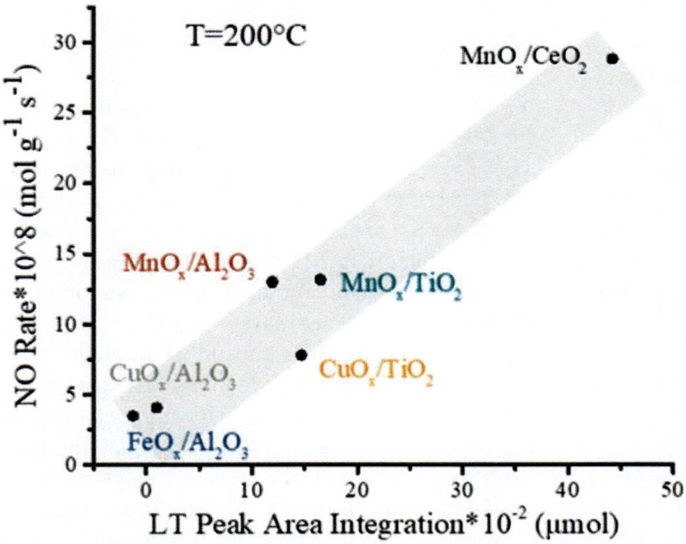

Fig. 2.29 Correlation between the LT-SCR rates (in terms of first-order NO rates) and the area of LT peaks deconvoluted from the TPD profiles

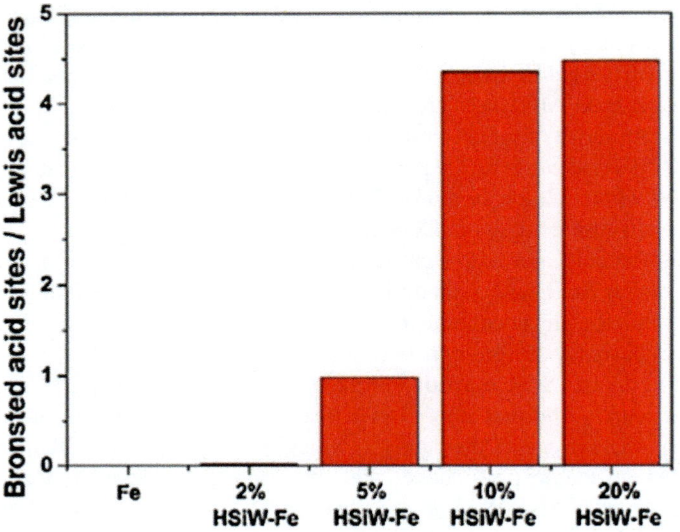

Fig. 2.30 Ratio of the number of Brønsted acid sites to Lewis acid sites over the series catalysts [62]

in the distributions of the catalyst surface acid sites: Brønsted acid sites become the dominant acid sites on catalysts with an HSiW content $\geq 10\%$. Consequently, doping with $H_4SiW_{12}O_{40}$ introduced an abundance of acid sites, promoting the adsorption of reductant NH_3 on the catalysts in high-temperature regions. A NO_x conversion of over 80% was obtained in the temperature region of 250–450 °C using catalysts with an $H_4SiW_{12}O_{40}$ content of 5 and 10%, as shown in Fig. 2.31a.

An increased occurrence of side reactions, including the oxidation of NH_3 to NO_x, is expected at higher temperatures, as this is closely related to the reactivity of catalyst surface oxygen species. H_2-TPR profiles and XPS results demonstrate that abundant reactive oxygen species are generated on the surface of the pure Fe support, which promotes the oxidation of NH_3 to NO_x at high temperatures. Thus, NH_3 is more prone to react with O_2 to produce NO_x than with NO_x to form harmless N_2 and H_2O over the pure Fe support. This additional NO_x increases its concentration in the flue gas and decreases the amount of reductant NH_3 to react with NO_x, leading to a negative NO_x conversion above 300 °C. Adding HSiW impedes the oxidation behavior of reductant NH_3, making most NH_3 species more prone to participate in the reaction with NO_x rather than being directly oxidized to NO_x in the higher temperature region over the $H_4SiW_{12}O_{40}$-decorated catalysts, which helps in explaining the enhanced high-temperature SCR activity of the high-HSiW content catalysts to some extent.

Overall, a balance of acid and redox properties allowed for the increased activity of the high-$H_4SiW_{12}O_{40}$ content catalysts. Thus, when designing catalysts to be suitable for a wide temperature window, it is necessary to balance the acid and redox properties.

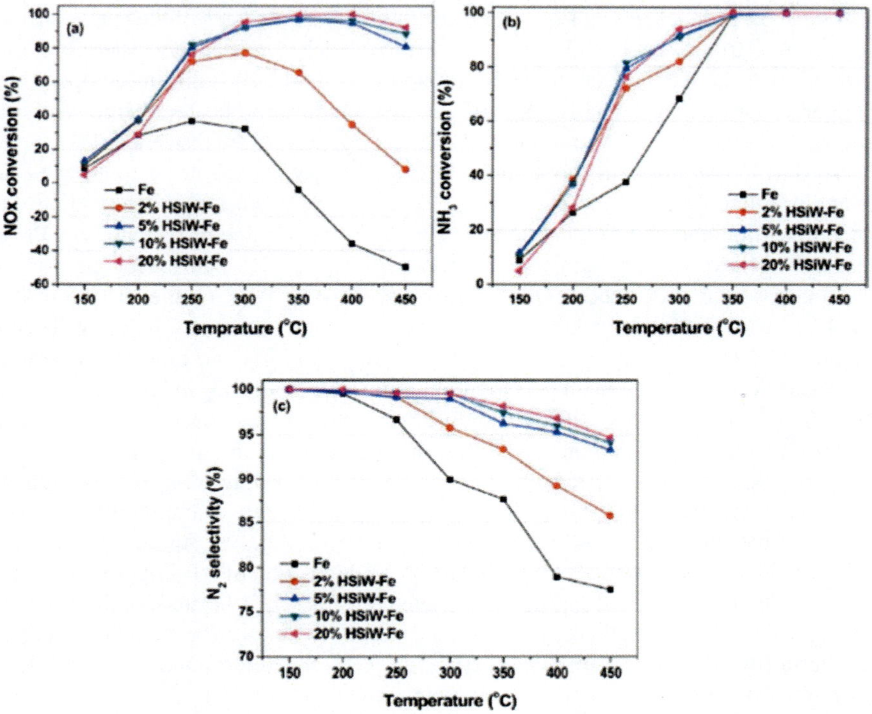

Fig. 2.31 **a** NH_3-SCR activity, **b** NH_3 conversion, and **c** N_2 selectivity of the series HSiW-Fe catalysts. Conditions: 500 ppm NO, 500 ppm NH_3, and 5% O_2 in N_2 at 0.5 L min^{-1}, 0.2 g catalyst, GHSV = 150,000 mLg^{-1} h^{-1} [62]

2.3.2.2 Effect of Ammonia Bisulfate on Low-Temperature SCR Catalyst

Flue gas from a coal-fired power plant contains sulphur dioxide (SO_2); SO_2 can be oxidized to sulphur trioxide SO_3, which further reacts with water and ammonia to form ammonium bisulfate (NH_4HSO_4, ABS) and ammonium sulfate ($(NH_4)_2SO_4$) in the SCR system, respectively, as follows:

$$NH_3 + SO_3 + H_2O \rightarrow NH_4HSO_4, \tag{2.23}$$

$$2NH_3 + SO_3 + H_2O \rightarrow (NH_4)_2SO_4. \tag{2.24}$$

The resulting $(NH_4)_2SO_4$ is a dry powder that can be removed by soot blowing and thus has little effect on the catalyst. However, the dew point of ABS is between 280 and 320 °C. Below the dew point, ammonia and sulfuric acid condense into ABS as described in the chemical Eq. (2.23), which blocks the surface of the catalyst [63–65] and the Brønsted acid sites, thus decreasing the deNO$_x$ efficiency. Therefore, the formation of the ABS inside the SCR system limits the catalyst minimum operating temperature [66]. Additionally, the ABS deposits in the air preheater, thereby fouling and corroding the air preheater and reducing its operational effectiveness [67].

To develop an efficient low-temperature SCR process capable of treating SO_2-containing flue gas, considerable efforts have been dedicated to understanding the decomposition and reaction behaviors of ABS on the surface of the catalyst. Prior researchers have reported that the ammonium ions of ABS can react with NO in the flue gas, which can accelerate the decomposition of ABS on the surface of the catalyst [68–70]. Synthesized birnessite catalysts [71] with fewer OH groups inhibited the NH_4HSO_3-forming reaction between SO_2 and NH_3 and thus demonstrated excellent SO_2 resistance in an investigation of the effects of SO_2 in low-temperature NH_3-SCR. Any deposited ABS on the γ-Fe_2O_3 surface was consumed by NO and O_2; as a result, there was increased deNO$_x$ activity after an initial decline caused by the formation of sulfate species on adsorption sites [72]. Furthermore, researchers studying the interaction between ABS and V_2O_5/TiO_2 catalysts clarified that amorphous ammonium sulfate salts deposited on the V/Ti catalyst surface decompose more readily than crystallite salts [73]. Given the deposition of ABS on the catalyst's surface, electrons around the Ti atoms deviated toward SO_4^{2-} because of the species' strong electronegativity, thereby rendering Ti atoms in an electron-deficient state, as shown in Fig. 2.32. Compared with crystallite ABS, the amorphous ABS deposited on V_2O_5/TiO_2 catalysts can act as a reductant and react with NO.

Researchers employing thermogravimetry to investigate the thermal decomposition of ABS formed on different metal oxides have demonstrated that single-component catalysts, such as CeO_2, TiO_2, Ni_2O_3, AC, and CrO_3, may have a high resistance to ABS [74]. Although this information can guide the design of ABS-resistant catalysts for low-temperature SCR, these thermogravimetric analyses employed dry air or N_2 as a background gas, which is considerably more inert than industrial flue gases. DFT was later used to study the anti-SO_2 poisoning mechanism on the surface of a MoO_3/V_2O_5 catalyst [75]. Here, SO_2 is converted into SO_3 on $V_2O_5(010)$ and transformed into ABS. The comprehensive formation, deposition, and reaction mechanism of ABS on the V_2O_5/TiO_2 catalyst's surface, characterized by XPS and TRM experiments [76], indicated that ABS is mainly formed in the

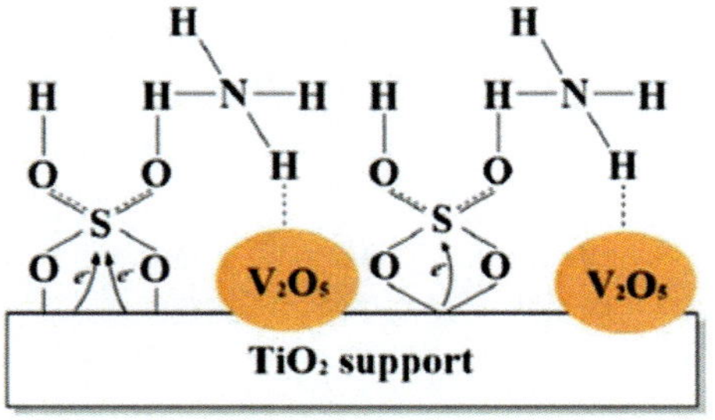

Fig. 2.32 Ammonium bisulfate deposition model on TiO_2 support [73]

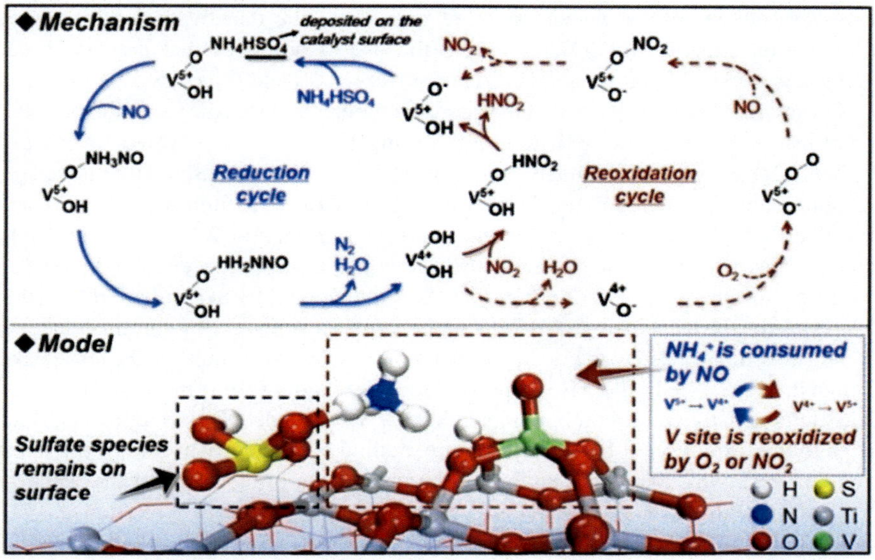

Fig. 2.33 Deposition and reaction mechanism of ammonium bisulfate on a V_2O_5 SCR catalyst [76]

gas phase through the nucleation of SO_3, H_2O, and NH_3, before depositing onto the surface of the catalyst, as summarized in Fig. 2.33. The decomposition of ABS on the surface of the V_2O_5/TiO_2 catalyst comprises two steps. First, NO is reduced by the NH_4^+ of ABS to form N_2 and H_2O by transferring an electron to the adjacent vanadium site. The reduced vanadium site is then reoxidized by either O_2 or NO_2. At low temperatures, owing to the weak reoxidizing ability of O_2, the reaction of ABS with NO in the NO/O_2 mixture is relatively slow. Here, the rate-determining step of the latter reaction is the reoxidation of the reduced vanadium (V^{4+}) sites; NO_2 is a better reoxidizing agent than O_2.

2.3.2.3 Optimizing SCR Catalysts with Higher Resistance to Ammonia Bisulfate

Measures can be taken to reduce the formation of ammonia bisulfate (ABS) and its resulting negative effects on catalysts activity, including maintaining an operating temperature above 300 °C to reduce the ABS deposition, replacing steel components of the air preheater with ceramic ones, changing the structure of the air preheater. The catalyst used during SCR can also be optimized for its resistance to ammonium bisulfate formation, possibly allowing lower temperatures.

Several strategies have been reported to address the deactivation of catalysts after the introduction of SO_2, including the use of additives that can boost the decomposition or inhibit the generation of surface ABS and $(NH_4)_4SO_4$.

The interaction between ABS and vanadium species and the role vanadium species plays in the reactivity of ABS with NO during SCR was investigated [69] using V_2O_5/TiO_2 catalysts. Here, the reactivity of ABS with NO increased with increasing vanadium content, owing to the formation of polymeric vanadium species and reactive electrophilic oxygen species on the catalysts. Introducing WO_3 to a V_2O_5/TiO_2 catalyst was also found to promote the decomposition of deposited ABS on the catalyst's surface [70]. WO_3 doping increases the amount of SO_2 released at lower temperatures during heating; ABS decomposition is promoted by the introduction of WO_3, as illustrated by the decrease in the temperatures of the weight loss peaks shown in Fig. 2.34a. In the ABS-deposited samples, adding WO_3 increased the electron cloud density around the S atoms in SO_4^{2-}, which aided the reduction of S^{6+} to S^{4+} and the formation of SO_2. Furthermore, adding WO_3 mildly promotes the reactivity of ABS with NO while inhibiting the production of N_2O during the reaction process. Overall, WO_3-promoted catalysts exhibited enhanced SO_2 resistance.

Examining the thermal decomposition of ABS on 21 different kinds of single-metal oxides demonstrated that its thermal stability markedly decreased when using an Al_2O_3 catalyst [77]. Similarly, it has been reported that doping with other metal elements (e.g., Al [78], Eu [79], Ho [80]) effectively improved poisoning resistance.

Owing to their increased reactivity of ABS with NO, Nb_2O_5- and Sb_2O_5-doped V_2O_5-WO_3/TiO_2 catalysts have demonstrated a better resistance than the unmodified catalyst to SO_2, as shown in Fig. 2.35 [81]. Adding Nb_2O_5 and Sb_2O_5 increases the electron density around the S atoms in SO_4^{2-}, which promotes the transformation of SO_4^{2-} to SO_2 at low temperatures. Adding promoters reduces the number of sulfate sites that accommodate NH_4^+ on the catalysts, resulting in a diminished amount of surface NH_4^+ during heating, as shown in Fig. 2.36. As a result, the decomposi-

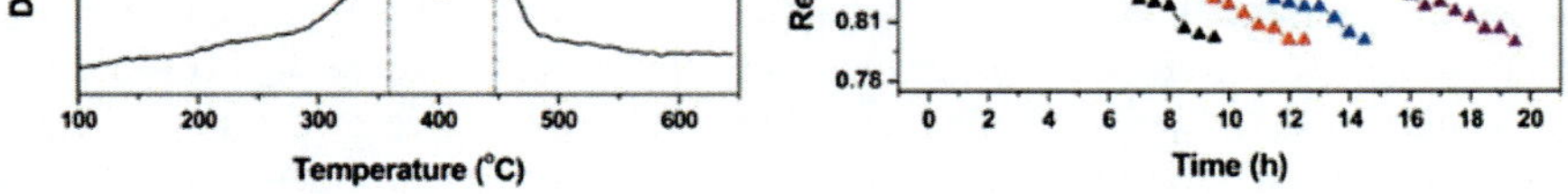

Fig. 2.34 **a** DTG profiles and **b** relative activity in the presence of SO_2 and H_2O at 250 °C during NH_3-SCR of NO over V/Ti, VW_5/Ti, $VW_{6.5}$/Ti, and VW_8/Ti [70]

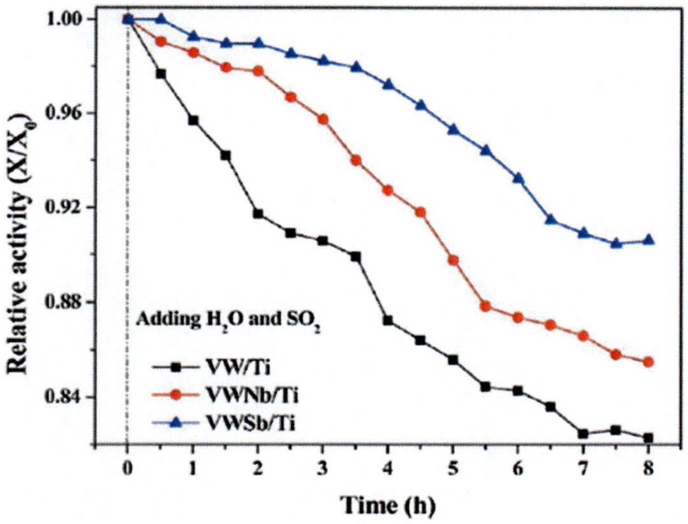

Fig. 2.35 Relative catalytic activity in the presence of SO_2 and H_2O for the SCR of NO with NH_3 over VW/Ti, VWNb/Ti, and VWSb/Ti at 250 °C. Conditions: 667 ppm NO and 667 ppm NH_3 with 5% O_2, 1000 ppm SO_2, and 10% H_2O and N_2 at 1.2 L/min, and 0.3 g of catalyst [81]

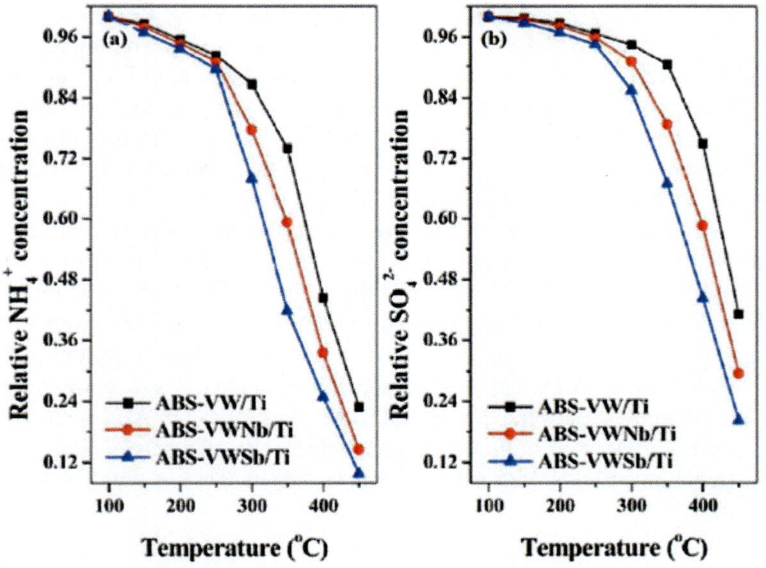

Fig. 2.36 Relative concentration of **a** NH_4^+ and **b** SO_4^{2-} on the series catalysts during heating [81]

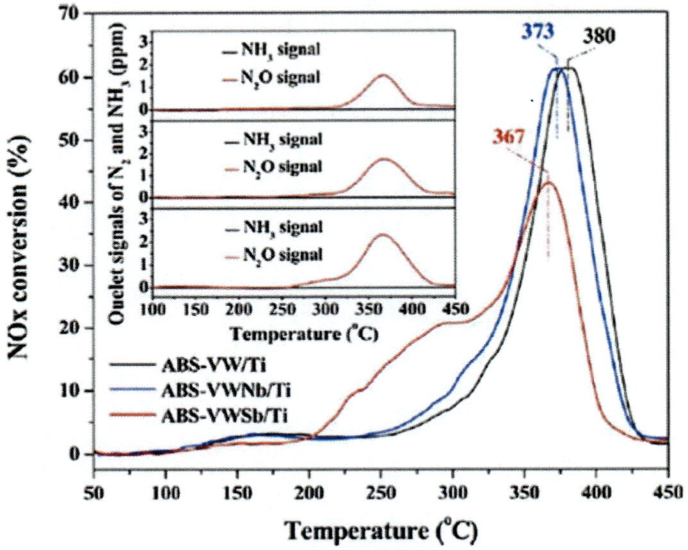

Fig. 2.37 TPSR profiles of ABS with NO. Conditions: 650 ppm NO with 5% O_2 and N_2 at a flow rate of 500 ml min^{-1} and 0.3 g samples [81]

tion of ABS is promoted on the promoters-modified catalysts. The improved reactivity of ABS with NO may be caused by the increased amount of V^{5+} and reactive electrophilic oxygen species on the Sb_2O_5- and Nb_2O_5-doped catalyst's surfaces.

Introducing Nb_2O_5 and Sb_2O_5 promotes the reactivity of ABS decomposition in varying degrees, as shown in Fig. 2.37. The scope of the relative NH^{4+} concentration is increased with the addition of Nb_2O_5. It is suggested that the higher NO_x conversion rate might relate to the slight improvement of the decomposition of ABS over the Nb_2O_5 containing catalyst surface, in the temperature region of 300–350 °C. However, only the trace NH^{4+} was detected at 200 min which is much quicker than the other two catalysts. Therefore, it is reaching the highest reactivity of ABS decomposition over the surface of the Sb-doped catalyst. Therefore, the increased reactivity of NH_4^+ in ABS with NO is vital to the increased NO_x conversion with the Sb-doped catalyst between 200 and 300 °C. Thus, NH_4^+ in ABS is the primary reductant during reactions with NO at lower temperatures, whereas NH_3 produced from the decomposition of ABS acts as the main reductant to react with NO at higher temperatures.

2.3.3 SO_2/SO_3 Conversion Inhibition

While treating SO_2-containing flue gases, the undesired oxidation of SO_2 occurs along with the NO reduction in the SCR unit as follows:

$$2SO_2 + O_2 \rightarrow 2SO_3. \quad (2.25)$$

The generated SO_3 forms sulfuric acid and ammonium bisulfate by reacting with water and unconverted ammonia in the flue gases, respectively, as shown below:

$$SO_3 + H_2O \rightarrow H_2SO_4, \quad (2.26)$$

$$H_2SO_4 + NH_3 \rightarrow NH_4HSO_4 \quad (2.27)$$

This leads to a pressure drop and downstream corrosion of the SCR unit, therefore, inhibiting SO_2 oxidation while maintaining sufficient NO removal efficiency is crucial.

The chemical components of the catalyst directly affect the oxidation rate of SO_2 over one catalyst. Although commercially used vanadium oxides enhance the deNO_x efficiency, they also improve the oxidation of SO_2. The oxidation rate of SO_2 over a V_2O_5/TiO_2 catalyst has a nearly linear relationship with the V_2O_5 loading. Moreover, adding WO_3, which typically acts as an SCR promoter, could also promote the adsorption of SO_2 and increase its oxidation rate.

Both the removal of NO_x and the oxidation of SO_2 are affected by the monolith structure of the catalyst. The rate of SO_2 oxidation has been reported to depend linearly on the wall thickness of the catalyst; several models describing reactors and their reactions in the catalyst monolith have been established to obtain insights into the effect of the catalyst's structure on SO_2 oxidation.

An introduction to the strategies for designing both chemical components and catalyst structures for SCR catalysts with a low SO_2/SO_3 conversion is thus discussed here.

2.3.3.1 Mechanism of SO_2 Oxidation Over SCR Catalyst

Fundamental studies on the mechanism behind SO_2 oxidation over V_2O_5/TiO_2-based catalysts have demonstrated that both V=O and V–OH groups are likely involved in the adsorption of SO_2 and its oxidation over V_2O_5/TiO_2 catalysts and that bridging Ti–O–V oxygen influences the adsorption and subsequent oxidation of SO_2 [82–84]. However, the detailed mechanism of SO_2 oxidation over the catalyst with different active component loadings requires further investigation.

XRD can be used to characterize the relationship between bulk crystalline phases and the loading of vanadium in the catalyst; patterns of V_2O_5/TiO_2 catalysts with different loadings of V_2O_5 are shown in Fig. 2.38 [85]. TiO_2 samples comprise a mixed phase of anatase and rutile TiO_2 in the ratio of 90:10. In each sample, the peaks representing anatase TiO_2 and rutile TiO_2 were detected. No changes in the anatase or rutile phases were observed in the XRD patterns after ammonium metavanadate loading and calcination, and this indicates that the crystalline phase of the TiO_2 support was not modified. Crystalline V_2O_5 was found only on the 6 V/Ti

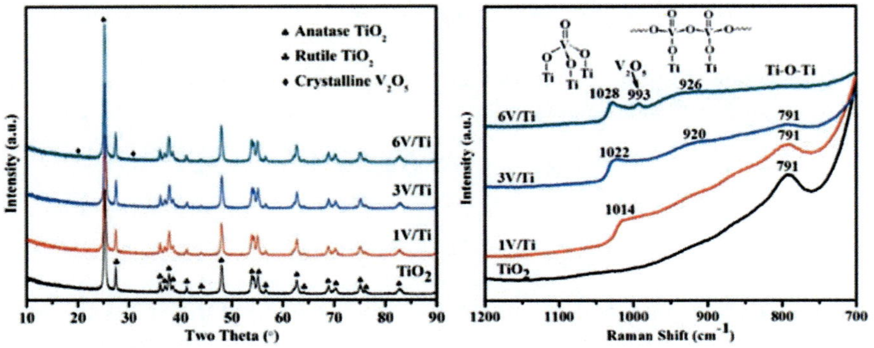

Fig. 2.38 XRD patterns and Raman spectra of catalysts with different V_2O_5 loadings [85]

samples. These results indicate that in all other samples, the vanadium oxide was highly dispersed and existed in an amorphous state on the TiO_2 support. And these results correspond to the surface density analyses of the V atom.

Raman spectra of the TiO_2, 1 V/Ti, 3 V/Ti, and 6 V/Ti samples from 700 to 1200 cm^{-1} under ambient conditions are shown in Fig. 2.38 [85]. The Raman band at 791 cm^{-1} corresponds to the Ti–O–Ti vibration of the TiO_2 support and diminished in intensity with increasing vanadium oxide content because of the increased light absorption by the yellow/orange color of vanadium oxide. The addition of vanadium oxide to the TiO_2 support resulted in three new Raman bands at approximately 1020, 993, and 920 cm^{-1}, attributable to the isolated vanadium oxide species, crystalline V_2O_5, and polymerized vanadium oxide, respectively. The Raman band of the isolated vanadium oxide species shifted from 1014 to 1028 cm^{-1} with increasing vanadium oxide content (from 1 to 6%). On the surface of the 3 V/Ti and 6 V/Ti, the Raman band corresponding to the polymerized vanadium oxide appeared at approximately 920 cm^{-1}; this band was not observed in the 1 V/Ti sample. At higher vanadium oxide loadings on TiO_2 (6 V/Ti), crystalline V_2O_5 was formed and gave rise to a strong Raman band at 993 cm^{-1}.

The results of SO_2-TPD conducted for V_2O_5/TiO_2 catalyst samples with different vanadium loading are summarized in Fig. 2.39 [85]. Pure TiO_2 showed three peaks centered at 150, 270, and 740 °C. The peaks at 150 and 270 °C separately represent the desorption of molecular SO_2 from the surface and the combination of hydroxyl and sulfur dioxide on the surface. The high-temperature peak represents the decomposition of stable sulfate compounds on the surface. Increasing the vanadium oxide content caused the area of the peaks at 150 and 270 °C to decrease; similarly, both the temperature and area of the high-temperature peak decreased with increasing vanadium oxide content. On the V_2O_5/TiO_2 surface, sulfate could only bond to the Ti–O site. After vanadium loading this was not possible because the Ti–O site was occupied as Ti–O–V. Thus, the area of the peak belonging to the sulfate decreased with increasing vanadium oxide coverage. A new peak at approximately 400–500 °C

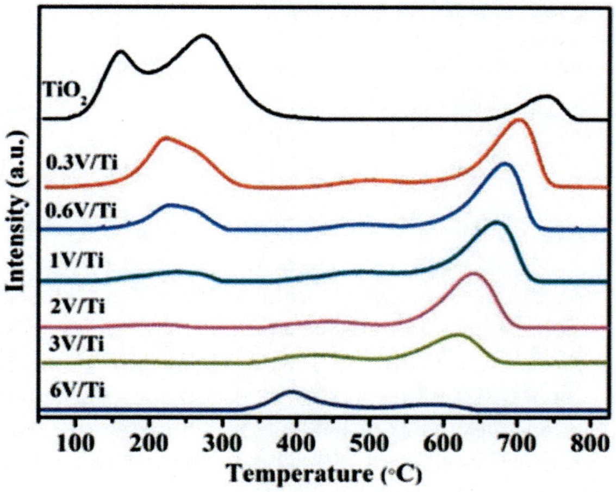

Fig. 2.39 SO_2-TPD profiles of the catalysts with different V_2O_5 loadings [85]

appeared for the V_2O_5/TiO_2 samples and increased with the increasing vanadium oxide content. This peak is likely associated with the decomposition of surface sulfite.

V_2O_5/TiO_2 catalysts with different vanadium loadings (0.6–6% V_2O_5) were then subjected to SO_2 at 300–400 °C to clarify the oxidation activity. The turnover frequency (TOF) of SO_2 oxidation over the V_2O_5/TiO_2 catalysts was calculated at low SO_2 conversion (lower than 20%). The TOF can be utilized to normalize SO_2 oxidation activity per surface vanadium atom and is expressed in molecules of SO_2 oxidized per surface vanadium oxide site per second; the TOF as a function of vanadium loading is shown in Fig. 2.40 [85]. Normalizing the reaction rate with

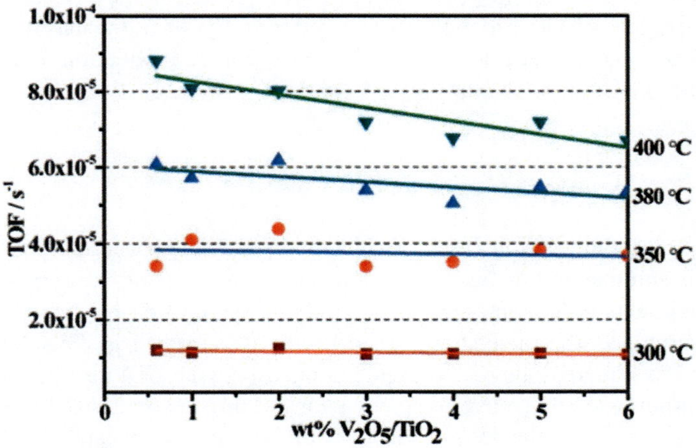

Fig. 2.40 Turnover frequency (TOF) of SO_2 oxidation over V_2O_5/TiO_2 catalysts as a function of V_2O_5 loading and reaction temperature [85]

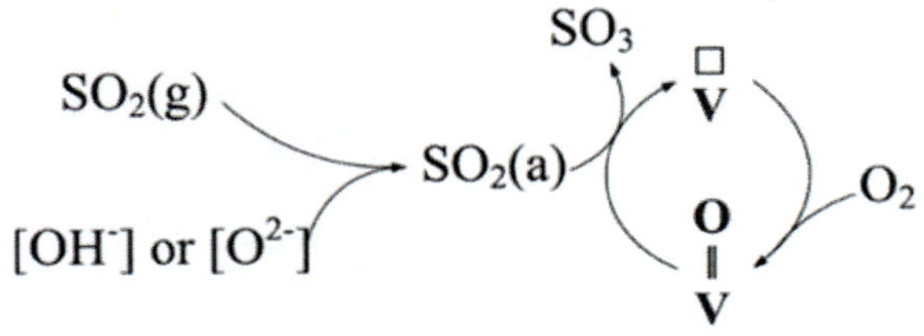

Fig. 2.41 Possible reaction mechanism for SO_2 oxidation by O_2 over V_2O_5/TiO_2 catalysts [85]

respect to the vanadium loading assumes that all the surface vanadium oxide species participate equally in the reaction.

At a V_2O_5 loading of 1 wt%, the magnitude of the TOF varied from 1×10^{-5} s^{-1} at 300 °C to 8.05×10^{-5} s^{-1} at 400 °C. The temperature drastically affected the reaction rate; the TOF increased with increasing temperature and roughly doubled with every increase of 50 °C. No obvious changes were observed in the TOF for SO_2 oxidation over different V_2O_5/TiO_2 catalysts at a constant temperature. Unlike during SCR, the surface reaction appears to be the rate determining step during SO_2 oxidation. The TOF of SO_2 oxidation is far lower than that of the SCR reaction (10^{-3} to 10^{-2} s^{-1}) over the V_2O_5/TiO_2 catalysts. The apparent activation energy of the V_2O_5/TiO_2 with a loading of 0.6–6% was 67.2 ± 2.8 kJ·mol^{-1}, implying that SO_2 oxidation over a V_2O_5/TiO_2 catalyst occurs at the same active sites regardless of V_2O_5 loading.

Isolated vanadium species on the V_2O_5/TiO_2 catalysts may have been dispersed on the TiO_2 surface via Ti–O–V bonds, which then form polymeric vanadium species through V–O–V bonds at the expense of the Ti–O–V bonds, thereby decreasing the amounts of Ti–O–V bonds and increasing the amounts of V–O–V bonds as the VO_x coverage increased. Although V=O and Ti–OH sites are crucial in the adsorption and oxidation of SO_2, no obvious changes were observed in the TOF and apparent activation energy of SO_2 oxidation over the different V_2O_5/TiO_2 catalysts. In particular, V=O plays a critical role during SO_2 adsorption and oxidation; one plausible reaction mechanism is shown in Fig. 2.41 [85].

2.3.3.2 Low SO_2/SO_3-Conversion SCR Catalysts

The effect of adding various metal oxides to V_2O_5/TiO_2-based commercial SCR catalysts is summarized in Fig. 2.42, obtained via a series of experiments bench-scale experiments at 350 and 400 °C. Here, 0.35 mmol of each oxide per gram of catalyst was added, the catalyst contained 1 wt% of V_2O_5, 0.1 g of catalyst powder was loaded in a quartz tube reactor with an inner diameter of 4 mm, and a gas flow of 1200 ml/min was conducted into the reactor. Adding the oxides of La, Zn, In, Al, Zr, Y, and Sn decreased the deNO_x efficiency of the commercial catalyst; the catalyst containing La_2O_3 had a near-zero deNO_x activity. Alternatively, adding the oxides of Co, Fe, Ge, Sb, Nb, Cu, Mn, Ce, Cr, W, and Mo, enhanced deNO_x activity. In particular, adding WO_3 and MoO_3 increased the deNO_x efficiency at 350 °C from

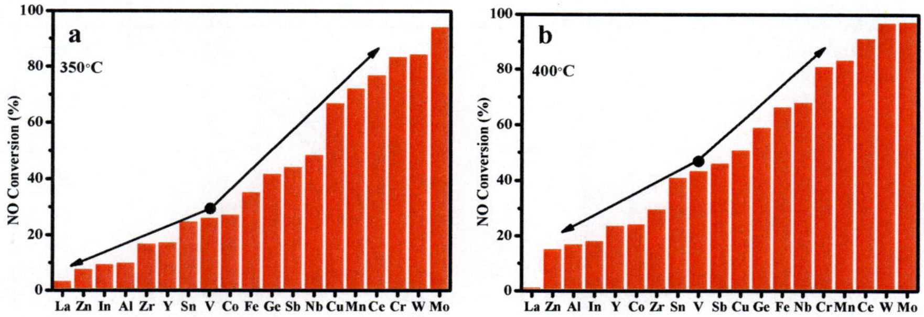

Fig. 2.42 Influence of metal oxide additives on the NO conversion of a 1 wt% V_2O_5/TiO_2 commercial SCR catalyst a: at 350 °C. b: at 400 °C

25.9 to 84.3 and 94.1%, respectively, and at 400 °C from 43.1 to 96.7 and 97.1%, respectively. Thus, adding WO_3 and MoO_3 can significantly improve the deNO_x efficiency of the V_2O_5/TiO_2 catalyst.

The effects of these additives on the catalyst's SO_2 oxidation rates were also demonstrated in a similar fashion by loading a 0.3-g catalyst into a quartz tube reactor with an inner diameter of 4 mm and a gas flow of 200 ml/mint. As summarized in Fig. 2.43, the addition of the oxides of Ge, Nb, Sb, Cr, Ce, Mn, W, and Mo resulted in higher SO_2 oxidation rates. At 400 °C, adding WO_3 and MoO_3 increased the SO_2 oxidation rate from 1.63 to 4.2 and 6.07%, respectively.

The precursors used for metal oxide additives also affect the resulting SCR and SO_2 oxidation. As an example, NO conversion and its temperature dependence subsequent to MoO_3 doping (5 wt%) using three precursors are shown in Fig. 2.44, where N-Mo represents ammonium heptamolybdate, Si–Mo represents silicomolybdic acid, and P-Mo represents phosphomolybdic acid. In all instances, NO conversion is improved and the temperature window is widened; at 350 °C,

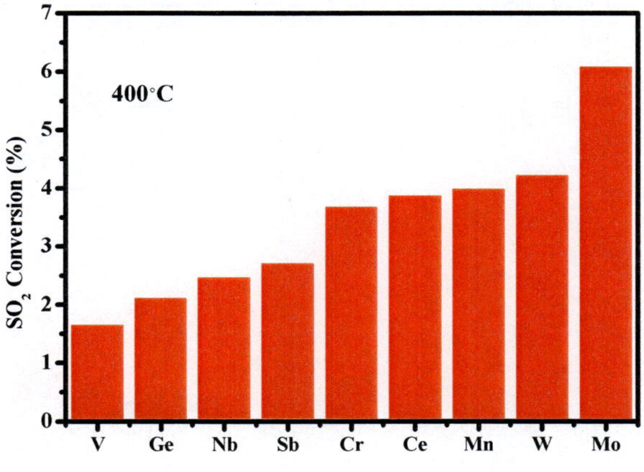

Fig. 2.43 Influence of metal oxide additives on the SO_2 oxidation of a 1 wt% V_2O_5/TiO_2 commercial SCR catalyst

Fig. 2.44 NO conversion with different MoO_3 precursors

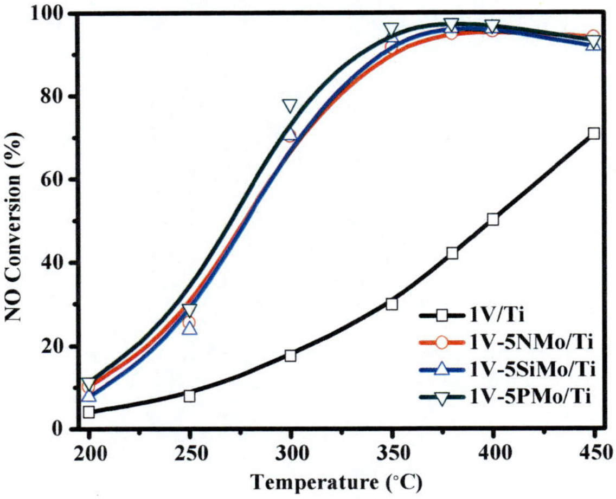

the NO removal efficiency is improved from 29.8 to 91.6% (N-Mo/Ti), 93.6% (Si-Mo/Ti), and 96.4% (P-Mo/Ti). However, using phosphomolybdic acid as the MoO_3 precursor provided the best deNO_x efficiency in the range of 200 °C-450°C.

Similarly, SO_2 oxidation and its temperature dependence after MoO_3 doping (5 wt%) using the same three precursors are shown in Fig. 2.45. In all instances, the SO_2 oxidation rate of the catalyst increased after doping. At 400 °C, the SO_2 oxidation of catalyst increased from 3.12 to 6.72% (N-Mo/Ti), 5.92% (Si-Mo/Ti) and 5.52% (P-Mo/Ti). The lowest increase in SO_2 oxidation was achieved with a precursor of phosphomolybdic acid.

Phosphomolybdic acid allows for the highest increase in NO conversion with the lowest increase in SO_2 oxidation; therefore, NO conversion and SO_2 oxidation at different levels of doping (3, 5, 7, and 10%) were then considered, as shown in

Fig. 2.45 SO_2 conversion with different MoO_3 precursors

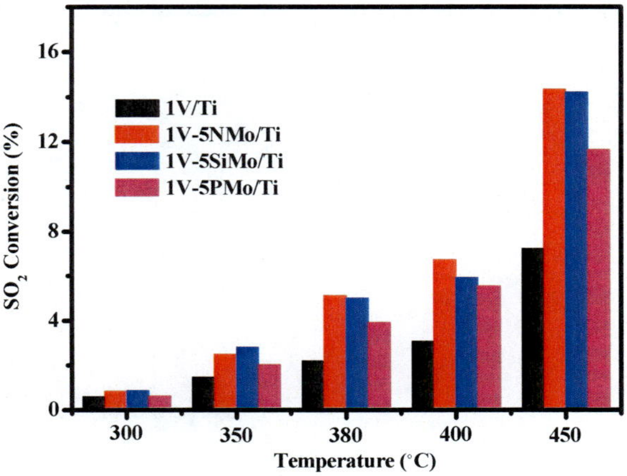

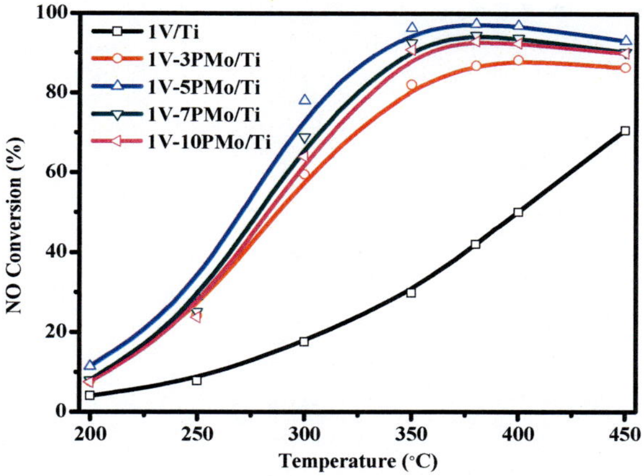

Fig. 2.46 NO conversion with varying MoO_3 precursor dosage

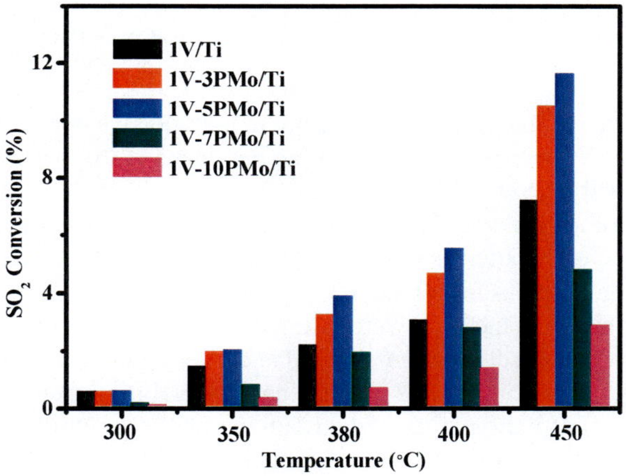

Fig. 2.47 SO_2 conversions with varying MoO_3 precursor dosage

Figs. 2.46 and 2.47, respectively. In all cases, the NO conversion efficiency was maximized at phosphomolybdic acid loading of 5 wt%. Similarly, the SO_2 oxidation is maximized at a precursor loading of 5 wt%. Further increases of this reduced the SO_2 oxidation rate below that obtained using the pure V_2O_5/TiO_2 catalyst. It can be concluded, higher doping of the phosphomolybdic acid precursor may help inhibit SO_2 oxidation while only mildly decreasing NO conversion.

2.3.3.3 SO_2 Oxidation in Monolith SCR Catalyst

The structure of the monolith SCR catalyst and the operating parameters affecting to the NO_x reduction and SO_2 oxidation processes. A dual-layer model of a monolith catalyst support structure, as schematically shown in Fig. 2.48, can help clarify the

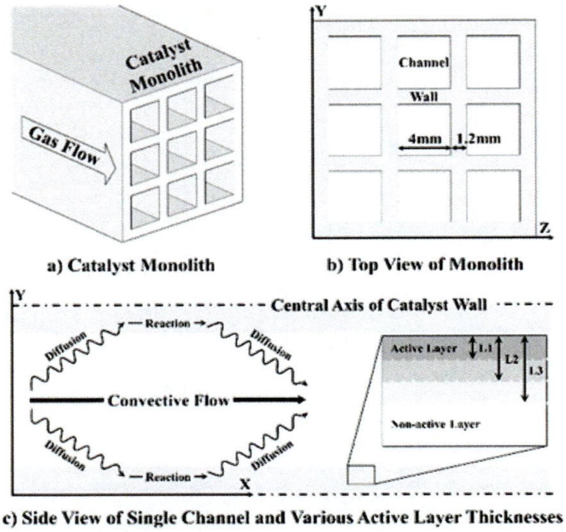

Fig. 2.48 Schematic of the monolith catalyst model [86]

role of the monolith's structure in the selective catalytic NO reduction and oxidation of SO_2 under typical operating parameters [86]. In this work, the modeled monolith had a length of 200 mm, having square channels with a 4-mm side length and 1.2-mm-thick walls. Using the model, the influences of operating parameters and active layer thicknesses were examined by assuming an active layer from 0.05 to 0.6 mm in thickness is reactive to NO_x reduction and SO_2 oxidation; the nonactive layers (i.e., the rest of the wall) were assumed inert but permeable. The porous medium approach, which describes the flow field and effects of chemical reactions simultaneously, was used to consider the external and internal mass transfer of the catalyst wall.

From the simulated distribution of NO cncentration within a transverse sector of a channel of the monolith, shown in Fig. 2.49 [86], the highest and lowest concentration of NO in the radial direction appears at the central axis and distributed along the wall, respectively. As SCR occurs at the wall, the NO concentration decreased gradually toward the wall and then sharply decreased inside the wall. The concentration of NO drastically decreased from 262 ppm at $x = 4.0$ mm to 2 ppm at $x = 4.2$ mm, indicating that SCR occurs mainly within a thin layer (0.2 mm) of the wall near the channel. Similar results can be obtained over other transverse sectors. Overall, the reaction rates of NO and NH_3 inside the wall are much higher than their diffusion rates; i.e., NO and NH_3 are consumed before diffusing to the wall center, indicating that most of the bulk catalyst contained in the wall is not involved in the reaction.

The simulated distribution of SO_2 and SO_3 concentration is shown for a similar transverse sector in Fig. 2.50 [86]. Here, the concentration of SO_3 increases gradually from the central axis until reaching the wall, where it increases sharply; at $y =$ 100 mm, the concentration increased from 14 to 62 ppm within the catalyst wall. These results suggest that the reaction rate of SO_2 oxidation inside the catalyst wall is slower than the diffusion rate of SO_2. The produced SO_3 diffuses from the

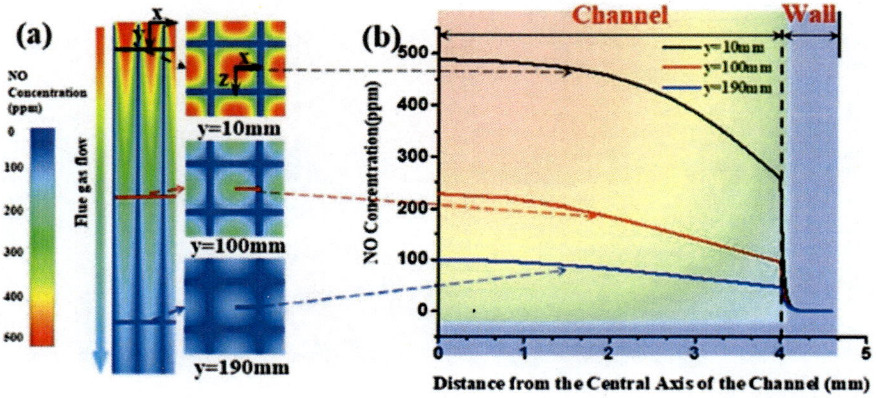

Fig. 2.49 Concentration distribution of NO in a transverse sector of the monolith (honeycomb) catalyst: **a** transverse sector profile and **b** NO concentration in the radial direction. Reaction conditions: $T = 340$ °C, GHSV = 8000 h^{-1}, [NO] = 500 ppm, [NH_3] = 500 ppm, and [SO_2] = 500 ppm [86]

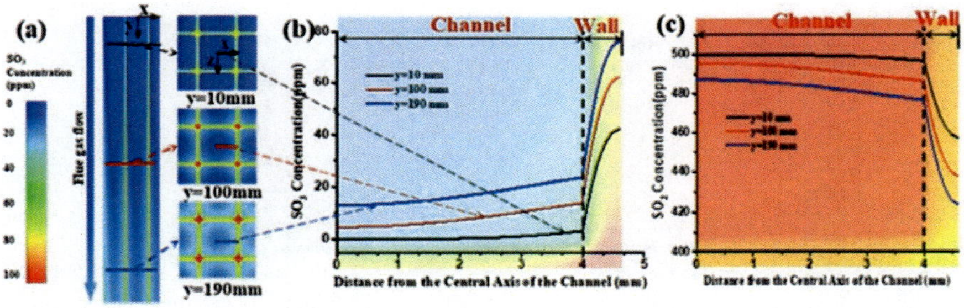

Fig. 2.50 Concentration distribution of SO_3 and SO_2 in a transverse sector of the monolith catalyst: **a** transverse sector profile; **b** SO_3 and **c** SO_2 concentration in the radial direction. Reaction conditions: $T = 340$ °C, GHSV = 8000 h^{-1}, [NO] = 500 ppm, [NH_3] = 500 ppm, [SO_2] = 500 ppm [86]

wall surface to the channel; a distinct SO_3 concentration gradient was formed in the channel, where the SO_3 concentration increased from 4 ppm at $y = 10$ mm to 14 ppm at position $y = 190$ mm.

To clarify the influence of active layer thickness under typical operating condition, seven active layer thicknesses (0.05, 0.10, 0.15, 0.20, 0.25, 0.30, and 0.60 mm) were studied under varying temperatures and GHSV: 320–400 °C at 20 °C intervals and 2000 h^{-1} to 10,000 h^{-1} at 2000 h^{-1} intervals, respectively. In all cases, the feed gas contained 1000 ppm of NO, NH_3 and SO_2. A summary of these 175 cases is shown in Fig. 2.51 [87], where each column represents a temperature and each row represents a gas velocity. Each case is represented by a pixel, the color of which represents

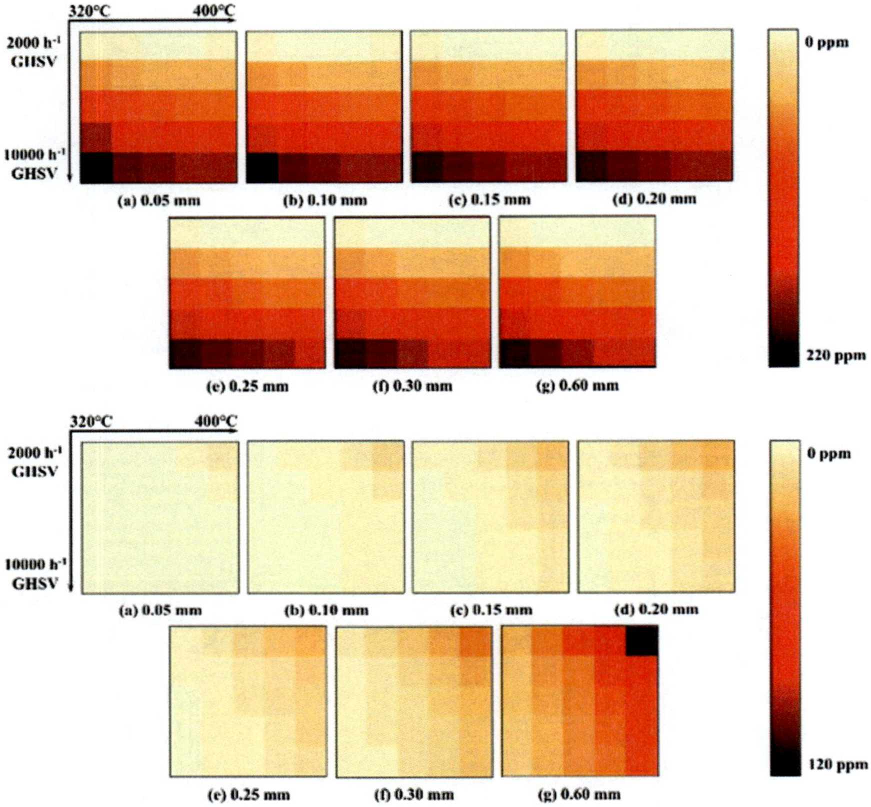

Fig. 2.51 Average concentration of NO (top) and SO_3 (bottom) in ppm at the outlet of the monolith with an active layer thickness of **a** 0.05, **b** 0.10, **c** 0.15, **d** 0.20, **e** 0.25, **f** 0.30, and **g** 0.60 mm [87]

the average concentration of NO (top) or SO_3 (bottom) in ppm at the outlet of the monolith.

Overall, the GHSV influenced the outlet concentration of NO in all scenarios, whereas the influence of temperature became more remarkable as GHSV decreased. At wall thicknesses below 0.15 mm, the thickness of the active layer did not influence the concentration as much as the GHSV and temperature did. The highest and lowest NO outlet concentrations (216.2 and 11.8 ppm, respectively) occurred at operating conditions of 320 °C, GHSV = 10,000 h^{-1}, and an active layer thickness of 0.05 mm, and 360 °C, GHSV = 2000 h^{-1}, and an active layer thickness of 0.60 mm, respectively. At low GHSVs (e.g., 2000 or 4000 h^{-1}), increasing the temperature did not increase the conversion of NO, particularly in scenarios with a thinner active layer (e.g., 0.05 and 0.10 mm). This phenomenon likely occurs owing to a decrease in the reactants' residence time in the active layer due to the increase in diffusion under high temperature.

Conversely, the outlet concentration of SO_3 was positively related to the thickness of the active layer, in accordance with prior work indicating that SO_2 oxidation occurs across the entire active layer. Both the GHSV and temperature affect the outlet SO_3 concentration significantly: The highest and lowest SO_3 outlet concentrations (120.7 and 0.468 ppm, respectively) occurred at operating conditions of 400 °C, GHSV = 2000 h^{-1}, and an active thickness of 0.60 mm, and 320 °C, GHSV = 10,000 h^{-1}, and an active layer thickness of 0.05 mm, respectively. Temperature and active layer thickness variations had a greater influence on SO_2 conversion than GHSV variations, indicating that decreasing the temperature and active layer thickness are better strategies than decreasing the GHSV.

Detailed distributions of NO and SO_3 within the catalyst wall at four positions within the monolith are plotted in Fig. 2.52 [87], including at the inlet of the monolith ($x = 20$ mm), the middle line of the monolith ($x = 100$ mm), the outlet of the monolith ($x = 200$ mm), and the central axis of the monolith ($y = 8.4$ mm). Here, the concentration of NO dropped sharply within a thin layer close to the wall surface and then declined gently; in all cases, especially 0.05 and 0.10 mm, the sharp decline is seen at the interface between the active and inactive layers. Thus, the sharp drop in NO concentration is caused by the catalytic reaction, whereas the gentle decline is attributed to the diffusion. At $x = 20$ mm, increasing the active layer above 0.15 mm caused only minor increases in NO reduction. Similarly, an active layer of 0.10 mm was adequate at $x = 200$ mm. This observation suggests that an optimal NO reduction strategy would include a catalytic monolith with a thicker active layer at the inlet than at the outlet.

Similar trends of SO_2 conversion are observable at all three positions. The concentration of SO_3 increased with an increase in active layer thickness until a thickness of 0.60 mm. At this point a large gap was observed between thicknesses of 0.30 and 0.60 mm, indicating that the SO_2 oxidation drastically increased when the active layer is thicker than 0.30 mm. Furthermore, the distribution of SO_3 concentration along with the central axis of the monolith was highly related to the active layer thickness in the catalyst wall. To maintain low sulfur conversion, the active layer should be thinner than 0.30 mm. Considering the mechanical strength that must be maintained by the monolith walls, a dual-layer design may be used to reduce SO_2 oxidation while maintaining a high conversion of NO.

2.3.4 Applications

By the end of 2017, the capacity of installed coal-fired units in China was 1106.04 GW; more than 90% of these units had been mounted with SCR systems using a total catalyst volume of over 900,000 m^3. As this leads to the annual production of nearly 300,000 m^3 of deactivated catalyst, significant research efforts have been devoted to improving industrial catalyst production and regenerating deactivated catalysts. Another crucial aspect of SCR is the uniform distribution of the NH_3 injected into the

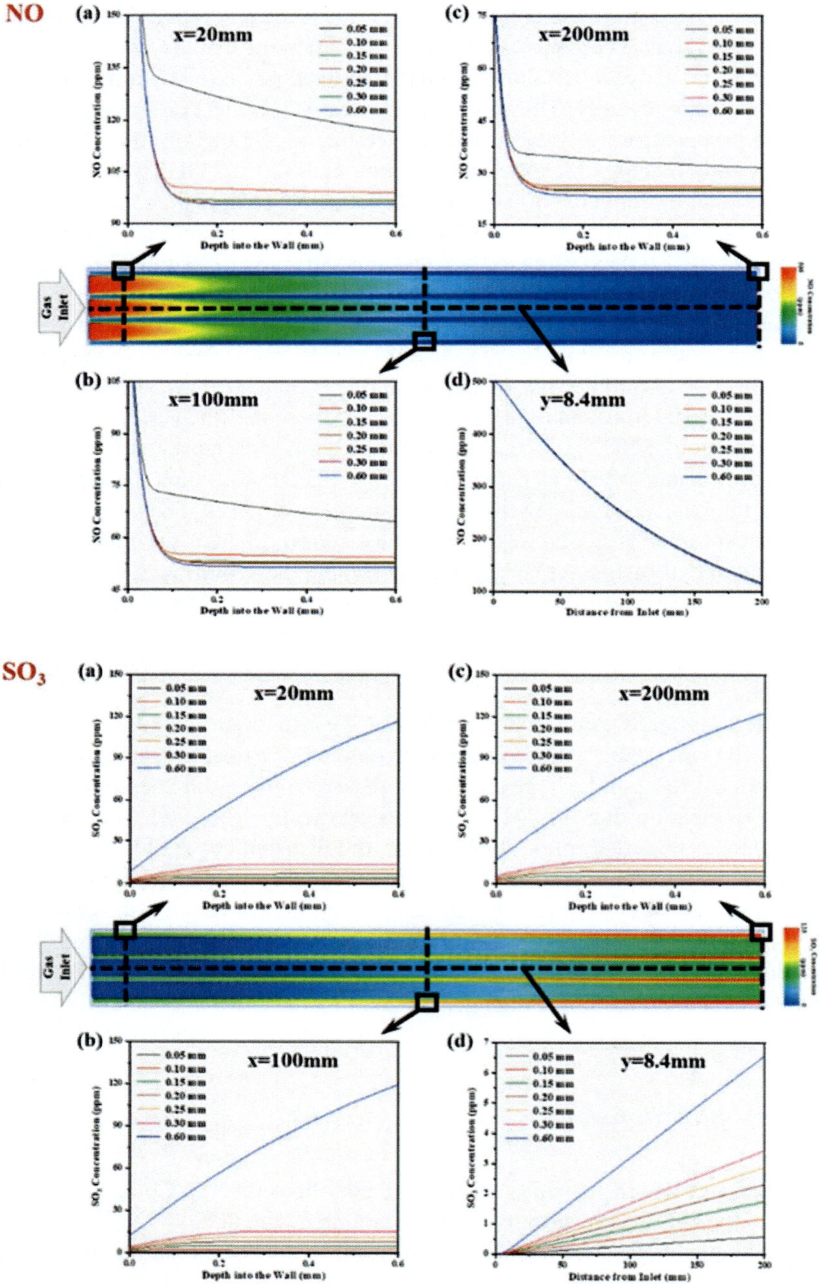

Fig. 2.52 Effect of active layer thickness on the simulated distribution of NO and SO_3 concentration: **a** in catalyst wall at $x = 20$ mm; **b** in catalyst wall at $x = 100$ mm; **c** in catalyst wall at $x = 200$ mm; and **d** along with the central axis of the monolith at $y = 8.4$ mm. Operating parameters: $T = 320$ °C, GHSV = 4000 h^{-1}, and [NO] = [NH_3] = [SO_2] = 1000 ppm [87]

flue gas. Several factors influence this distribution, including the injection method and mixing process which is determined by the reactor structure.

2.3.4.1 Industrial Preparation of SCR Catalysts

In coal-fired power plants, the SCR system is usually downstream of the economizer, where the high-ash flue gas is fed. To facilitate operation and maintain a low pressure drop, monolith catalysts are usually used, i.e., a single structure comprising inorganic oxides or metals with uniformly sized and parallel channels. Some common monolith designs, including a square channel monolith, a parallel plate monolith, and corrugated plate packing, are shown in Fig. 2.53. Among these catalyst modules, a square channel monolith comprises supports, active components, promoters, and additives to help with extrusion, a parallel plate monolith uses a metal plate as a carrier over which active components are impregnated and then sintered, and corrugated plate packing uses corrugated fiberglass or a ceramic board as the catalyst carrier.

A typical preparation process of a square channel V_2O_5–WO_3/TiO_2 monolith, including batching, kneading, filtering, extrusion, drying, calcination, and cutting, is shown in Fig. 2.54.

Batching and kneading: First, an ammonium metavanadate solution, ammonium metatungstate, and deionized water are added to TiO_2 and mixed thoroughly to ensure that ammonium metavanadate and ammonium metatungstate are uniformly impregnated onto the surface of TiO_2. A pore-forming agent and structural promoter are then added to the mixture, as is a binder to ensure viscosity and strength. The uniformly mixed materials then sit for 24 h.

Filtration and extrusion: The paste is filtered to remove any impurities that might inhibit extrusion and then pugged and degassed to improve the density, strengthen the catalysts, and prevent cracking of large holes caused by air during extrusion. The paste is then extruded by applying pressure that forces it to pass through a hole with a specific form and dimension.

Square channel monolith   Parallel plate monolith   Corrugated plate packing

Fig. 2.53 General appearances of monolith catalysts

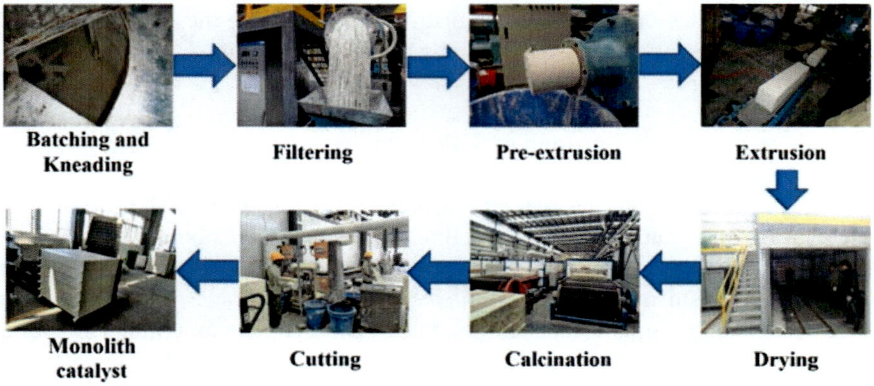

Fig. 2.54 Preparation process of commercial monolith catalysts

Drying: The catalyst is then dried to remove excess moisture mostly at 100–110 °C. It is worth noticing that drying process may causes large holes in the bulk phase formed by the rapid evaporation of water, thereby improving the size and volume of the pores; but on contrary, it also may cause the formed catalyst to rupture, with the mechanical strength decreasing. It is considered that the proper temperature is based on the fully removed of moisture, of which upper limit depends on the resistance of catalysts against the structure destruction.

Calcination: After drying, the catalyst is calcined in a muffle furnace. In this process, it must make sure that the precursors are completely decomposed while the structure of the catalyst should not be excessively destroyed to prevent the loss of specific surface area. In detail, compared to the proper temperature the additives may not completely decompose in lower temperature and the catalyst is sintered with specific surface area decreasing at higher temperature. Furthermore, too fast temperature rising may cause catalyst cracking, so that it is widely accepted that temperature rising rate at 10 °C per minute is safe for destruction.

According to different flue gas parameters and components, the monolith production can be varied by changing the extrusion molds, production parameters, and catalyst component precursors; three monolith examples with varying pore numbers, pore sizes, and wall thicknesses are presented in Fig. 2.55.

2.3.4.2 Regeneration of Deactivated SCR Catalysts

Catalysts used during SCR inevitably become inactive after long-term service (three years for average); an example deactivated catalyst, plugged with fly ash, is shown in Fig. 2.56. Catalyst deactivation is mainly caused by one of the following four mechanisms:

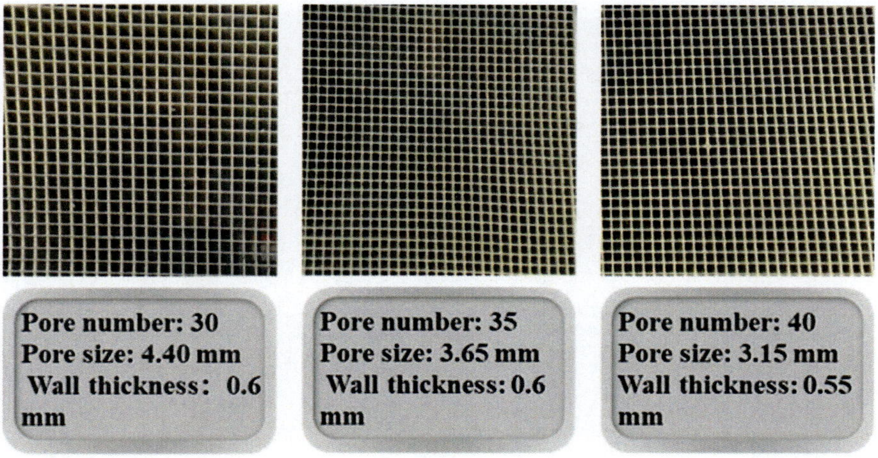

Fig. 2.55 Series monolith catalyst products prepared for different flue gas condition such as gas concentration, sulfur content and dust concentration

Fig. 2.56 Deactivated SCR catalyst. Left: catalyst plugged by fly ash. Right: scanning electron microscopy (SEM) image of a deactivated catalyst

- thermal sintering, which caused by the high flus gas temperatures within the reactor, will lead to a permanent loss of catalyst activity due to a change in the pore structure of the catalyst
- fouling or sticking, which is caused by ammonium bisulfate, fly ash, and other particulate matter in the flue gas
- poisoning, which caused by flue gas elements, such as calcium oxide, potassium, sodium, arsenic, chlorine, fluorine, and lead, will deactivate the catalysts by diffusing into the pores and occupying the active sites
- erosion, which caused by particulate matter impingement and high-velocity gas, will lead to the loss of active components.

Fig. 2.57 Regeneration process of deactivated SCR catalyst

At the end-of-life, the catalyst can be either regenerated or replaced. Of these deactivation mechanisms, only thermal sintering is irreversible and causes permanent damage. Catalyst regeneration is thus the priority, particularly considering the cost and pollution caused by replacing deactivated catalysts.

Regeneration of deactivated SCR catalysts commonly includes physical and chemical cleaning treatments as well as a reloading of active components to compensate for any active components losses during operation and the physical and chemical treatments. A summary of this process is presented in Fig. 2.57 and is discussed below.

Before regeneration, the cause of deactivation is analyzed via physical and chemical characterization and activity tests. Physical changes to the catalyst may include blockage, wear, and a decrease in mechanical strength; chemical changes include a decreased SCR selectivity and increased SO_2/SO_3 conversion rate. To clarify the cause of deactivation and thus determine the required regeneration measures must be taken, parameters such as the specific surface area, crystal structure, element composition, particle morphology are used to determine the presence of pore blockage, poisoning, or sintering. Regeneration measures then include purging, physical and chemical cleaning, reloading, and calcination/drying, as discussed below.

Purging: As SCR catalysts are usually employed for flue gases with a high concentration of fly ash, ash commonly accumulates on the surface of the catalyst and blocks the pores, causing a decrease in catalyst activity. Before purging, large pieces of fly ash accumulated on the upwind side of the catalyst module are removed. The catalyst is then put into a purging chamber and purged with compressed air. Although purging removes most of the fly ash caught in the pores, heavily clogged catalysts require additional treatment.

Physical cleaning: After purging, the catalyst is put into a water-filled physical cleaning tank, where ultrasound or aeration is used to clean the surface of the catalyst and remove fly ash from the pores. During ultrasound treatments, a kind of corrosion

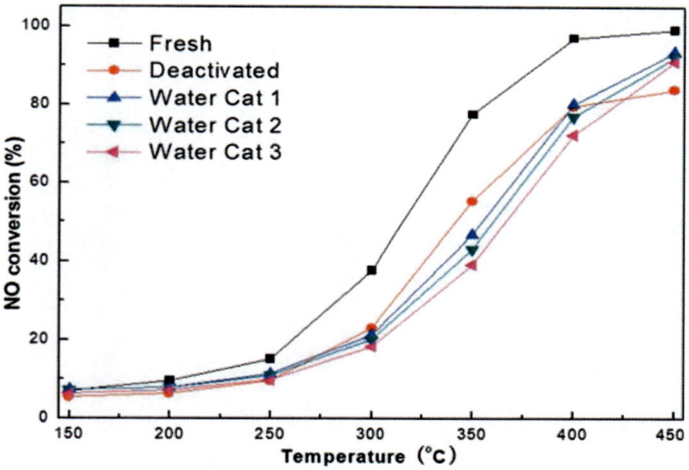

Fig. 2.58 Effect of physical cleaning on catalyst activity

occurred on the catalyst surface under the unstable fluid conditions which is called cavitation, causes the fly ash to disperse, emulsify, and peel. During aeration, high-pressure air is fed into the water to repeatedly impact the blocked pores. However, as physical cleaning only removes fly ash on the surface and cannot effectively remove toxic elements, the catalyst activity does not significantly improve, as shown in Fig. 2.58.

Chemical cleaning: Chemical cleaning is then used to remove the poisoning elements in the catalyst micropores and thus ensure the restoration of the catalyst's active sites. According to the cause of deactivation, additives such as inorganic acids, organic chelating agents, and penetrants are used in conjunction with aeration to ensure that the additives can enter the micropores and react with the poisons. Subsequent to the 15–30 min treatment, the catalyst is rinsed to wash away any lingering poison and treatment solution left on the surface. An example of the effect of poison removal from a deactivated catalyst by chemical cleaning is shown in Fig. 2.59.

Reloading treatment: After chemical cleaning, active components are reloaded to help in recovering the catalyst activity. Active components are uniformly and effectively loaded on the surface of the catalyst via a two-step process, in which the catalyst is immersed in a solution with organic coupling agent under ultrasonic conditions, and then transferred to a solution dissolved with precursors of active components.

Drying and calcination: The catalyst is subsequently dried and calcined at high temperatures in a tunnel kiln. During calcination, organic matter is oxidized to CO_2 and H_2O, and active component precursors are decomposed. The uniform distribution of velocity and temperature fields in the kiln are crucial in catalyst regeneration by promoting the reloading of active components and inhibiting catalyst cracking.

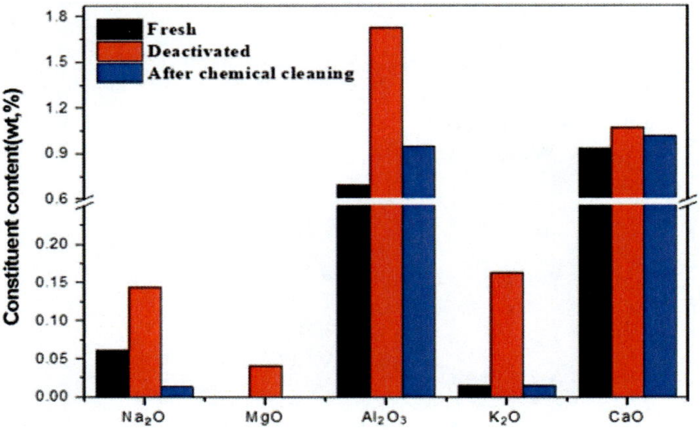

Fig. 2.59 Removal of poisoning elements via chemical cleaning

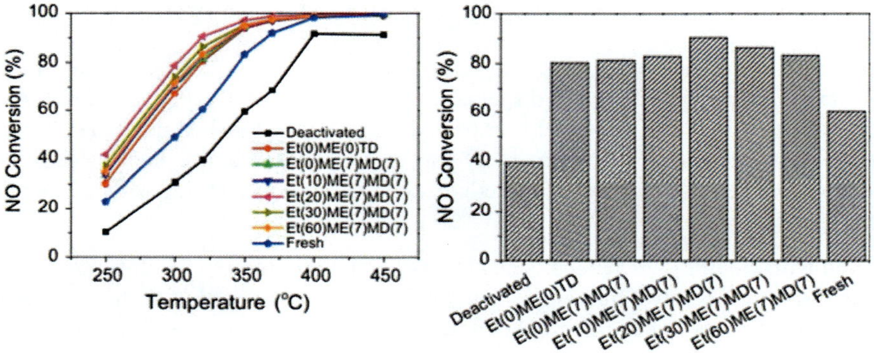

Fig. 2.60 NO conversion of regenerated catalysts with different organical solutions during the regeneration process. Left: 250–450 °C. Right: 320 °C

The resulting NO conversion of some typical regenerated catalysts is shown in Fig. 2.60. In some cases, such as the samples washed with organic solutions, the regenerated catalysts perform better than the fresh sample owing to the active component loaded in the reloading process.

2.3.4.3 Integrated Optimization of SCR System

Achieving high de-NO_x efficiency requires that NH_3 and NO_x are well mixed; reactor design and ammonia injection thus play pivotal roles. Relative standard deviation, C_v, is usually used for evaluation, as shown below:

$$C_v = \sigma_v / \overline{x} \times 100\%, \tag{2.28}$$

$$\sigma_v = \sqrt{\sum (x_i - \bar{x})^2 / (n - 1)}, \tag{2.29}$$

where $\bar{x}$ represents a mean value, x_i represents a single value, n represents the sample size, and σ_v is the standard deviation.

Deviations in the NH_3/NO_x molar ratio at the inlet of the first layer of catalyst, velocity, temperature, and incident angle are calculated accordingly. Improper design, installation, and maintenance of the SCR system can cause these parameters to surpass limitations, which then increases the ammonia slip and corrosion and blockage of the air preheater. Moreover, the ammonia distribution and flow characteristics of flue gas greatly influence the utilization of the catalyst, and consequently, the $deNO_x$ efficiency. However, owing to space, layout, and cost constraints, the flue length and mixing distance are always insufficient. Expansion of gases and bending structure of reactor also decrease the mixing efficiency of NH_3 and NO_x, thereby causing an uneven distribution of NH_3 and NO_x in flue gas. Improving $deNO_x$ thus requires the SCR system to be optimized, especially the reactor structure and NH_3 injection method.

Optimizing the SCR reactor: Due to space constraints, there is commonly a section of flue with variable cross-sectional area before the ammonia injection grid (AIG). Since this sudden change to cross-sectional area causes an uneven air velocity distribution, the flue structure and deflector must be optimized to meet the demand of flue rectifying.

An example arrangement of the deflectors in a variable flue cross section is shown in Fig. 2.61. Here, the rapid spatial expansion of the first cross section (left) affects the flow field in the system. In this case, large deflectors are preferable. Conversely, the cross section on the right shows less structural change, as there is only an inclined wall on one side. Accordingly, small deflectors are fixed to promote the uniform distribution of the flow field while maintaining a low pressure drop.

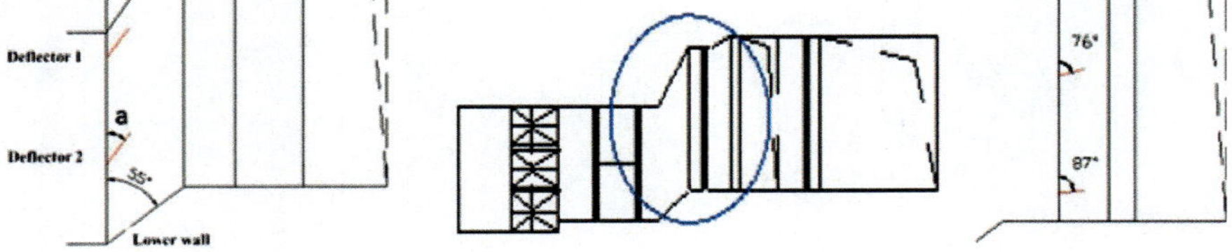

Fig. 2.61 Example deflector arrangement in variable cross-section flue

Three other typical variable cross sections and their deflector design schemes are shown in Fig. 2.62. Their velocity distributions before the AIG inlet, shown in Fig. 2.63, had a corresponding C_v of 5.93, 6.3, and 14.3%, respectively. Overall, the distribution of velocity was clearly affected by structure; the velocity distribution worsened with the dramatic change of cross section and related spatial variation, as evidenced by the increasing C_v values, despite the addition of deflectors in schemes 2 and 3.

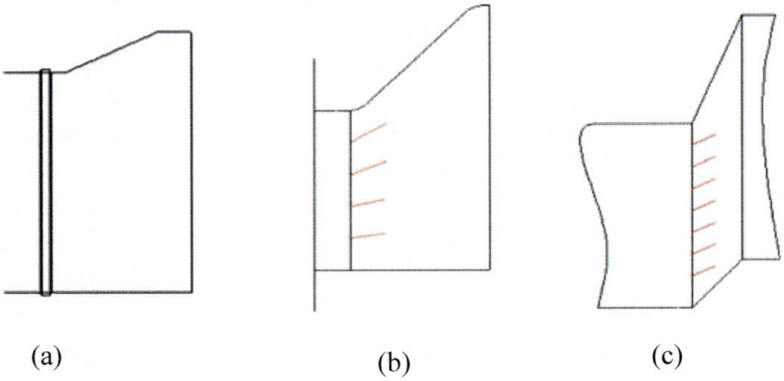

Fig. 2.62 Typical variable cross sections and related schemes

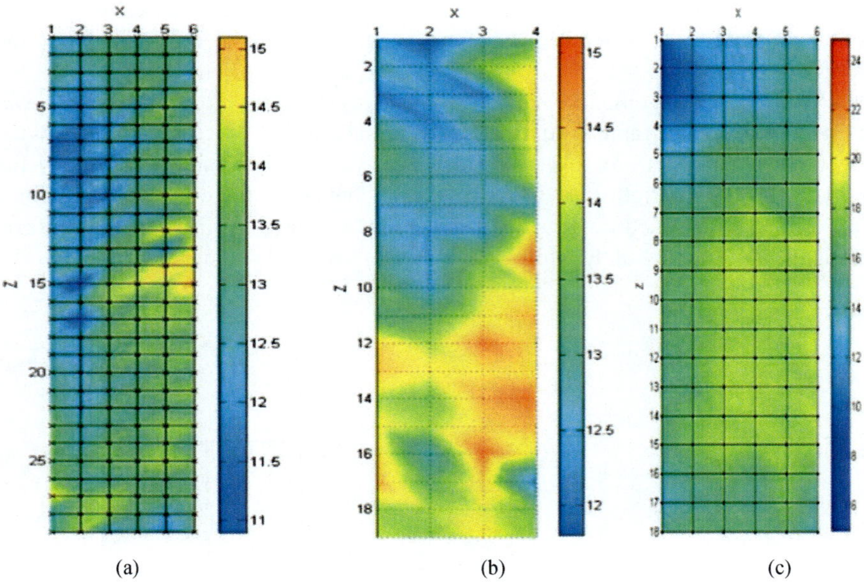

Fig. 2.63 Velocity distribution before AIG inlet of schemes

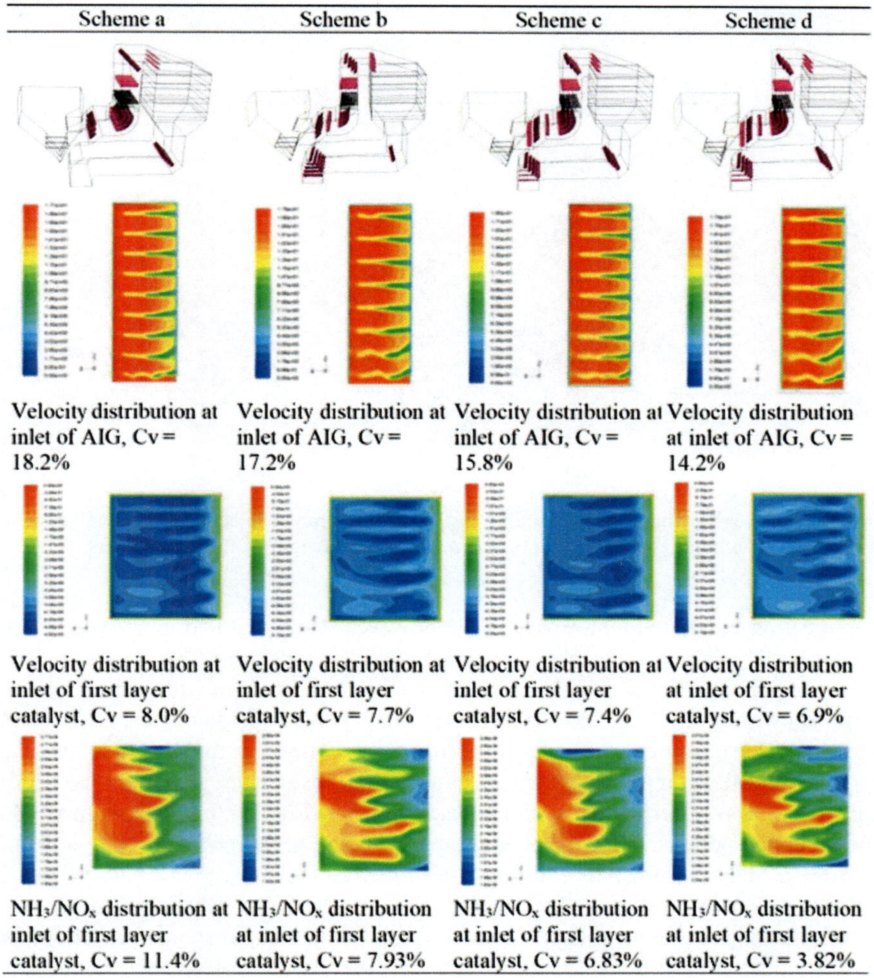

Fig. 2.64 An example of the integrated design process results of an SCR reactor

An example of the integrated design results of an SCR reactor is shown in Fig. 2.64. Four deflector schemes, a, b, c and d, are compared. Here, the C_v of velocity distribution decreased from 14.2% at the AIG inlet to 6.9% at the first layer catalyst inlet, as well as NH_3/NO_x distribution, which shows a low value of 3.82% at the first layer catalyst inlet.

Enhancing ammonia injection and mixing: Conventional ammonia injection grids are equipped with many small nozzles and thus require a high investment cost and a complex control system. Further, the small nozzles can become easily clogged and make precise control of the jet flow difficult. Static mixers installed downstream from

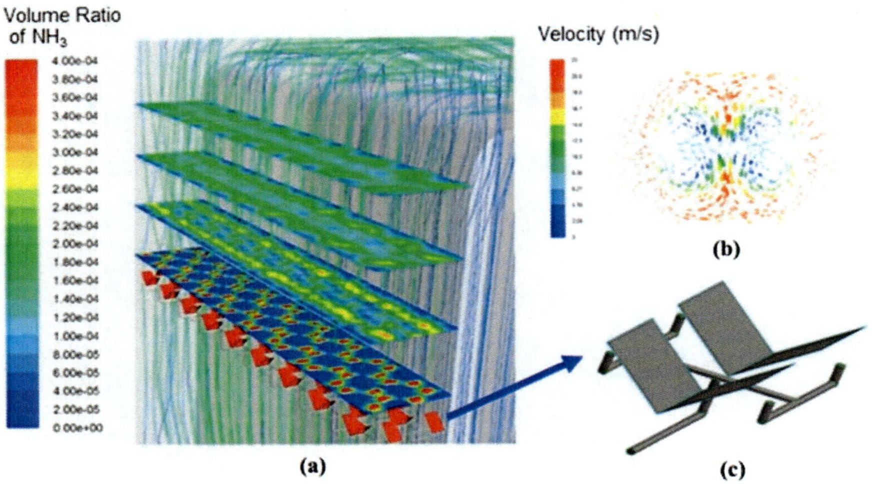

Fig. 2.65 Novel NH_3 injection and mixing system, **a** the ammonia injection and mixing system, **b** the manifold of nozzles converges to the main tube, **c** the structure of the static mixers

the injection grid help to improve mixing; however, these mixers often employ large blades with a high coverage area, causing a large pressure drop.

A recently developed ammonia injection and mixing system with a low pressure drop, shown in Fig. 2.65a, comprises static mixers and related nozzles. Each static mixer is made of folded plates with specific angles. Upstream from the mixer, nozzle groups, typically comprising four nozzles and two folded plates, are fixed. The manifold of nozzles converges to the main tube, as shown in Fig. 2.65b. This design allows for accurate control in a local area and rapid response to variations in NO_x concentrations in the flue gas. After passing the mixer, as shown in Fig. 2.65c, jet flows form symmetrical vortices, which enhance heat and mass transfer, facilitate mesoscopic and microscopic mixing processes, shorten the mixing distance, and allow for uniform distribution of NH_3 in the flue gas. By employing this system, researchers have demonstrated a reduction of C_v into 2.14% in the NH_3/NO_x distribution located at the inlet of the first catalyst layer, as shown in Fig. 2.66.

2.3.4.4 Applications

Poison-resistant catalyst: As discussed in Sect. 2.3.1, flue gas contains a variety of elements (e.g., alkali/alkaline earth metals, As, and Pb) that lead to catalyst deactivation and thus challenge the long-term stable operation of the SCR system. Researchers have thus developed a poison-resistant catalyst; a long-term pilot-scale stability test (5000 h) demonstrated its poison resistance, as shown in Fig. 2.67. Other developed

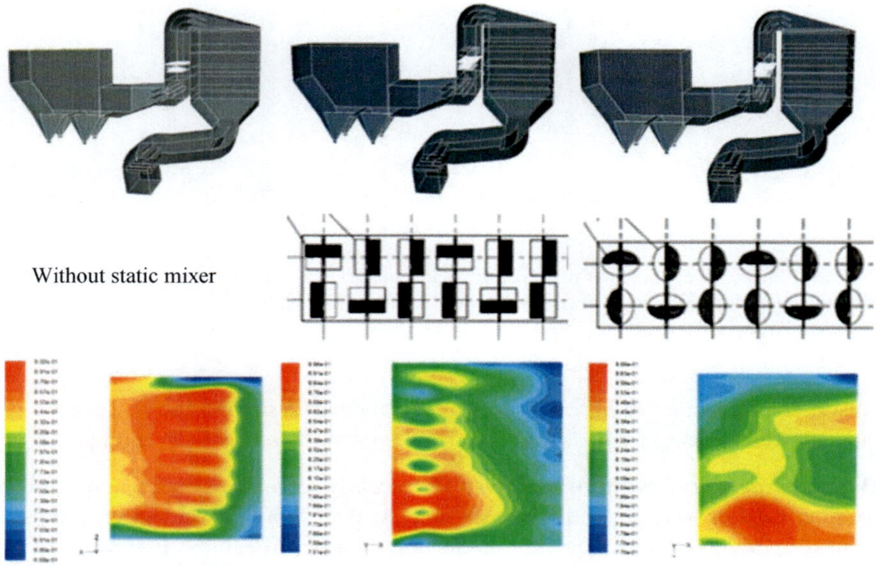

NH3/NOx distribution at inlet of the first layer catalyst C_v = 5.88%

NH3/NOx distribution at inlet of the first layer catalyst C_v = 3.75%

NH3/NOx distribution at inlet of the first layer catalyst C_v = 2.14%

Fig. 2.66 Effect of static mixer on the NH_3/NO_x distribution

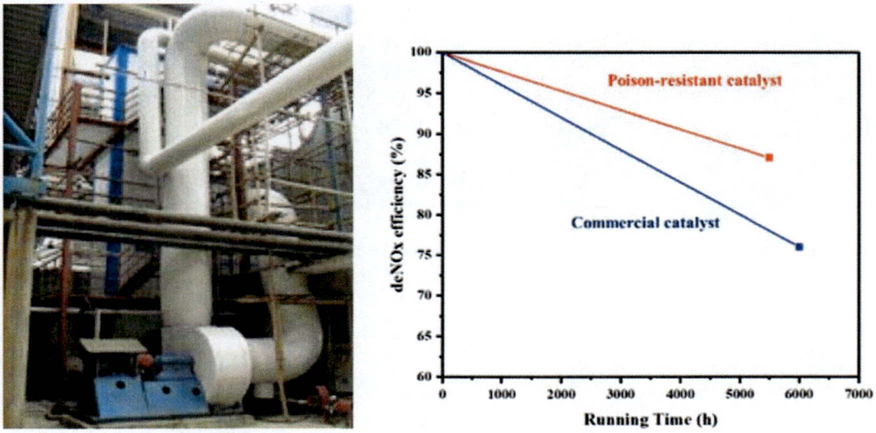

(a) (b)

Fig. 2.67 Pilot-scale experimental study on the stability of q (**a**) Pilot-scale experimental platform; (**b**) Long-term stability test results of the anti-poison catalyst

(a) (b) (c)

Fig. 2.68 Full-scale applications of poison-resistant catalysts: (**a**) Ca content in ash > 20%, deNO$_x$ efficiency > 86%; (**b**) S content of coal > 3%, deNO$_x$ efficiency > 90%; and (**c**) alkali metal content of coal > 3%, deNO$_x$ efficiency > 80%

poison-resistant catalysts have also been used in industry, as shown in Fig. 2.68, where they allowed for long-term efficient and stable operation of the SCR equipped.

Wide temperature window catalyst: A recently developed wide temperature window catalyst was applied to the SCR system of a 330-MW coal-fired power plant (see Fig. 2.69) with flue gas parameters as shown in Table 2.1. Here, the flue

Fig. 2.69 330-MW coal-fired unit employing a wide-temperature window catalyst for SCR

Table 2.1 Parameters of SCR inlet flue gas

Item	Unit	Parameters				
		100% BMCR	100% THA	75% THA	50% THA	35% BMCR
Flue gas volume	m^3/h	2,617,072	2,441,301	162,165	999,889	816,234
Volume flow-rate	m^3/h	1,112,623	1,049,162	736,972	484,150	416,594
Mass flow-rate	kg/h	1,511,492	1,425,280	1,001,172	657,715	565,940
Temperature	°C	365	365	333	297	268

Table 2.2 Test results under different load condition

Item		Unit	Design value	Measured value
100% load	deNO_x efficiency	%	≥ 85	85.31
	Ammonia slip	ppm	≤ 3	2.45
30% load	deNO_x efficiency	%	≥ 85	85.05
	Ammonia slip	ppm	≤ 3	2.58

Fig. 2.70 Practical application of the low SO_2/SO_3 conversion SCR catalyst

gas temperature fluctuates within the range of 268–365 °C. At different load conditions, NO_x removal efficiency remained above 85%, as shown in Table 2.2, thus demonstrating the catalyst's adaptability to temperature.

Low SO_2/SO_3 conversion catalyst: A recently developed low SO_2/SO_3 conversion SCR catalyst was applied to the 330-MW thermal power unit shown in Fig. 2.70, for which the concentration of SO_2 at the inlet of the SCR system was approximately 700 mg/m^3. The catalyst allowed for a higher deNO_x efficiency from 72.7 to 82.11% and a lower SO_2/SO_3 conversion rate from 0.48 to 0.17%, as summarized in Table 2.3.

Regeneration of deactivated catalyst: A redesigned manufacturing line for regenerating deactivated SCR monoliths is shown in Fig. 2.71; an example of a plate-type catalyst module before and after regeneration is shown in Fig. 2.72. After regeneration, the monolith had less dust accumulation and looked considerably cleaner. Furthermore, by comparing the concentration of poisoning elements and the specific surface area before and after catalyst regeneration, as shown in Tables 2.4 and 2.5, respectively, regeneration reduced the presence of poisoning elements and increased the specific surface area and pore volume of the catalyst.

The regenerated catalysts were then used for SCR; the results, summarized in Table 2.6, demonstrate that regeneration allowed a near-full recovery to fresh catalyst performances. At 100% BMCR (382 °C) and 35% BMCR (275 °C), the deNO_x

Table 2.3 Results from the 330-MW scale test of a low SO_2/SO_3 conversion catalyst

Item	Unit	Test results	
		Commercial catalyst	Low SO_2/SO_3 conversion catalyst
$C_{N}O$ at inlet	mg/m^3	432.9	425
$C_{N}O$ at outlet	mg/m^3	117.8	76
deNO$_x$ efficiency	%	72.7	82.11
$C_{S}O_2$ at inlet	mg/m^3	739	692.73
SO_3 conversion	%	0.48	0.17
NH_3 slip	ppm	12.49	1.24
Temperature at inlet	°C	364	362.7

(a) (b)

Fig. 2.71 Example of production line for deactivated catalyst regeneration

(a) (b)

Fig. 2.72 **a** Deactivated and **b** regenerated SCR catalyst

Table 2.4 Presence of poisoning elements in deactivated and regenerated catalyst

Catalyst	Na_2O	P_2O_5	SO_3	K_2O	CaO	Fe_2O_3
	%	%	%	%	%	%
Deactivated	0.329	0.106	9.23	0.303	0.597	5.43
Regenerated	0.094	0.0336	0.489	0.051	0.0416	1.23

Table 2.5 Specific surface area and pore volume of catalyst samples

Catalyst	specific surface area (m^2/g)	pore volume (cm^3/g)
Deactivated	50	0.2523
Regenerated	57	0.3009

Table 2.6 SCR performance of the regenerated catalyst

Temperature	deNO_x efficiency	Ammonia slip	The conversion rate of SO_2/SO_3
°C	%	ppm	%
382	80.8	1.9	0.48
275	80.2	0.6	–

efficiency of the regenerated catalyst was > 80%, the NH_3 slip was < 3 ppm, and the conversion rate of SO_2/SO_3 was 0.48% at 382 °C.

Further, the developed regeneration process can also be used for secondary regenerations (i.e., after deactivation following regeneration) with only minor performance reduction, as shown in Fig. 2.73. This regeneration process therefore helps reduce catalyst waste and investment costs.

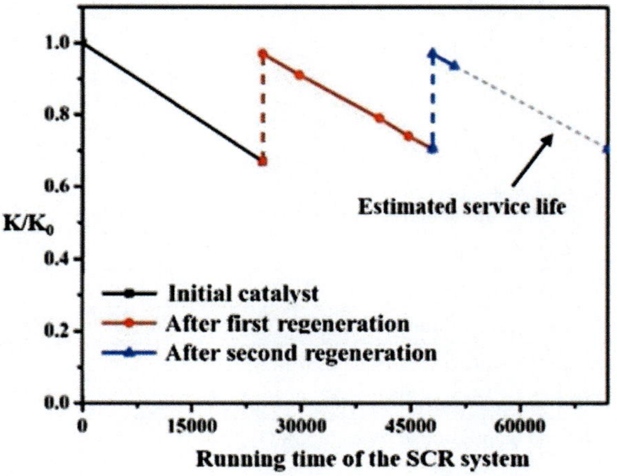

Fig. 2.73 Estimated performance of the catalyst after a second regeneration. K/K_0 represents the ratio of on-line activity to initial activity

(a) (b)

Fig. 2.74 **a** Schematic and **b** photograph of key components of the ammonia injection and mixing system

Design and optimization of the SCR reactor: Using the methodology and technology introduced in Sect. 2.3.4.3, an optimized SCR system was applied to a 150-MW thermal power unit with flue gas parameters, as shown in Table 2.7. Here, the unit was equipped with two sets of SCR reactors, each with three layers of catalysts,

Table 2.7 Flue gas parameters at the inlet of deNO$_x$ device

Item	Unit (W/V)%	Data
Composition of flue gas at economizer outlet (excess air coefficient of 1.22)		Coal
CO_2	%	13.98
O_2	%	3.54
N_2	%	73.87
SO_2	%	0.16
H_2O	%	8.45
Flue gas volume and temperature of economizer outlet at different boiler loads		
Item (deNO$_x$ device inlet)		BMCR
Wet flue gas rate Nm^3/h (Coal)		495,106
Flue gas temperature °C (Coal)		364
SO_2 concentration (Dry base, 6%O_2)		4148.7
Dust concentration (Dry base, 6%O_2)		40.36
NO$_x$ concentration (Dry base, 6%O_2)		400

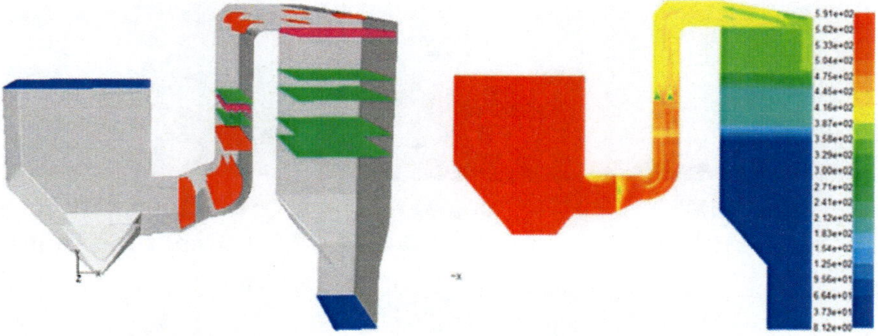

(a) (b)

Fig. 2.75 **a** Diagram of the optimized 150-MW SCR reactor and **b** simulated pressure field distribution

located between the economizer outlet and the air preheater, including liquid NH_3 storage and preparation system. At the NH_3 injection and mixing site, the flue is 7890 × 1400 mm^2 and the reactor is 7890 × 5070 mm^2. The designed NH_3 injection and mixing system and a photograph of some key components are shown in Fig. 2.74a, b, respectively.

Moreover, the flue at the bottom of the economizer, flue of the SCR inlet, SCR reactor, DC grid, SCR catalyst, and SCR outlet and the flue at the inlet of the air preheater were also optimized Fig. 2.75a. A numerical simulation and cold model indicate that the designed NH_3 injection and mixing system has excellent mixing performance, evidenced by the low relative standard deviation of NH_3/NO_x molar ratio concentration at the inlet of the first layer catalyst ($C_v < 5\%$) and velocity ($C_v < 10\%$). Furthermore, the NH_3 injection mixing device demonstrated a relatively low pressure drop because both the numerical simulation and cold state physical model had a pressure drop below 90 Pa, as shown in Fig. 2.75b.

The performed 168 h tests show that the NO_x concentration at the inlet and outlet of the SCR system was reduced from 304.3 to 16.64 mg/m^3, respectively; furthermore, NH_3 slipping was rarely detected. The system thus demonstrated deNO_x efficiencies of $> 90\%$.

These discussed methods for industrial catalyst production, deactivated catalyst regeneration, and integrated system optimization have been broadly adopted for NO_x of the flue gas of a variety of thermal power plants and industrial facilities; some are shown in Fig. 2.76. In each of the facilities shown here, NO_x emissions could be as low as 50 mg/m^3, well within the ultra-low emission standards set for coal-fired power plants in China.

Fig. 2.76 Applications of the SCR technology in installed at various capacities and industries

2.4 Simultaneous Control of Multiple Pollutants

Flue gas contains NO_x and a variety of other pollutants, including heavy metals and VOCs. To reduce their environmental influence and comply with the increasingly strict emission standards without significant added cost or space requirements, industrial enterprises have aimed to develop multi-pollutant control technologies that allow for the simultaneous control of heavy metals, VOCs, and NO_x.

2.4.1 Simultaneous Control of Mercury

Mercury emissions have become a worldwide concern owing to the bioaccumulation of mercury and its considerable toxicity and propagable properties [88]. Mercury is typically present in coal-fired flue gas as particle-bound (Hg^P), oxidized (Hg^{2+}), and elemental (Hg^0) mercury. Existing air pollution control devices (APCDs) generally can only capture Hg^P and Hg^{2+}; efforts have thus recently been made toward the development of technologies capable of Hg^0 capture in coal-fired power plants.

Post-combustion Hg^0 removal technologies, such as catalytic oxidation and sorbent injection, have been widely investigated. As the cost of sorbent injection technology is significantly high, researchers have focused on catalytic oxidation of Hg^0 to Hg^{2+} and subsequent absorption in the desulfurizing absorption tower process. Catalytic oxidation can be performed using HBr or HCl as an oxidizing agent, as shown in the following equations [89]:

$$2HCl_{(ad)} + M_xO_y \rightarrow M_xO_{y-1} + H_2O + 2Cl^* \quad (2.30)$$

$$HgO - M_xO_{y-1} + 2Cl^* \rightarrow HgCl_{2(g)}\uparrow + M_xO_y \quad (2.31)$$

$$Hg^0_{(g)} + 2Cl^* \rightarrow HgCl_{2(g)} \quad (2.32)$$

However, the concentration of HBr or HCl in the flue gas is usually lower than the required level. As there is a substantial amount of O_2 (approximately 3–10%) in the flue gas, efforts have thus been made to develop catalysts to enable the oxidation of Hg^0 using O_2 as the oxidizing agent.

In order to guide the design of mercury oxidation catalyst, the reaction mechanism has been studied. Generally, Hg^0 oxidation may occur via three possible mechanisms: the MvK, LH, and ER mechanisms [90]. In the MvK mechanism, the gaseous mercury molecules are oxidized to Hg^{2+}, leading to the reduction of metal oxides active sites and the formation of catalyst surface vacancies, which would then be replaced by gas phase oxygen [91]. In the LH mechanism, O_2 molecules in the gas phase adsorb and dissociate into the O* species on the surface of the catalyst, and Hg^0 molecules adsorb on the active sites without dissociation at the same time, then Hg^0 molecules are oxidized into Hg^{2+} by O* species. In the ER mechanism, only Hg^0 molecule adsorbs on the surface of the catalyst, then it directly reacts with the O_2 molecules in the gas phase [92].

Qualitative studies have helped clarify the Hg^0 oxidation mechanism. For example, the number of surface-active sites of the catalyst is considered to be crucial to the Hg^0 removal capacity, as is supported by first-principle calculations [93, 94]. Additionally, DFT was used to clarify the interactions between Hg^0 and O_2 on the (110) surface of an IrO_2 catalyst [95]. Of the five stable Hg^0 adsorption configurations identified, the most stable adsorption position was at the top of a fivefold coordinated Ir atom (denoted as $Ir_{cus\text{-}top}$). An in-depth analysis of the interactions between the Hg^0 atom and O atom on the IrO_2 (110) surface showed that the adsorption energy of O is higher than that of Hg^0 on the $Ir_{cus\text{-}top}$. Moreover, the pre-adsorption of O atoms has a positive effect on the adoption of Hg^0, and the adsorption was identified as chemisorption, as shown in Fig. 2.77. More importantly, since the adsorption energy of Hg^0 is much lower after pre-adsorption of O atom, the LH mechanism was determined to be the most probable reaction mechanism, as shown in Fig. 2.78. Here, the (110) surface of IrO_2 can strongly bind the reactants (Hg and O atoms) and release HgO.

Building from the assumption of the LH mechanism, researchers have aimed to develop catalysts for Hg^0 oxidation, such as by employing the Sabatier principle to a suite of candidate materials [96]. Here, the synergistic effect between Mo–O and Ce–O was demonstrated to allow for excellent catalytic activity with a low required activation energy. First-principle calculations and experimental evaluations helped clarify the synergistic effect of these catalysts based on a structure–activity relationship.

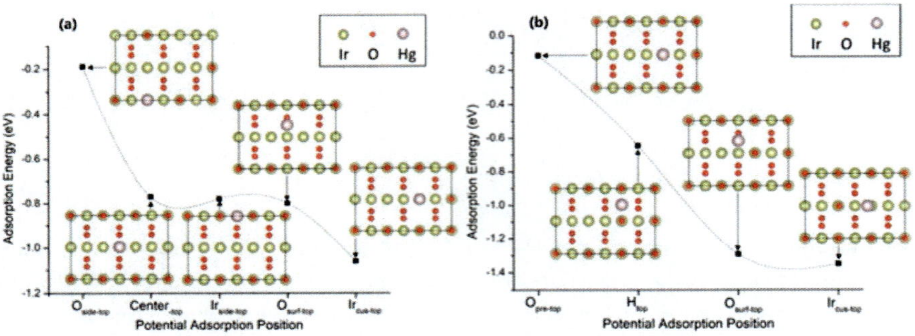

Fig. 2.77 Characteristic Hg adsorption energy as a function of various potential adsorption positions obtained using DFT calculation on the (110) surface of (**a**) an IrO_2 catalyst and (**b**) an IrO_2 catalyst with one pre-adsorbed O atom [95]

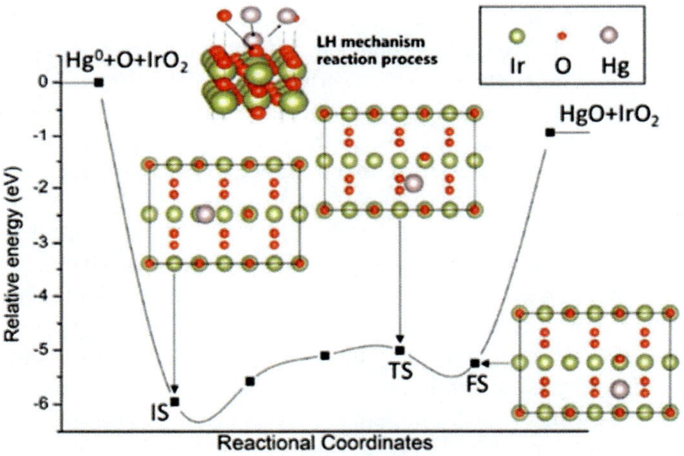

Fig. 2.78 Adsorption pathways and energy profiles of Hg^0 adsorption process on the IrO_2 (110) surface [95]

With Hg^0, CeO_2 has a relatively strong bond strength, whereas MoO_3 has a relatively weak bond strength. Pre-doping MoO_3 results in the transformation of CeO_2 clusters to an additional exposed CeO_2 (111) surface, which leads to a increase of Bronsted acid sites. Moreover, there is an overlap between the Hg^0 desorption region of MoO_3 and the Hg^0 adsorption region of CeO_2 (with adjusted optimum bond strength with Hg^0), which contributes to the catalytic reaction of Hg^0 by O_2, as shown in Fig. 2.79. Thus, the mechanism of Hg^0 catalytic oxidation was clarified and is shown in Fig. 2.80. A CeO_2 and MoO_3 bimetallic sample may thus enhance catalytic performance by the intermediate bond strength achieved via the synergistic effects of the low and high strength of the Hg^0-MoO_3 and Hg^0-CeO_2 bonds, respectively.

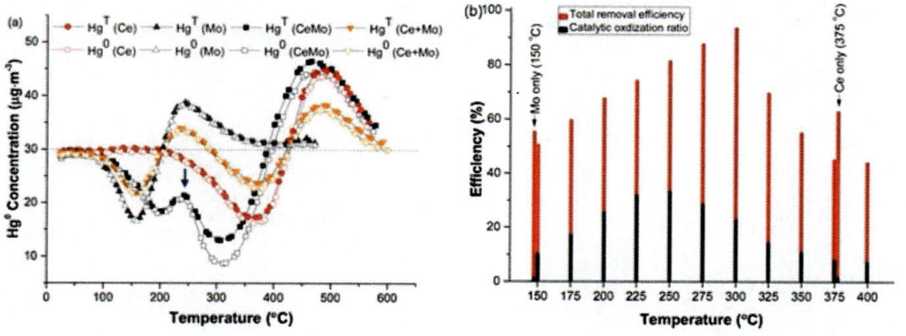

Fig. 2.79 **a** Integrated dynamic and **b** steady-state analyses of Hg^0 oxidation catalysts; HGT = total amount of mercury [96]

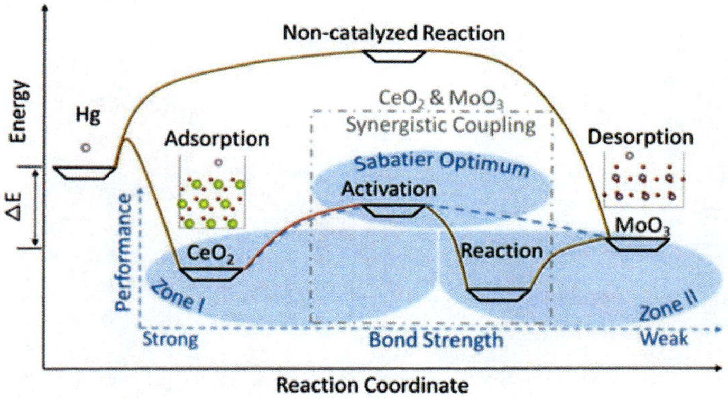

Fig. 2.80 Schematic representation of the Sabatier principle and rational design of CeO_2-MoO_3 bimetallic catalyst for catalytic oxidation of Hg^0 [96]

MoO_3 has also been used to adjust the δ-MnO_2 nanosheet for catalytic oxidation of Hg^0 to Hg^{2+} by exploring Mo–Mn catalysts through a combination of experimental and computational approaches. Here, a series of catalysts with varying Mn and Mo loading (1–10%) were synthesized and characterized. Among them, the 1.25Mo_2Mn catalyst exhibited the highest Hg^0 removal performance of approximately 98% (oxidation ratio > 50%, the rest is removed by adsorption) with excellent stability over 10 h at 250 °C, as shown in Fig. 2.81 [94].

The proposed reaction mechanism for this Hg^0 oxidation on MoO_3-adjusted δ-MnO_2 nanosheet is shown in Fig. 2.82. In general, the synergistic interaction between MoO_3 and the surface of the defective δ-MnO_2 nanosheet in the 1.25Mo_2Mn catalyst allowed for a higher conversion of Hg^0 to Hg^{2+}. This was achieved by the MoO_3-adjusted properties of activation energy, bond strength, Brønsted acid sites, and defects, which were favorable for the catalytic oxidation of Hg^0. In addition, the desorption step, which was defined as the rate-determining step, was promoted by

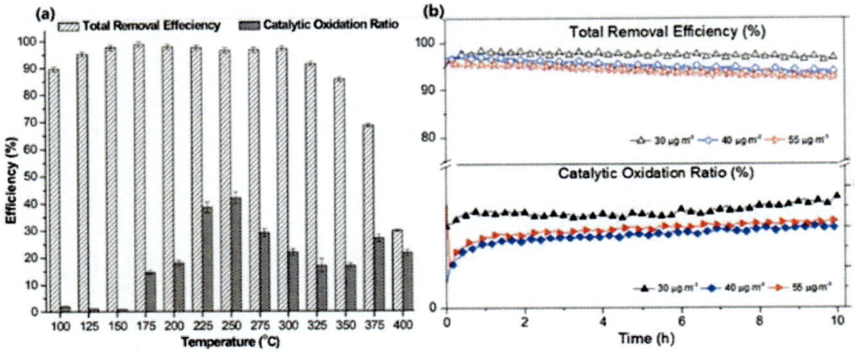

Fig. 2.81 Steady-state analysis of catalytic oxidation of Hg^0 over a $1.25Mo_2Mn$ catalyst: (**a**) removal efficiency and oxidation ratio and (**b**) effect of Hg^0 concentration [94]

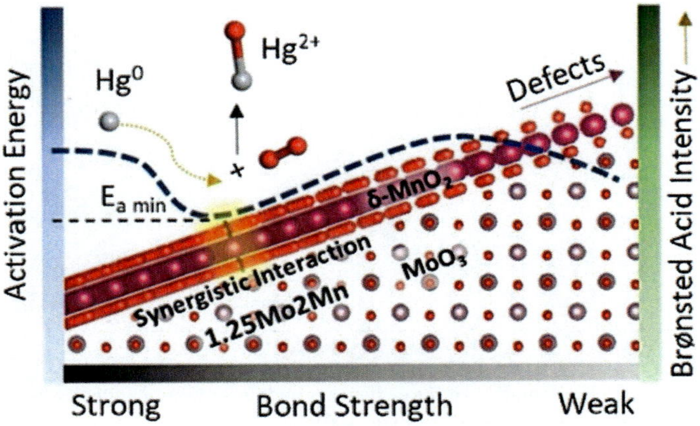

Fig. 2.82 Proposed mechanism for MoO_3-adjusted δ-MnO_2 nanosheet for Hg^0 catalytic oxidation [94]

the introduction of MoO_3, which led to a significant reduction in desorption energy [94].

The oxidation of Hg^0 to Hg^{2+} via a V_2O_5-$WO_3(MoO_3)/TiO_2$ catalyst has been demonstrated to have decent performance; however, the oxidation efficiency is significantly affected by HCl concentration [97]. Here, lab-scale tests at 300 °C demonstrated that the homogeneous oxidation of Hg^0 was negligible and the Deacon reaction was not observed using an untreated V_2O_5-$WO_3(MoO_3)/TiO_2$ catalyst. Only a small amount of Hg^0 was adsorbed onto the surface of the catalyst, as shown in Fig. 2.83. However, after adding HCl into the reactor, the adsorption behavior of Hg^0 changed drastically; HCl caused the desorption of mercury from the catalyst. These results demonstrate that HCl adsorbs on the surface of the catalyst in a pure N_2 environment and competes with weekly adsorbed Hg^0 on the active sites, as shown

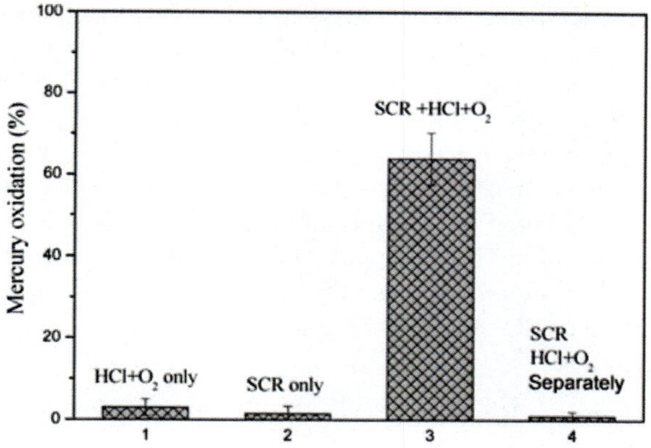

Fig. 2.83 Oxidation of Hg^0 by O_2 and HCl over a commercial V_2O_5-based SCR catalyst at 300 °C [98]

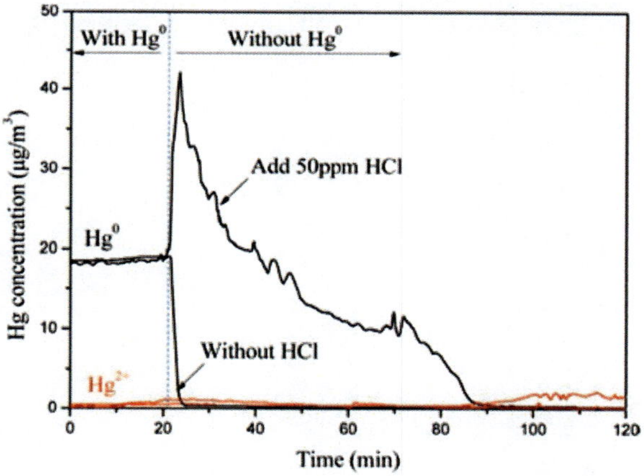

Fig. 2.84 Release of mercury from the surface of a commercial V_2O_5-based catalyst by adding HCl at 300 °C [98]

in Fig. 2.84. Furthermore, the adsorption of HCl onto the catalyst was confirmed via XPS and FTIR, allowing reaction pathways to be deduced. The V_2O_5 sites on the SCR catalyst are thus favorable for the adsorption of HCl, thereby allowing the activated Cl to react with adjacent Hg^0. Therefore, the mercury oxidation on the V_2O_5-based SCR catalyst can be regarded as a heterogeneous reaction according to the LH mechanism [98].

In a system using subbituminous coal or lignite, the Hg^0 oxidation efficiency is relatively low owing to the fairly low concentration of HCl in the flue gas. Impregnating commercial V_2O_5/TiO_2 catalyst with metal oxides may enhance mercury oxidation in such a system. Transition metals have been widely used for mercury oxidation due to their low cost, better oxidation performance and wide reaction temperature (80–500 °C). A series of transition metal oxides, e.g., MnO_x, CeO_x, FeO_x, CuO, V_2O_5, NiO_x, and CoO_x, have thus been examined for simultaneous NO_x and mercury conversion with emphasis on a lower working temperature and broader temperature window. Among them, ruthenium oxide (RuO_2) has shown to be promising by allowing the minimum HCl concentration required for Hg^0 oxidation to be reduced. With only 5 ppm of HCl in the simulated gases, the Hg^0 oxidation efficiency over a RuO_2-doped SCR catalyst was over 90% [99].

Cr_2O_3/TiO_2 has also been highlighted for its excellent mercury oxidation in flue gas generated by coal combustion [100]. Further, a Cr_2O_3-modified V_2O_5/TiO_2 catalyst exhibited improved catalytic activity for both Hg^0 oxidation and NO reduction, as shown in Figs. 2.85 and 2.86. Here, the presence of NO_x, O_2, and HCl promoted Hg^0 oxidation whereas SO_2 inhibited it likely owing to the competitive adsorption between SO_2 and Hg^0. H_2O also inhibited mercury oxidation [101].

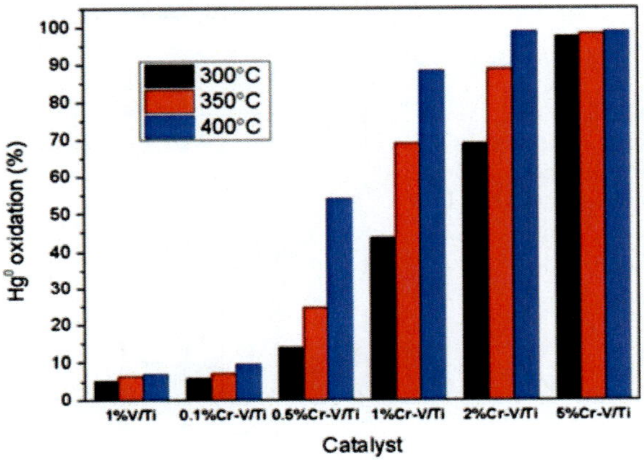

Fig. 2.85 Activity of Hg^0 oxidation from 300 to 400 °C over Cr_2O_3-modified V_2O_5/TiO_2 catalysts with different Cr contents [101]

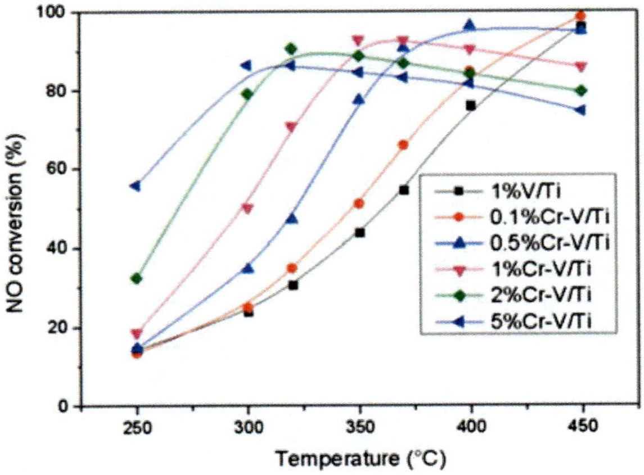

Fig. 2.86 NO conversion as a function of temperature over Cr_2O_3-modified V_2O_5/TiO_2 catalysts with different Cr contents [101]

2.4.2 Simultaneous Control of Volatile Organic Compounds

VOCs are one of the most common precursors of ozone, photochemical smog, and haze. Owing to strict emission control regulations, technologies for removing dilute VOCs with high efficiency are in great demand. Catalytic conversion is an effective method to control emissions from flue gas from coal-fired sources. Most catalysts, however, have mainly focused on a single pollutant type, making it difficult to reduce emissions of multiple pollutants simultaneously.

Catalytic oxidation of VOCs using noble metals and transition metal oxides has been investigated in recent decades. Noble metals are more active at low temperatures than transition metal oxides, but their high cost and poor resistance to halogens have limited their application. Some transition metal oxides, such as copper oxide and cerium oxide, have also been employed. Copper oxide catalysts possess high activity for catalytic oxidation of VOCs, whereas cerium oxide catalysts show outstanding oxygen storage capacity and reducibility of Ce^{4+}/Ce^{3+} compared to other fluorite-type oxides. In one study, a series of cobalt-cerium spinel oxides catalysts were prepared via a citric acid complex method with different Co/Ce ratios [102]. Among them, the $Co_2Ce_1O_x$ catalyst exhibited the highest activity at low temperatures with a T_{50} value of 220 °C, as shown in Fig. 2.87. Cerium ions could thus enter into the spinel-type lattice of cobalt oxides and cause considerable lattic distortion. The activated Co–O bonds promote propane (C_3H_8) activation and enhance te reducibility of active Co^{3+} species. Furthermore, a high Co^{3+} content and oxygen mobility were also observed in the Co-Ce spinel oxide catalysts.

Cu-Ce binary metal oxide catalysts have also been demonstrated to possess activity comparable to supported noble metal catalysts in acetone oxidation, as the transformation between Ce^{4+} and Ce^{3+} becomes easier when other transition metal ions are introduced to the ceria lattice. To demonstrate this phenomenon, a series of

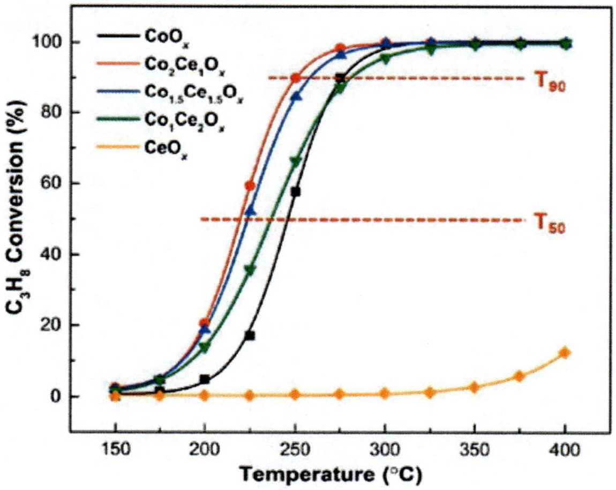

Fig. 2.87 Comparison of Co-Ce spinel oxide catalysts via total propane (C_3H_8) oxidation [102]

$CuCeO_x$ nanofiber catalysts with different Cu/(Cu + Ce) molar ratios were synthesized using an electrospinning method and evaluated for acetone oxidation at different temperatures (100–280 °C) with a GHSV of 79,000 ml g^{-1} h^{-1} [103]. Overall, the nanofiber catalysts possessed better catalytic performance than catalysts prepared by urea-nitrate combustion and sol–gel methods (Fig. 2.88).

V_2O_5/TiO_2 catalysts have also been studied for their catalytic performance during NH_3-SCR, chlorinated VOCs oxidation and photocatalysis. Furthermore, the redox properties of V_2O_5/TiO_2 catalysts may be modified because of the electronic interaction between TiO_2 and VO_x species. A controllable strategy was proposed to fabricate novel hierarchical V_2O_5/TiO_2 nanofiber catalysts, shown via SEM in Fig. 2.89

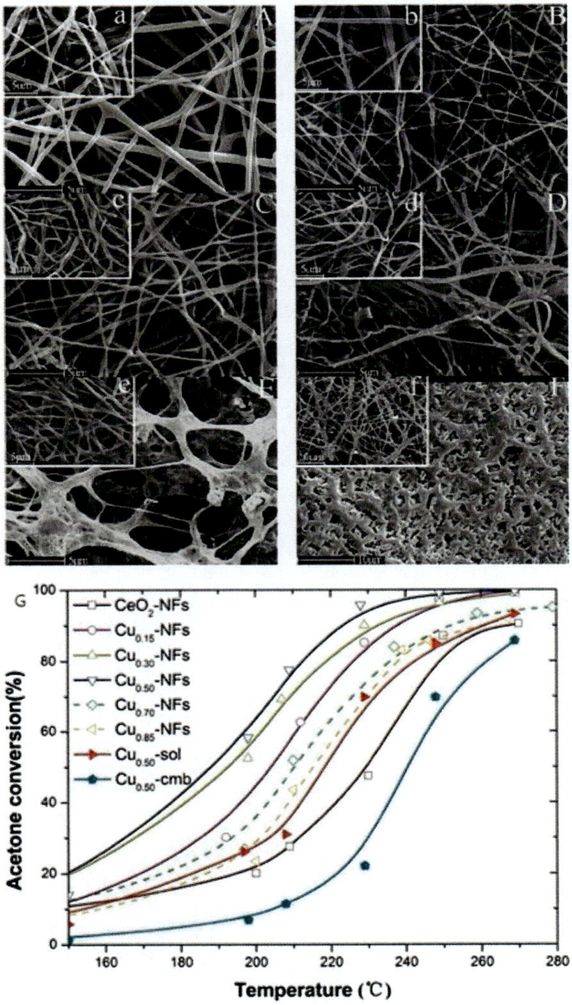

Fig. 2.88 SEM images of as-spun nanofibers: **a** CeO_2, **b** $Cu_{0.15}$, **c** $Cu_{0.30}$, **d** $Cu_{0.50}$, **e** $Cu_{0.70}$, **f** $Cu_{0.15}$; calcinated nanofibers: (A) CeO_2, (B) $Cu_{0.15}$, (C) $Cu_{0.30}$, (D) $Cu_{0.50}$, (E) $Cu_{0.70}$, (F) $Cu_{0.15}$, (G) Conversion of acetone [103]

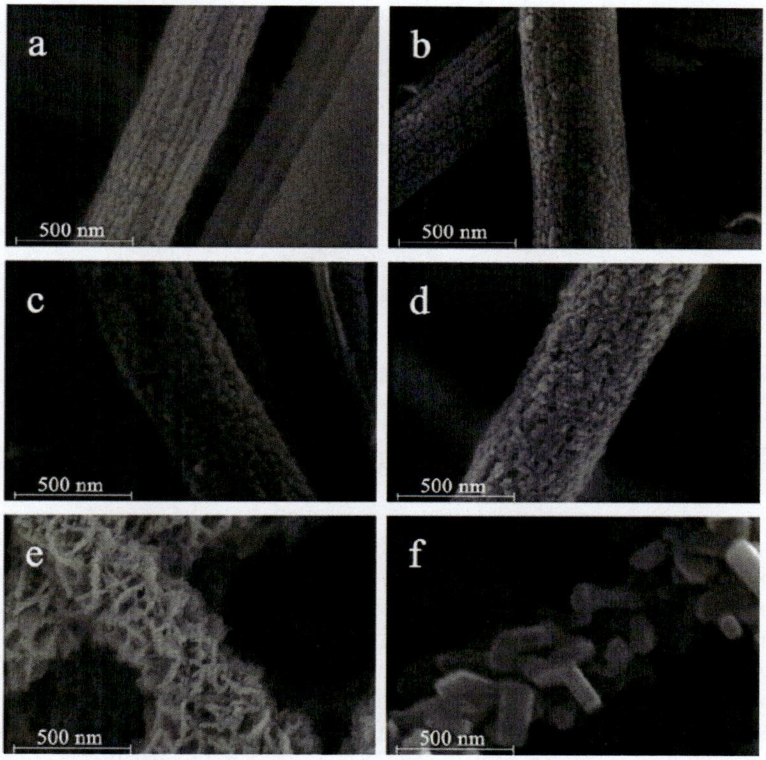

Fig. 2.89 **a–e** SEM images of hierarchical nanofibers before calcination with a hydrothermal time of **a** 3 h, **b** 6 h, **c** 12 h, **d** 18 h, **e** 24 h, and **f** SEM image of calcinated V_2O_5/TiO_2 nanofibers with a hydrothermal time of 24 h [104]

[104]. The catalysts, which comprised primary TiO_2 nanofibers and secondary V_2O_5 nanoparticles, were manufactured combining electrospinning and hydrothermal growth methods. The controllable synthesis processes and possible formation mechanisms were also demonstrated via a series of time-dependent experiments. The hierarchical V_2O_5/TiO_2 nanofiber catalysts showed excellent performance during acetone oxidation with a T_{90} temperature of 300 °C.

MnO_x/TiO_2 nanofiber catalysts synthesized via a combined electrospinning and hydrothermal growth technique have also demonstrated a capacity to oxidize acetone, as shown in Fig. 2.90 [105]. This synthesized nanomaterial comprising primary TiO_2 nanofibers and secondary MnO_x nanoneedles showed outstanding performance on low-concentration acetone oxidation with a T_{90} temperature of 290 °C.

To simultaneously remove VOCs and NO_x from flue gas, researchers have employed layered mixed catalysts, wherein different pollutants may occur at different active sites of the catalyst. To remove hydrocarbons (HCs) and NO_x, some researchers have investigated the catalytic selective reduction of NO using hydrocarbons as the

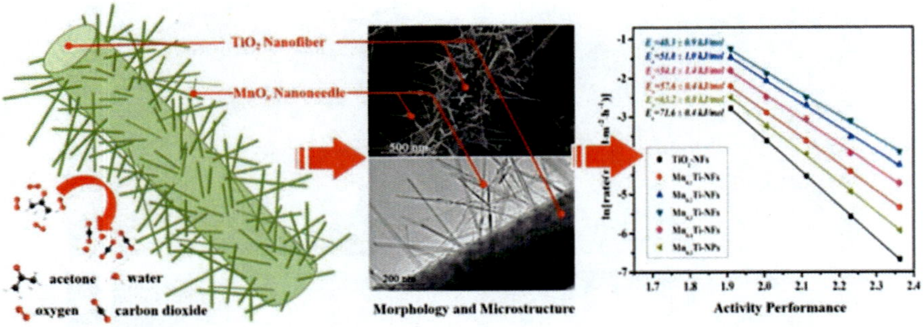

Fig. 2.90 Novel hierarchical MnO_x/TiO_2 composite nanofiber fabricated by combining the electrospinning technique and hydrothermal growth method [105]

reducing agents. However, HCs with strong reducibility are needed to get the required reaction activity. For example, at 290 °C and a C_3H_8 space velocity of 47,000 h^{-1}, the propylene oxidation of Cu/Y catalysts with a Cu loading of 16 wt.% using propylene as the reducing agent was close to 100%, the NO conversion rate was 73%, and the N_2 selectivity reached 98% [106]. Although the conversion rate of NO began to decrease at higher temperatures, increasing the O_2 concentration promoted the oxidation of NO to NO_2, significantly increasing $deNO_x$.

Researchers have aimed to optimize the distribution of active components, bed thickness, particle size, and other structural parameters of Co-based layered mixed catalysts used for multi-pollutant removal. Further, an analysis of the interaction mechanism in these catalysts between NO, NH_3, and C_3H_8 showed that the two catalysts were mixed into an upper and lower layer, comprising V_1/Ti and $Co_{0.9}Ce_{0.1}O_x$ catalysts, respectively. This mixed structure allowed the excellent redox performance of the Co-based catalyst and the high $deNO_x$ efficiency to be maintained. Here, the T_{50} temperature of C_3H_8 oxidation was as low as 227 °C at a space velocity of 90,000 ml g^{-1} h^{-1}.

To optimize the catalysts layout, the $Co_{0.9}C_{e0.1}O_x$ and V_1/Ti catalysts were mixed in three layouts with the volume ratio of 1:1, as shown in the upper left of Fig. 2.91. In mode 1, the two layers are homogenously blended (physically). In mode 2, the upper layer comprises the V_1/Ti catalyst and the lower layer comprises the $Co_{0.9}C_{e0.1}O_x$ catalyst; these layers are switched in mode 3. C_3H_8 and NO were introduced to examine the mixed catalyst's ability to remove multiple pollutants.

Although each mode demonstrated high C_3H_8 oxidation between 275 °C-350°C, as shown in Fig. 2.91, mode 2 allowed for the highest C_3H_8 oxidation. The T_{50} temperature of mode 2 was 245 °C, which was 35 and 20 °C lower than that of mode 1 and 3, respectively. At 275 °C, the C_3H_8 removal efficiency of mode 2 reached 90%, whereas modes 1 and 3 achieved 59 and 71%, respectively. The $Co_{0.9}Ce_{0.1}O_x$ catalyst thus exhibits excellent C_3H_8 and NH_3 oxidation activity. During oxidation, the metal sites are ready to capture reactant gas molecules with electron accumulations on these sites, leading to their reduced valence. The adjacent lattice oxygen sites can assist in the cleavage of the chemical bond of C_3H_8 as the oxygen atom provider.

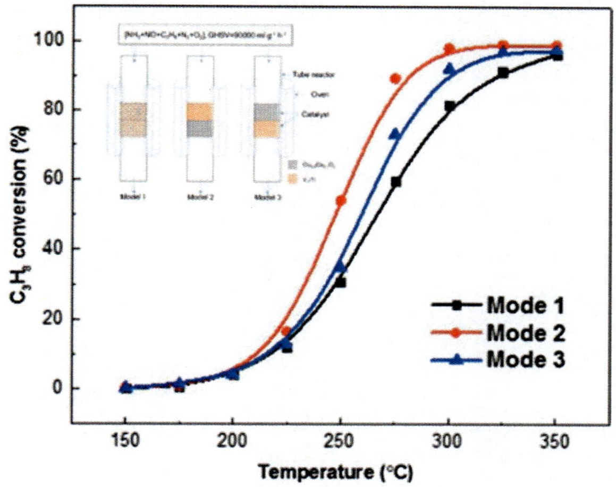

Fig. 2.91 C_3H_8 conversion resulting from the mixed $1V_1/Ti$ and $Co_{0.9}Ce_{0.1}O_x$ catalysts in modes 1, 2, and 3

When C_3H_8 and NH_3 are present in the reactive space, the competitive adsorption of these molecules on the metal sites is observed, the removal efficiency of both C_3H_8 and NH_3 decreased as a result. The C_3H_8 oxidation ability of the V/Ti catalyst is inadequate; however, the abundant strong acid sites of the V/Ti catalyst are favorable for NH_3 adsorption. Using mode 2, the $Co_{0.9}Ce_{0.1}O_x$ catalyst with its strong oxidation capacity is located just downwind of the flue gas, where NH_3 is trapped by strong acid sites of the V/Ti catalyst to reduce C_3H_8. The competitive adsorption between C_3H_8 and NH_3 thus allows for the increased catalytic C_3H_8 oxidation present in mode 2.

The effect of catalyst layer arrangement has a much greater influence on the conversion of NO, as clear from Fig. 2.92. Here, mode 2 also provided the highest deNO_x efficiency; in the reaction temperature range of 175–325 °C, the NO conversion rate was above 85%. However, the deNO_x activity of modes 1 and 3 fluctuated greatly and demonstrated a valley-like curve. Below 175 °C, the NO conversion of all modes increased to approximately 80%. With further temperature increases, the deNO_x activity of modes 1 and 3 dropped rapidly; the NO conversion rate of mode 1 was nearly zero at 225–275°C, where ammonia is directly oxidized to NO. With the decrease of reducing agent and more NO is generated, the total NO conversion turned to be quite low or even negative. At the temperatures range between 275 and 325 °C, the deNO_x activity of modes 1 and 3 increased slightly, and the NO conversion of mode 1 was approximately 30–40%.

To clarify this large variation in deNO_x of the mixed catalyst modes, the resulting N_2O and NO_2 concentrations are shown in Figs. 2.93 and 2.94, respectively. Using mode 2, the generated N_2O and NO_2 remained low throughout the entire temperature range, indicating that most NO is reduced to N_2. However, a large amount of N_2O was produced when using modes 1 and 3 in the temperature range of 225–275 °C, indicating that these side reactions were favored. At temperatures range between 275 and 325 °C, the NO conversion rate increased, as a result of further oxidization of

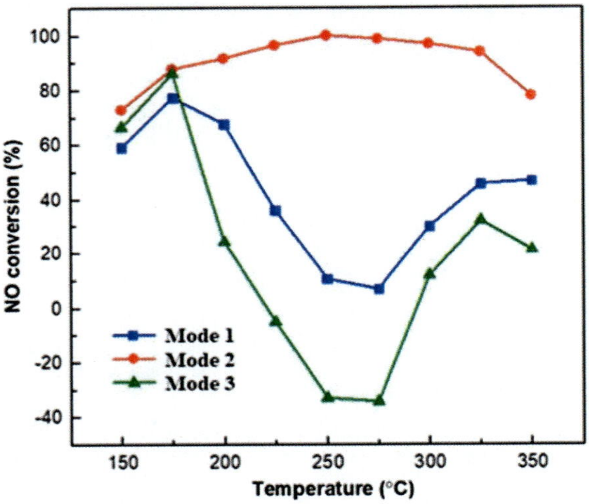

Fig. 2.92 NO conversion resulting from the mixed $1V_1/Ti$ and $Co_{0.9}Ce_{0.1}O_x$ catalysts in modes 1, 2, and 3

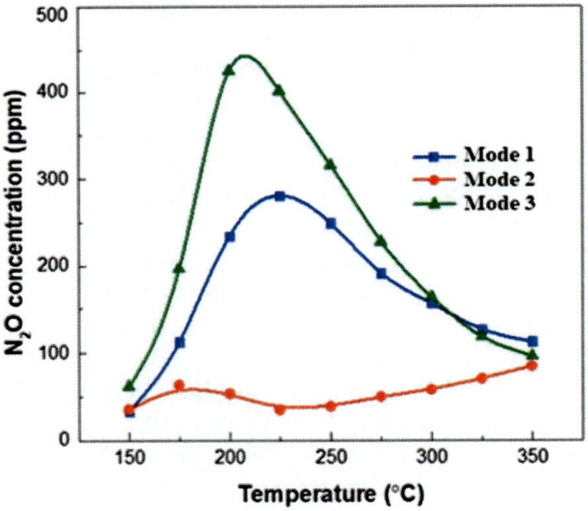

Fig. 2.93 N_2O concentration from the mixed $1V_1/Ti$ and $Co_{0.9}Ce_{0.1}O_x$ catalysts in modes 1, 2, and 3

NO to NO_2 at high temperature, instead of being reduced to N_2 in the SCR reactions. Overall, the layouts used in modes 1 and 3 are not conducive to the reduction and removal of NO.

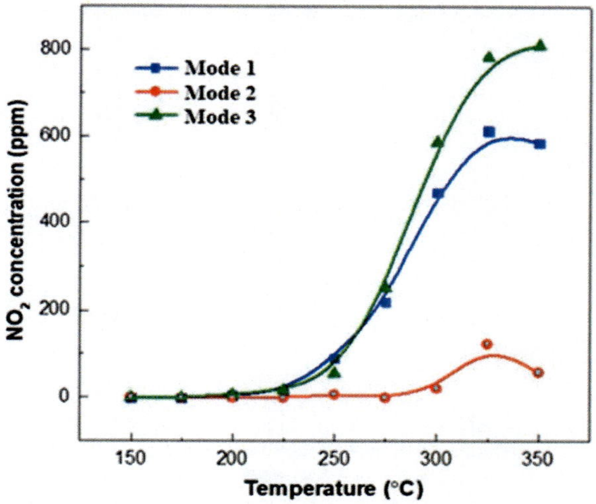

Fig. 2.94 NO_2 concentration from the mixed $1V_1/Ti$ and $Co_{0.9}Ce_{0.1}O_x$ catalysts in modes 1, 2, and 3

2.4.3 Applications

A 1000 MW coal-fired thermal plant, shown in Fig. 2.95a, was equipped with a multi-pollutant control SCR system; results are summarized in Table 2.8. An oxidation efficiency of Hg^0 above 71% with a high $deNO_x$ activity (50.7 m/h) was maintained.

In another application of a multi-pollutant control SCR system, the flue gas was treated of the 220 t/h thermoelectric power unit shown in Fig. 2.95b burning coal,

(a) (b)

Fig. 2.95 Multi-pollutant (NO_x and Hg) control SCR system applied to **a** a 1000-MW coal-fired power unit and **b** a 220 t/h thermoelectric power unit burning coal, waste, and sludge

Table 2.8 Test results of catalyst with synergistic pollutants control capability

Catalytic activity (m/h)	Ammonia slip (*10^{-6})	SO_2/SO_3 conversion (%)	Hg^0 oxidation efficiency (%)
50.7	25.1	0.671	71.4

waste, and sludge, which leads to a complex flue gas composition. Here, the NO_x and Hg emissions were controlled to be within 50 mg/m^3 and 2 μg/m^3, respectively.

References

1. Busca G, Lietti L, Ramis G et al (1998) Chemical and mechanistic aspects of the selective catalytic reduction of NO_x by ammonia over oxide catalysts: a review. Appl Catal B 18(1):1–36
2. Madia G, Koebel M, Elsener M et al (2002) Side reactions in the selective catalytic reduction of NO_x with various NO_2 fractions. Ind Eng Chem Res 41(16):4008–4015
3. Yates M, Martín JA, Martín-Luengo MÁ et al (2005) N_2O formation in the ammonia oxidation and in the SCR process with V_2O_5-WO_3 catalysts. Catal Today 107–108:120–125
4. Ravishankara AR, Daniel JS, Portmann RW (2009) Nitrous oxide (N_2O): the dominant ozone-depleting substance emitted in the 21st century. Science 326(5949):123
5. Giraud F, Couble J, Geantet C et al (2018) Experimental microkinetic approach of De-NO_x by NH_3 on $V_2O_5/WO_3/TiO_2$ catalysts. 5. Impacts of the NH_3-H_2O coadsorption on the coverage of sulfated TiO_2-based solids. J Phys Chem C 122(43):24619–24633
6. Roduit B, Wokaun A, Baiker A (1998) Global kinetic modeling of reactions occurring during selective catalytic reduction of NO by NH_3 over vanadia/Titania-Based catalysts. Ind Eng Chem Res 37(12):4577–4590
7. Han LP, Cai SX, Gao M et al (2019) Selective catalytic reduction of NO_x with NH_3 by using novel catalysts: state of the art and future prospects. Chem Rev 119(19):10916–10976
8. Anstrom M, Topsøe N-Y, Dumesic JA (2003) Density functional theory studies of mechanistic aspects of the SCR reaction on vanadium oxide catalysts. J Catal 213(2):115–125
9. Topsøe N-Y (1994) Mechanism of the selective catalytic reduction of nitric oxide by ammonia elucidated by in situ on-line fourier transform infrared spectroscopy. Science 265(5176):1217–1219
10. Du X, Gao X, Hu W et al (2014) Catalyst design based on DFT calculations: metal oxide catalysts for gas phase no reduction. J Phys Chem C 118(25):13617–13622
11. Topsoe NY, Dumesic JA, Topsoe H (1995) Vanadia-titania catalysts for selective catalytic reduction of nitric-oxide by ammonia: I.I. Studies of active sites and formulation of catalytic cycles. J Catal 151 (1):241–252
12. Qu R, Gao X, Cen K et al (2013) Relationship between structure and performance of a novel cerium-niobium binary oxide catalyst for selective catalytic reduction of NO with NH_3. Appl Catal B 142–143:290–297
13. Qu R, Peng Y, Sun X et al (2016) Identification of the reaction pathway and reactive species for the selective catalytic reduction of NO with NH_3 over cerium-niobium oxide catalysts. Catal Sci Technol 6(7):2136–2142
14. Liu F, He H, Zhang C et al (2011) Mechanism of the selective catalytic reduction of NO_x with NH_3 over environmental-friendly iron titanate catalyst. Catal Today 175(1):18–25
15. Ciardelli C, Nova I, Tronconi E et al (2004) A "nitrate route" for the low temperature "Fast SCR" reaction over a V_2O_5-WO_3/TiO_2 commercial catalyst. Chem Commun 23:2718–2719
16. Forzatti P, Nova I, Tronconi E (2009) Enhanced NH_3 selective catalytic reduction for NO_x abatement. Angewandte chemie 121(44):8516–8518

17. Hu W, Zou R, Dong Y et al (2020) Synergy of vanadia and ceria in the reaction mechanism of low-temperature selective catalytic reduction of NO_x by NH_3. J Catal 391:145–154
18. Du XS, Gao X, Qiu KZ et al (2015) The reaction of poisonous alkali oxides with vanadia SCR catalyst and the afterward influence: a DFT and experimental study. J Phys Chem C 119(4):1905–1912
19. Du XS, Yang GP, Chen YR et al (2017) The different poisoning behaviors of various alkali metal containing compounds on SCR catalyst. Appl Surf Sci 392:162–168
20. Peña DA, Uphade BS, Smirniotis PG (2004) TiO_2-supported metal oxide catalysts for low-temperature selective catalytic reduction of NO with NH_3: I. Evaluation and characterization of first row transition metals. J Catal 221(2):421–431
21. Guo RT, Wang QS, Pan WG et al (2014) The poisoning effect of Na and K on Mn/TiO_2 catalyst for selective catalytic reduction of NO with NH_3: a comparative study. Appl Surf Sci 317:111–116
22. Du XS, Gao X, Qu R et al (2012) The influence of alkali metals on the Ce-Ti mixed oxide catalyst for the selective catalytic reduction of NO_x. ChemCatChem 4(12):2075–2081
23. Zhou A, Yu D, Yang L et al (2016) Combined effects Na and SO_2 in flue gas on Mn-Ce/TiO_2 catalyst for low temperature selective catalytic reduction of NO by NH_3 simulated by Na_2SO_4 doping. Appl Surf Sci 378:167–173
24. Boxiong S, Yan Y, Shen BX, Yao Y, Chen JH et al (2013) Alkali metal deactivation of Mn-CeO_x/Zr-delaminated-clay for the low-temperature selective catalytic reduction of NO_x with NH_3. Microporous Mesoporous Mater 180:262–269
25. Shen BX, Deng LD, Chen JH (2013) Effect of K and Ca on catalytic activity of Mn-CeO_x/Ti-PILC. Front Environ Sci Eng 7(4):512–517
26. Du XS, Tang JY, Gao X et al (2016) Molecular transformations of arsenic species in the flue gas of typical power plants: a density functional theory study. Energy Fuels 30(5):4209–4214
27. Hu W, Gao X, Deng Y et al (2016) Deactivation mechanism of arsenic and resistance effect of SO_4^{2-} on commercial catalysts for selective catalytic reduction of NO_x with NH_3. Chem Eng J 293:118–128
28. Gao X, Du X-s, Fu Y-c et al (2011) Theoretical and experimental study on the deactivation of V_2O_5 based catalyst by lead for selective catalytic reduction of nitric oxides. Catal Today 175(1):625–630
29. Jiang Y, Gao X, Zhang YX et al (2014) Effects of $PbCl_2$ on selective catalytic reduction of NO with NH_3 over vanadia-based catalysts. J Hazard Mater 274:270–278
30. Zhang S, Liu SJ, Hu WS et al (2019) New insight into alkali resistance and low temperature activation on vanadia-titania catalysts for selective catalytic reduction of NO. Appl Surf Sci 466:99–109
31. Du X, Gao X, Fu Y et al (2012) The co-effect of Sb and Nb on the SCR performance of the V_2O_5/TiO_2 catalyst. J Colloid Interface Sci 368(1):406–412
32. Peng Y, Li J, Si W et al (2015) Ceria promotion on the potassium resistance of MnO_x/TiO_2 SCR catalysts: An experimental and DFT study. Chem Eng J 269:44–50
33. Du XS, Wang XM, Chen YR et al (2016) Supported metal sulfates on Ce-TiO_x as catalysts for NH_3-SCR of NO: High resistances to SO_2 and potassium. J Ind Eng Chem 36:271–278
34. Hu W, Zhang Y, Liu S et al (2017) Improvement in activity and alkali resistance of a novel V-Ce(SO_4)(2)/Ti catalyst for selective catalytic reduction of NO with NH_3. Appl Catal B-Environ 206:449–460
35. Kustov AL, Kustova MY, Fehrmann R et al (2005) Vanadia on sulphated-ZrO_2, a promising catalyst for NO abatement with ammonia in alkali containing flue gases. Appl Catal B 58(1):97–104
36. Jiang S, Li T, Zheng J et al (2020) Unveiling the remarkable arsenic resistance origin of alumina promoted cerium-tungsten catalysts for NH_3-SCR. Environ Sci Technol 54(22):14740–14749
37. Xu D, Wu W, Wang P et al (2020) Boosting the alkali/heavy metal poisoning resistance for NO removal by using iron-titanium pillared montmorillonite catalysts. J Hazardous Mater 399:122947

38. Busca G, Lietti L, Ramis G et al (1998) Chemical and mechanistic aspects of the selective catalytic reduction of NO_x by ammonia over oxide catalysts: a review. Appl Catal B-Environ 18(1–2):1–36
39. Reddy BM, Khan A, Yamada Y et al (2003) Structural characterization of CeO_2-TiO_2 and V_2O_5/CeO_2-TiO_2 catalysts by Raman and XPS techniques. J Phys Chem B 107(22):5162–5167
40. Chen L, Li J, Ge M (2009) Promotional effect of Ce-doped V_2O_5-WO_3/TiO_2 with low vanadium loadings for selective catalytic reduction of NO_x by NH_3. J Phys Chem C 113(50):21177–21184
41. Chen L, Li J, Ablikim W et al (2011) CeO_2-WO_3 mixed oxides for the selective catalytic reduction of NO_x by NH_3 over a wide temperature range. Catal Lett 141(12):1859–1864
42. Gao X, Jiang Y, Fu YC et al (2010) Preparation and characterization of CeO_2/TiO_2 catalysts for selective catalytic reduction of NO with NH_3. Catal Commun 11(5):465–469
43. Gao X, Jiang Y, Zhong Y et al (2010) The activity and characterization of CeO_2-TiO_2 catalysts prepared by the sol-gel method for selective catalytic reduction of NO with NH_3. J Hazard Mater 174(1–3):734–739
44. Gao X, Du XS, Cui LW et al (2010) A Ce-Cu-Ti oxide catalyst for the selective catalytic reduction of NO with NH_3. Catal Commun 12(4):255–258
45. Song H, Liu S, Zhang M et al (2018) A comparative study of the NH_3-SCR reactions over an original and Sb-modified V_2O_5-WO_3/TiO_2 catalyst at low temperatures. Energies 11(12):3339
46. Qi G, Yang RT (2003) A superior catalyst for low-temperature NO reduction with NH_3. Chem Commun 7:848–849
47. Smirniotis PG, Peña DA, Uphade BS (2001) Low-temperature selective catalytic reduction (SCR) of NO with NH_3 by using Mn, Cr, and Cu oxides supported on Hombikat TiO_2. Angew Chem Int Ed 40(13):2479–2482
48. Zhang R, Liu N, Luo Z et al (2014) A remarkable catalyst combination to widen the operating temperature window of the selective catalytic reduction of NO by NH_3. ChemCatChem 6(8):2263–2269
49. Qu RY, Gao X, Cen KF et al (2013) Relationship between structure and performance of a novel cerium-niobium binary oxide catalyst for selective catalytic reduction of NO with NH_3. Appl Catal B-Environ 142:290–297
50. Liu ZM, Zhang SX, Li JH et al (2014) Novel V_2O_5-CeO_2/TiO_2 catalyst with low vanadium loading for the selective catalytic reduction of NO_x by NH_3. Appl Catal B-Environ 158:11–19
51. Lin F, Wang QL, Zhang JC et al (2019) Mechanism and kinetics study on low-temperature NH_3-SCR over manganese-cerium composite oxide catalysts. Ind Eng Chem Res 58(51):22763–22770
52. Hu W, Zhang Y, Liu S et al (2017) Improvement in activity and alkali resistance of a novel V-$Ce(SO_4)_2$/Ti catalyst for selective catalytic reduction of NO with NH_3. Appl Catal B 206:449–460
53. Liu ZM, Yi Y, Li JH et al (2013) A superior catalyst with dual redox cycles for the selective reduction of NO_x by ammonia. Chem Commun 49(70):7726–7728
54. Chang HZ, Chen XY, Li JH et al (2013) Improvement of activity and SO_2 tolerance of Sn-modified MnO_x-CeO_2 catalysts for NH_3-SCR at low temperatures. Environ Sci Technol 47(10):5294–5301
55. Lai JK, Wachs IE (2018) A perspective on the selective catalytic reduction (SCR) of NO with NH_3 by supported V_2O_5-WO_3/TiO_2 catalysts. ACS Catal 8(7):6537–6551
56. Tronconi E, Nova I, Ciardelli C et al (2007) Redox features in the catalytic mechanism of the "standard" and "fast" NH_3-SCR of NO_x over a V-based catalyst investigated by dynamic methods. J Catal 245(1):1–10
57. Nova I, Ciardelli C, Tronconi E et al (2006) NH_3-SCR of NO over a V-based catalyst: low-T redox kinetics with NH_3 inhibition. AIChE J 52(9):3222–3233
58. Hu W, Zou R, Dong Y et al (2021) Mechanism and enhancement of the low-temperature selective catalytic reduction of NO_x with NH_3 by bifunctional catalytic mixtures. Indus Eng Chem Res 60(18):6446–6454

59. Sun Q, Gao ZX, Chen HY et al (2001) Reduction of NO_x with ammonia over Fe/MFI: reaction mechanism based on isotopic labeling. J Catal 201(1):89–99
60. Ruggeri MP, Selleri T, Colombo M et al (2014) Identification of nitrites/HONO as primary products of NO oxidation over Fe-ZSM-5 and their role in the Standard SCR mechanism: a chemical trapping study. J Catal 311:266–270
61. Selleri T, Ruggeri MP, Nova I et al (2016) The low temperature interaction of NO + O_2 with a commercial Cu-CHA catalyst: a chemical trapping study. Top Catal 59(8–9):678–685
62. Ye D, Qu R, Zhang Y et al (2018) Investigating the role of $H_4SiW_{12}O_{40}$ in the acidity, oxidability and activity of $H_4SiW_{12}O_{40}$-Fe_2O_3 catalysts for the selective catalytic reduction of NO with NH_3. Mol Catal 448:177–184
63. Li C, Shen M, Yu T et al (2017) The mechanism of ammonium bisulfate formation and decomposition over V/WTi catalysts for NH_3-selective catalytic reduction at various temperatures. Phys Chem Chem Phys 19(23):15194–15206
64. Magnusson M, Fridell E, Ingelsten HH (2012) The influence of sulfur dioxide and water on the performance of a marine SCR catalyst. Appl Catal B 111:20–26
65. Ma SC, JIn X, Sun YX (2010) Formation mechanism and control of ammonium bisulfate in SCR flue gas denitrification process. Therm Power Gener 39(8):12–17 (in Chinese)
66. Thogersen J, Slabiak T, White N (2010) Ammonium bisulphate inhibition of SCR catalysts. Haldor Topsoe Inc., Frederikssund, Denmark
67. Zheng L, Zhou M, Huang Z et al (2016) Self-protection mechanism of hexagonal WO_3-based $DeNO_x$ catalysts against Alkali poisoning. Environ Sci Technol 50(21):11951–11956
68. Wang X, Du X, Zhang L et al (2018) Promotion of NH_4HSO_4 decomposition in NO/NO_2 contained atmosphere at low temperature over V_2O_5-WO_3/TiO_2 catalyst for NO reduction. Appl Catal A 559:112–121
69. Qu R, Ye D, Zheng C et al (2016) Exploring the role of V_2O_5 in the reactivity of NH_4HSO_4 with NO on V_2O_5/TiO_2 SCR catalysts. RSC Adv 6(104):102436–102443
70. Ye D, Qu R, Song H et al (2016) Investigation of the promotion effect of WO_3 on the decomposition and reactivity of NH_4HSO_4 with NO on V_2O_5-WO_3/TiO_2 SCR catalysts. RSC Adv 6(60):55584–55592
71. Fang X, Liu Y, Chen L et al (2020) Influence of surface active groups on SO_2 resistance of birnessite for low-temperature NH_3-SCR. Chem Eng J 399:125798
72. Yu Y, Tan W, An D et al (2021) Insight into the SO_2 resistance mechanism on γ-Fe_2O_3 catalyst in NH_3-SCR reaction: a collaborated experimental and DFT study. Appl Catal B: Environ 281:119544
73. Ye D, Qu R, Song H et al (2016) New insights into the various decomposition and reactivity behaviors of NH_4HSO_4 with NO on V_2O_5/TiO_2 catalyst surfaces. Chem Eng J 283:846–854
74. Pang C, Zhuo Y, Qin Y (2018) Research on the NH_4HSO_4 decomposition activity of different support-guidance to the development of NH_4HSO_4-resistant SCR catalysts for NO_x abatement. In: 2018 2nd international conference on green energy and applications (ICGEA), 24–26 March 2018, 254–262
75. Chai YX, Zhang GZ, He H et al (2020) Theoretical study of the catalytic activity and anti-SO_2 poisoning of a MoO_3/V_2O_5 selective catalytic reduction catalyst. ACS Omega 5(42):26978–26985
76. Wang X, Du X, Liu S et al (2020) Understanding the deposition and reaction mechanism of ammonium bisulfate on a vanadia SCR catalyst: A combined DFT and experimental study. Appl Catal B: Environ 260:118168
77. Fan ZY, Shi JW, Niu CH et al (2020) The insight into the role of Al_2O_3 in promoting the SO_2 tolerance of MnO_x for low-temperature selective catalytic reduction of NO_x with NH_3. Chem Eng J 398:125572
78. Li J, Guo J, Shi X et al (2020) Effect of aluminum on the catalytic performance and reaction mechanism of Mn/MCM-41 for NH_3-SCR reaction. Appl Surf Sci 534:147592
79. Yu C, Hou D, Huang B et al (2020) A MnO_x@Eu-CeO_x nanorod catalyst with multiple protective effects: Strong SO_2-tolerance for low temperature $DeNO_x$ processes. J Hazardous Mater 399:123011

80. Zhang Y, Li G, Wu P et al (2020) Effect of SO_2 on the low-temperature denitrification performance of Ho-modified Mn/Ti catalyst. Chem Eng J 400:122597
81. Ye D, Qu R, Zheng C et al (2018) Mechanistic investigation of enhanced reactivity of NH_4HSO_4 and NO on Nb- and Sb-doped VW/Ti SCR catalysts. Appl Catal A 549:310–319
82. Kamata H, Ohara H, Takahashi K et al (2001) SO_2 oxidation over the V_2O_5/TiO_2 SCR catalyst. Catal Lett 73(1):79–83
83. Dunn JP, Koppula PR, Stenger G et al (1998) Oxidation of sulfur dioxide to sulfur trioxide over supported vanadia catalysts. Appl Catal B 19(2):103–117
84. Dunn JP, Stenger HG, Wachs IE (1999) Molecular structure-reactivity relationships for the oxidation of sulfur dioxide over supported metal oxide catalysts. Catal Today 53(4):543–556
85. Ji P, Gao X, Du X et al (2016) Relationship between the molecular structure of V_2O_5/TiO_2 catalysts and the reactivity of SO_2 oxidation. Catal Sci Technol 6(4):1187–1194
86. Zheng C, Xiao L, Qu R et al (2019) Numerical simulation of selective catalytic reduction of NO and SO_2 oxidation in monolith catalyst. Chem Eng J 361:874–884
87. Xin Q, Hua Z, Fu Y et al (2020) Investigation on optimal active layer thickness and pore size in dual-layer NH_3-SCR monolith for low SO_2 oxidation by numerical simulation. Fuel 279:118420
88. Zhao H, Ezeh CI, Ren W et al (2019) Integration of machine learning approaches for accelerated discovery of transition-metal dichalcogenides as Hg^0 sensing materials. Appl Energy 254:113651
89. Li H, Feng S, Liu Y et al (2017) Binding of mercury species and typical flue gas components on ZnS (110). Energy Fuels 31(5):5355–5362
90. Suarez Negreira A, Wilcox J (2013) DFT study of Hg oxidation across Vanadia-Titania SCR catalyst under flue gas conditions. J Phys Chem C 117(4):1761–1772
91. Widmann D, Behm RJ (2018) Dynamic surface composition in a Mars-van Krevelen type reaction: CO oxidation on Au/TiO_2. J Catal 357:263–273
92. Yang S, Qi F, Xiong S et al (2016) MnO_x supported on Fe-Ti spinel: a novel Mn based low temperature SCR catalyst with a high N_2 selectivity. Appl Catal B 181:570–580
93. Zhao H, Mu X, Zheng C et al (2019) Structural defects in 2D MoS_2 nanosheets and their roles in the adsorption of airborne elemental mercury. J Hazard Mater 366:240–249
94. Zhao H, Ezeh CI, Yin S et al (2020) MoO_3-adjusted δ-MnO_2 nanosheet for catalytic oxidation of Hg^0 to Hg^{2+}. Appl Catal B: Environ 263:117829
95. Zhao H, Liu S, Li W et al (2019) Mechanism of Hg^0 and O_2 interaction on the IrO_2 (110) surface: a density functional theory study. Energy Fuels 33(2):1354–1362
96. Zhao H, Yin S, Lu L et al (2020) Catalytic oxidation of Hg^0 with O_2 induced by synergistic coupling of CeO_2 and MoO_3. J Hazardous Mater 381:121037
97. Cao Y, Chen B, Wu J et al (2007) Study of mercury oxidation by a selective catalytic reduction catalyst in a pilot-scale slipstream reactor at a utility boiler burning bituminous coal. Energy Fuels 21(1):145–156
98. He S, Zhou J, Zhu Y et al (2009) Mercury oxidation over a Vanadia-based selective catalytic reduction catalyst. Energy Fuels 23(1):253–259
99. Yan N, Chen W, Chen J et al (2011) Significance of RuO_2 modified SCR catalyst for elemental mercury oxidation in coal-fired flue gas. Environ Sci Technol 45(13):5725–5730
100. Kamata H, Ueno S-i, Sato N et al (2009) Mercury oxidation by hydrochloric acid over TiO_2 supported metal oxide catalysts in coal combustion flue gas. Fuel Process Technol 90(7):947–951
101. Song H, Zhang M, Yu J et al (2018) The effect of Cr addition on Hg^0 oxidation and NO reduction over V_2O_5/TiO_2 catalyst. Aerosol Air Qual Res 18(3):803–810
102. Zhang S, Liu SJ, Zhu XC et al (2019) Low temperature catalytic oxidation of propane over cobalt-cerium spinel oxides catalysts. Appl Surf Sci 479:1132–1140
103. Qin R, Chen JH, Gao X et al (2014) Catalytic oxidation of acetone over $CuCeO_x$ nanofibers prepared by an electrospinning method. RSC Adv 4(83):43874–43881
104. Zhu XC, Chen JH, Yu XN et al (2015) Controllable synthesis of novel hierarchical V_2O_5/TiO_2 nanofibers with improved acetone oxidation performance. RSC Adv 5(39):30416–30424

105. Zhu XC, Zhang S, Yu XN et al (2017) Controllable synthesis of hierarchical MnO_x/TiO_2 composite nanofibers for complete oxidation of low-concentration acetone. J Hazard Mater 337:105–114
106. Habib HA, Basner R, Brandenburg R et al (2014) Selective catalytic reduction of NO_x of ship diesel engine exhaust gas with C_3H_6 over Cu/Y Zeolite. ACS Catal 4(8):2479–2491

Chapter 3
Pollutant Control by Electric Methods

3.1 Introduction

PM is a primary pollutant that comprises fine solid particles or liquid droplets suspended in the atmosphere or flue gas. The main sources of PM emissions are combustion, vehicle exhaust, petroleum and chemical industries, construction, and dust [1]. Particles approximately 10 μm or smaller in diameter (PM_{10}) are classified as inhalable particles; those smaller than 2.5 μm ($PM_{2.5}$) are fine inhalable particles. Particles smaller than 1 μm in diameter ($PM_{1.0}$) can permeate into the lower respiratory tract and lungs, causing respiratory problems, nonfatal heart attacks, cancer, and premature deaths [2, 3]. To accurately assess the risks of PM emissions, the size, mass, number concentrations, and morphology and chemical composition (e.g., organic, metal, and salt contents) must be considered [4].

PM emitted from stationary sources can be classified into filterable PM (FPM) and condensable PM (CPM). Sulfur trioxide (SO_3) is an important component of CPM and causes corrosion and other damages to coal-fired power plants (CFPPs) [5–7]. Representative industrial cases of flue gas emissions in some CFPPs are shown in Fig. 3.1a [8, 9]. Furthermore, many other industries contribute to PM emissions across a wide temperature spectrum, as shown in Fig. 3.1b. Solving urgent public health problems such as the removal and prevention of PM emissions represents a tremendous scientific and economic challenge. Cost-effective methods for achieving more efficient and prudent emission control are thus under development [10–12].

ESPs are a cost-effective technology used to remove PM in CFPPs wherein an electric field is formed between a discharge electrode (connected to high voltage source) and a collecting electrode (normally grounded) to ionize the flue gas [13–15]. The high voltage source generates corona discharge at the discharge electrode and ionizes the gas molecules to release free electrons and ions [16, 17]. By applying this electric field to the flue gases, particles in flue gases collide with the negative ions formed by free electrons and become charged. The charged particles migrate from the discharge electrode to the collecting electrode across the gas stream under the influence of the electrostatic field, as summarized in Fig. 3.2.

X. Gao et al., *Multi-Pollutant Control for Flue Gases*, Advanced Topics in Science and Technology in China 63, https://doi.org/10.1007/978-981-16-1518-4_3

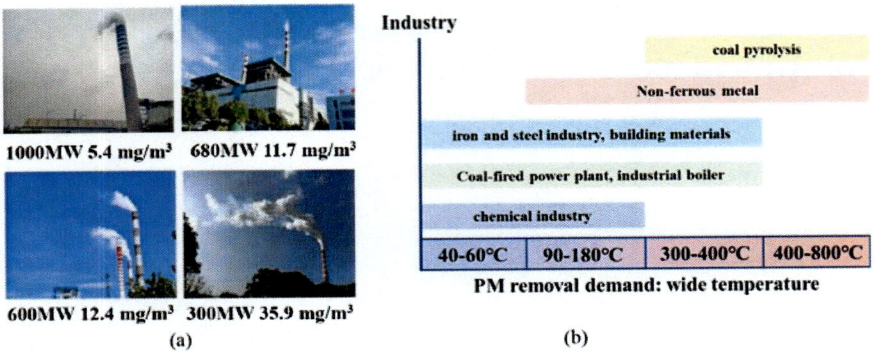

Fig. 3.1 **a** Flue gas exhaust and emission SO_3 concentration in some coal-fired power plants (CFPPs); **b** Temperature range of particulate matter (PM) removal in various industries

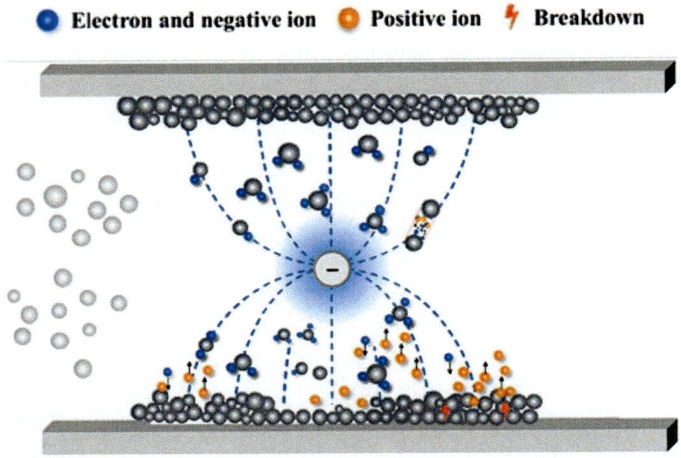

Fig. 3.2 Principle of electrostatic precipitation

In recent years, more stringent PM emission standards have been proposed in many countries for controlling the emission of fine particles; in some industries, PM emissions must now remain below 10 mg/m^3 [18–20]. However, certain complex emission conditions, detailed below, have restricted the removal efficiency of PM in different industries.

High particle concentration and ultrafine particle size: In some industrial applications, ESPs often operate under extreme conditions. For example, in a cement kiln, the mass concentration can reach 200 g/m^3; after a wet desulfurization tower, the number concentration can reach 10^9 cm^{-3}. At these high particle concentrations, the removal efficiency deteriorates rapidly, particularly for submicron (i.e., PM ranging from 0.1 to 1 μm in diameter) and ultrafine particles [21]. The corona current can be reduced by more than 95%, and the flashover frequency can increase by more than twice under the same operating voltage with the increase in ultrafine particle concentration.

Wide temperature range: Conventional ESPs generally operate below 200 °C. As discussed in previous chapters, the flue gas in commercial applications can have a much broader range because the temperature of wet flue gas after desulfurization ranges 40–55 °C whereas that of flue gas exiting a waste heat boiler of a glass kiln and nonferrous metallurgy facilities can reach over 400 and 600 °C, respectively. With this temperature variation, PM may exist as a solid, liquid (e.g., SO_3 aerosol and other condensable particles), or liquid–solid mixture (e.g., as a mixture of liquid drops and gypsum/dust, fine particles, and condensable particles).

Complex particle/atmosphere components: ESPs may induce the formation of complex particles caused by crushing, condensation, collision, and adhesion. The particle resistivity, dielectric constant, and morphology of the flue gas varies in the thermoelectric power, steel, chemical engineering, nonferrous, and other industries. In particular, some flue gas compositions or conditions, such as high SO_2 concentration (e.g., 15% in nonferrous metallurgy industry) and high humidity (above 40%), can decrease ESP efficiency [22].

Researchers have thus aimed to develop methods and modify existing devices to allow ESPs to operate under such complex conditions and achieve higher removal efficiency. Methods allowing for the simultaneous conversion/removal of other pollutants in the electrostatic field have also received considerable attention. Therefore, in this chapter, the removal of PM and the conversion of other gaseous pollutants through an applied external electrostatic field is introduced. To accomplish this, the fundamentals of PM control, including particle characterization, mechanisms, and primary influencing factors, are discussed in detail. Several PM removal enhancement methods are proposed, including particle growth enhancement, conditioning particle properties, a precharger, and electrode optimization. Electric multi-pollutant control methods, such as NO_x/SO_x conversion through plasma discharge, capturing heavy metals, and electrostatic field enrichment, are also discussed.

3.2 Fundamentals of Particulate Matter (PM) Control

3.2.1 Physicochemical Characterizations of PM

PM control requires understanding some fundamental concepts and properties. Particle characterizations from different industrial sources vary and can significantly affect the particle behavior, further influencing PM emission control. In particular, particle size distribution, chemical composition, morphology, and resistivity are critical in achieving efficient particle control.

3.2.1.1 Particle Size Distribution

Particle size is one of the most important parameters used to characterize particle behavior. Particle size is usually measured in microns (μm, 1×10^{-6} m) and allows particles to be classified into many categories based on their diameter, as shown in Fig. 3.3. Most particle properties, including migration velocity and particle charging capabilities, depend on particle size. Furthermore, most particle types cover a wide range of sizes; a 100-fold range between the smallest and largest particles of an aerosol is common. Not only do aerosol properties depend on particle size, but the nature of the laws governing these properties can change with the particle size. Particles in the atmosphere typically have diameters ranging from 0.01 to 100 μm, whereas larger particles typically settle out of the air. Following metrics are commonly used to describe the particle size.

Median number particle size: Widely used in particle science, it refers to the fact that when the total number of particles of various diameter smaller than a certain diameter accounts for 50% of the number of all particles, this diameter is called the median number diameter.

Median mass particle size: Also commonly used, this value, also known as the median mass diameter, represents the average diameter particle by mass. More specifically, particles with a smaller diameter account for 50% of the total particle mass (i.e., the total mass of all particles of different particle sizes).

Aerodynamic diameter: This metric is used to associate the behavior of an irregularly shaped particle to an ideal (i.e., spherical with a density of 1 g/cm^3) particle. The actual particle size can be replaced with an equivalent particle size, regardless

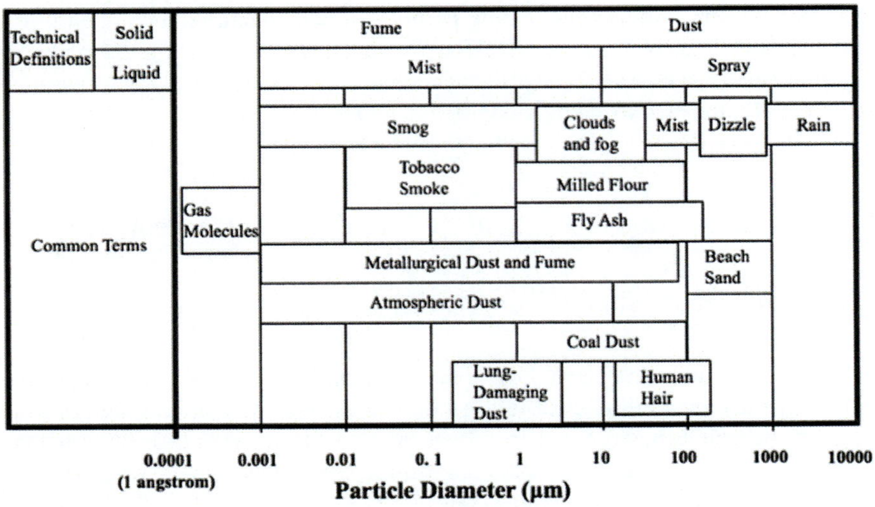

Fig. 3.3 Particle size distribution for different categories [23]

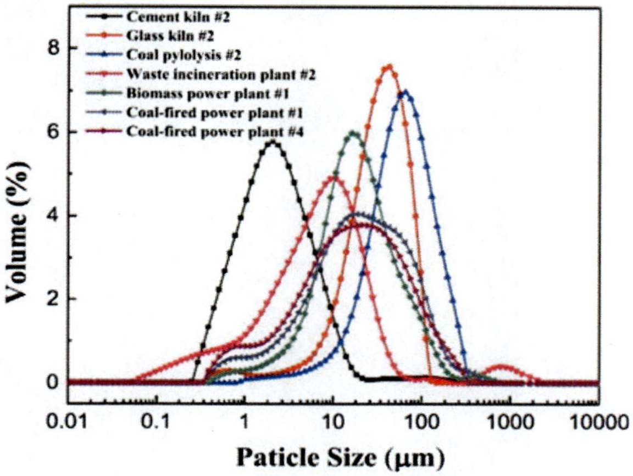

Fig. 3.4 Particle size distribution from different industries [24]

of its shape, size, or density, given that the particle's sedimentation velocity (v_s) in the air is the same as the sedimentation velocity of an idealized particle of that size.

Stokes diameter: Under the assumption that the diameter of a spherical dust particle of the same density has the same settling velocity as a certain nonspherical dust particle in still air, the Stokes diameter can be used as the equivalent particle size of nonspherical dust particles and measured by the liquid sedimentation method. The Stokes diameter (d_s) can be calculated as

$$d_s = \left(\frac{18\mu v_s}{g(\rho_c - \rho)}\right)^{\frac{1}{2}}, \tag{3.1}$$

where v_s represents the settling velocity of particles, m/s; μ represents the dynamic viscosity of the flue gas, Pa s; g is the gravitational constant, m/s^2; ρ_c is the particle density, kg/m^3; and ρ is the air density, kg/m^3.

Log-normal distributions, such as those shown in Fig. 3.4, are commonly used to detail particle size distributions because mass is concentrated in a small number of large particles. On this scale, the particle size–volume fraction curve is usually a bell-shaped curve. The median particle size from biomass power plants, waste incineration plants, coal gasifiers, and glass kilns is commonly in the range of 10–40 μm, similar to those from CFPPs.

3.2.1.2 Particle Morphology

Particle morphology is another parameter that strongly affects particle behavior. As shown in Fig. 3.5 [24], the morphology and heterogeneity of particles from various industries vary significantly. Overall, different particle formation mechanisms result in different particle morphologies. The particles from CFPPs are more spherical,

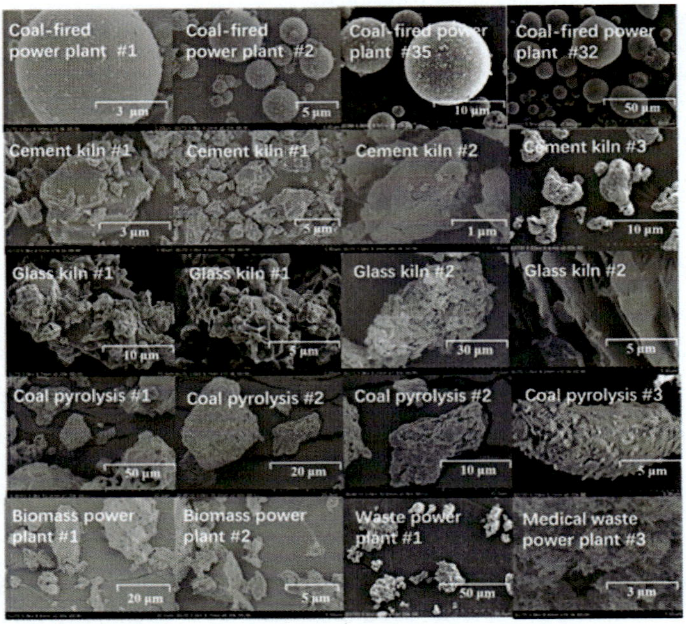

Fig. 3.5 Morphology of particles from various industries [24]

whereas those from cement kilns are irregular with some fine particles attached to large gaps on the particles. Most particles from glass kilns are loose and flocculent, although some were stacked as dense irregular blocks (layered or flocculent). Particles emitted from coal pyrolysis are porous and large owing to high carbon content of carbon. Temperature is a key factor in controlling the morphology of particles formed through solid-to-particle processes. During pyrolysis, the temperature can reach the melting point of fly ash, melting the particles and causing residual ash particles to be smooth and spherical [25–27]. Furthermore, the collision between particles during particle movement and transmission can cause particles to agglomerate and thus form new appearances. Although particle morphology specifics can vary by industry, most are nearly spherical. Therefore, researchers have commonly assumed a spherical morphology of particles to further study particle behavior.

3.2.1.3 Particle Chemical Composition

Although the chemical composition of particles can vary by industry, particles from most industries commonly contain Al, Si, Ca, Mg, Fe, Na, K, and Ti. A breakdown by weight of the chemical components of fly ash obtained from various industries in China, shown in Table 3.1, shows that fly ash mainly comprises SiO_2, Al_2O_3, Fe_2O_3, CaO, MgO, K_2O, and Na_2O. In particular, fly ash from the sampled CFPP mainly

Table 3.1 Chemical composition of typical fly ashes from different industries. CFPP = coal-fired power plant [24]

Source	SiO_2 (%)	Al_2O_3 (%)	Fe_2O_3 (%)	CaO (%)	MgO (%)	K_2O (%)	Na_2O (%)	Li_2O (%)
CFPP #1	45.04	23.57	8.99	12.84	1.39	1.44	1.100	0.0450
CFPP #2	46.65	23.89	8.07	12.86	1.43	1.49	1.120	0.0450
CFPP #3	42.33	24.65	8.99	12.79	1.45	1.43	1.090	0.0450
CFPP #4	43.90	28.3	5.34	5.56	0.631	1.3	0.570	0.0423
CFPP #5	45.84	17.76	8.09	8.99	6.13	1.41	2.170	0.0460
CFPP #30	42.55	21.97	9.77	6.47	8.12	0.987	4.670	0.0091
CFPP #31	51.02	21.87	5.29	13.9	1.80	1.32	2.070	0.0102
CFPP #32	37.35	28.14	4.05	19.99	2.48	0.53	2.080	0.0386
CFPP #33	44.95	29.54	3.22	13.60	1.34	0.738	0.918	0.0289
CFPP #34	53.65	20.33	6.98	4.79	3.85	1.52	6.030	0.0112
CFPP #35	41.99	27.47	3.51	16.62	1.76	0.53	0.623	0.0402
CFPP #36	36.85	32.87	3.00	17.90	1.67	0.347	0.575	0.0605
Biomass power plant #1	70.47	4.01	1.07	4.81	0.17	4.34	1.120	0.0198
Biomass power plant #2	28.02	11.35	4.30	30.43	8.36	5.19	1.820	0.0107
Waste incineration plant #1	7.67	2.37	1.13	48.82	3.51	3.23	10.890	0.0054
Waste incineration plant #2	24.23	14.19	3.10	23.42	8.73	2.51	7.290	0.0161
Cement kiln #1	14.57	3.87	1.59	42.45	1.35	0.684	0.240	0.0027
Cement kiln #2	11.28	4.24	1.93	43.05	2.10	0.491	0.122	0.0027
Cement kiln #3	5.95	5.44	0.78	63.45	1.79	0.088	0.332	0.0027
Glass kiln #1	11.20	5.49	1.85	13.80	1.04	1.23	8.310	0.0070
Glass kiln #2	2.59	1.93	1.80	14.50	2.04	0.166	15.300	0.0070
Coal pyrolysis #1	19.43	25.73	23.41	9.42	4.66	0.07	0.310	0.0102
Coal pyrolysis #2	39.73	17.28	3.04	1.61	1.15	0.41	0.300	0.0123
Coal pyrolysis #3	39.73	17.28	3.04	1.61	1.15	0.41	0.300	0.0037

comprises SiO_2 and Al_2O_3; SiO_2, Al_2O_3, Fe_2O_3 contents were 35–55%, 10–45%, and 2–10%, respectively. A few CFPPs had Fe_2O_3 content exceeding 10%.

During coal processing, the distribution and partitioning of minor and trace elements in the ash are a function of both the chemistry and mineral assemblages of the feed coal and the processing methods used, such as pulverization, boiler type and size, combustion parameters, combustion efficiency, and configuration of the ash collection system. Other fuels additions during coal combustion, such as petroleum coke or tire-derived fuel, can significantly alter the ash chemistry. The presence and composition of heavy metals and other trace elements can also vary by industry and combustion conditions. Overall, variations in the feed coal and combustion mode cause variations in the emitted particle composition, viscosity, and resistivity, which then impact how these particles can be removed.

3.2.1.4 Particle Resistivity

Particle resistivity (ρ_d), a key factor influencing particle removal via ESPs, is related to the particle size distribution, chemical composition, temperature, sulfur content of the burnt fuel, and other factors. A particle's resistivity represents the resistance of a particle layer with an area of 1 cm^2 and a thickness of 1 cm and can be calculated as follows:

$$\rho_d = \frac{V}{j\delta}, \tag{3.2}$$

where V is the voltage of the particle layer, V; j is the current density, A/m^2; and δ is the thickness of deposited particles, m.

As a particle's behavior can be significantly affected by its resistivity in the external electric field, particle resistivity is a key factor in the efficient and stable operation of ESPs. Inside the ESP, under the effect of the electric field, the charged particles can deposit on the collection plate, where form the deposited particles. High resistivity deposited particles can break down more easily, which will have an adverse effect on particle migration, whereas the re-entrainment of deposited particles with a low resistivity can be more easily occurred, further increasing the particle emission concentration.

In particular, the temperature of particles entrained in fly ash significantly influences their resistivity, as shown in Fig. 3.6, where a maximum resistivity of most samples was observed at approximately 373–473 K. Two mechanisms influence fly ash resistivity: surface resistivity and volume resistivity. At low temperatures, particles are dominated by surface conduction (conduction is the inverse of resistance), wherein specific resistivity increases with temperature. At temperatures of 373–473 K, both surface and volume conductivities reach a minimum, causing the resistivity to peak. The internal conductivity of particles plays a distinguishing role

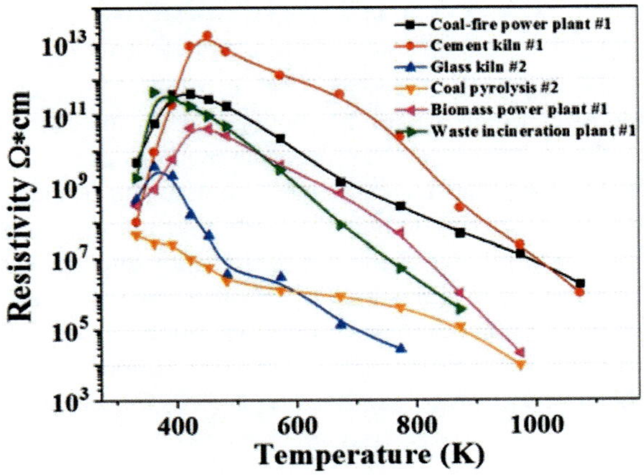

Fig. 3.6 Particle resistivity versus temperature

during these rising temperatures. The volume conductivity continues to increase with temperature, thereby causing the overall resistivity to decrease.

The influence of chemical composition on the specific resistivity of ash [24], as shown in Fig. 3.7. In particular, they found that the fly ash resistivity is strongly affected by the presence of Fe, K, and Na. Furthermore, the resistivity decreased with increasing Fe, K, Na, and Li content, whereas it increased with increasing Ca and Mg content. Si and Al had a weak influence on resistivity. Overall, the elemental sensitivity of resistivity was $N_{K+Na+Li} > N_{Fe} > N_{Ca+Mg} > N_{Al+Si}$. Several models have been proposed to predict particle resistivity, as summarized in Table 3.2.

3.2.2 Mechanisms of PM Control in Electric Field

Electrostatic precipitation involved the separation of particles from flue gases via an electric field. During operation, an ESP charges particles and then imposes an electric field on the region through which the particle-laden gas flows, exerting an attractive force on the charged particles that causes them to migrate to the oppositely charged electrode perpendicular to the direction of gas flow. This separation process can be divided into four essential steps: corona discharge, particle charging, particle migration, and particle deposition. However, in most applications, these steps all occur in one operational section.

3.2.2.1 Corona Discharge

Stable corona discharge is the basis of electrostatic precipitation. Corona discharge is caused by the generation of charged ions and a high-intensity electric field, allowing ions to attach to the surface of particles and charge the particles. The charged particles then migrate to the dust collection electrode and are captured under the force of the electric field [30].

Negative corona discharge, detailed in Fig. 3.8, is widely used in electrostatic precipitation. Here, the discharge region can be divided into ionization, attachment, and drift regions according to the strength of the electric field. In the ionization region, an uneven electric field is formed near the tip of the discharge electrode with a high-voltage power supply. At a high enough electric field intensity, high-speed electrons are released from the tip of the discharge electrode and ionized via collision with surrounding gas molecules, thereby generating new high-speed electrons and corresponding positive ions. The positive ions move toward the discharge electrode under the force of the electric field, whereas the newly generated electrons accelerate until they collide with and ionize the gas molecules. This process is known as an electron avalanche. In the attachment region, low-speed electrons that not energetic enough to ionize molecules attach to gas molecules to form negative ions. In the drift region, the negative ions continue to move toward the ground pole under the force of the electric field. As they do so, they can collide with particles to charge

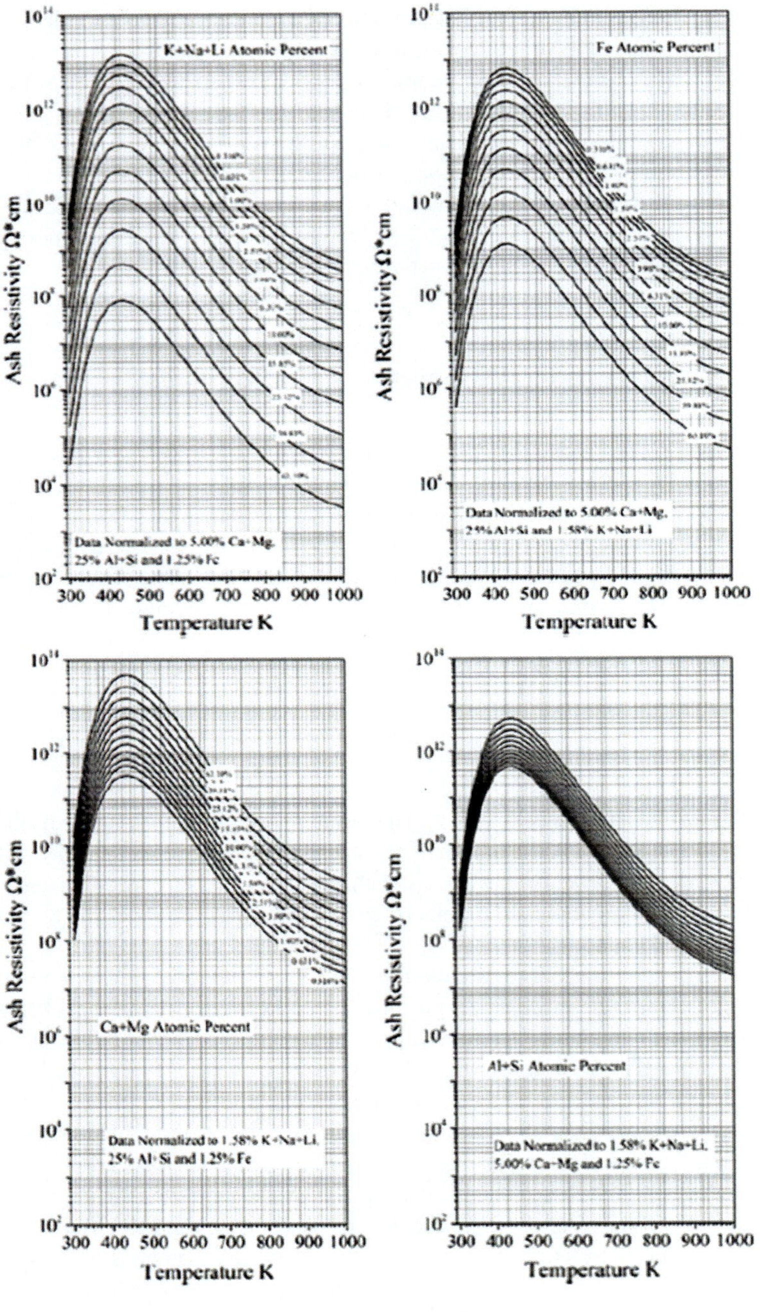

Fig. 3.7 Particle resistivity varying with temperature and particle composition, including K + Na + Li, Fe, Ca + Mg, Al + Si [24]

Table 3.2 Particle resistivity prediction models of typical fly ashes

Author	Model
Bickelhaupt et al. [28]	$\rho_s = 11.9857 - 2.2334\log X + \min(0, 0.056 - 0.056Z) - 0.03E - 7.662 \times 10^{-4}W - 3.2092 \times 10^{-4}We^{\frac{2303.3}{T}}$, $\rho_v = 2.0682 - 1.8916\log X - 0.9696\log Y + 1.2370\log\left(\frac{Z}{2.5}\right) - 0.03E + \frac{4334.5}{T}$, $\rho_a = 25.65278 - 0.371201C_{SO_3} - \frac{5667.313}{T} - 0.03E$, $\rho = \frac{1}{\rho_s} + \frac{1}{\rho_v} + \frac{1}{\rho_a}$.
Candra et al. [29]	$\rho_s = \exp\left[27.59774 - 2.233348\ln X - 0.0073895W - 0.069078E - \frac{0.00073895W(\exp)^{2303.3}}{T}\right]$, $\rho_v = \exp\left[-1.8916\ln X - 0.9696\ln Y + 1.237\ln Z + 3.62876 - 0.069078E + \frac{9980.58}{T}\right]$, $\rho_a = \exp\left[59.0677 - 0.854721CSO_3 - \frac{13049.47}{T} - 0.069078E\right]$, $\rho = \frac{1}{\rho_s} + \frac{1}{\rho_v} + \frac{1}{\rho_a}$.
Zheng et al. [24]	$\lg\rho = 2023.96 + [-1.22\lg N_{K+Na+Li} - 0.81(\lg N_{K+Na+Li})^2] + [0.31\lg N_{Si+Al} + 0.11(\lg N_{Si+Al})^2] + [0.75\lg N_{Ca+Mg} + 0.14(\lg N_{Ca+Mg})^2] + [-0.82\lg N_{Fe} - 0.61(\lg N_{Fe})^2] - 45801\frac{1}{T} - 1203\lg T + 182(\lg T)^2$

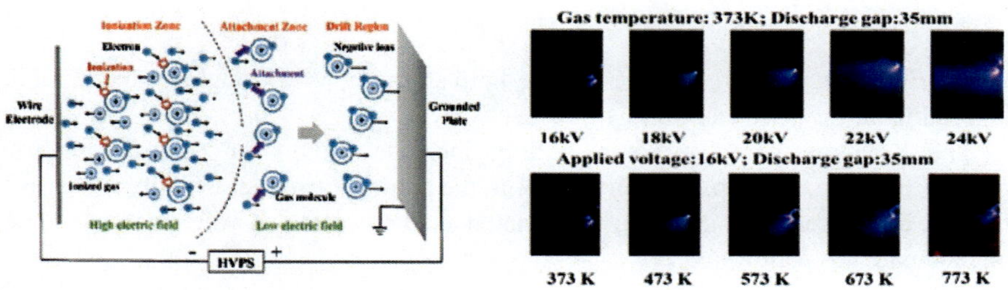

Fig. 3.8 Negative corona discharge process [31, 32]

them before reaching the grounded electrode. During discharge, the brightness and length of the corona light on the surface of the wire electrode increase as the applied voltage increases. The corona light gradually extends toward the grounded plate. If the voltage increases to reach the breakdown voltage, a spark breakdown can occur. The intensity of the corona discharge becomes stronger as the temperature increases under the same applied voltage; more gas molecules are ionized because Townsend's first ionization coefficient increases with the increase in temperature. The bright light around the wire is known as the gas molecule ionization zone, which extends until the bright corona light reaches the grounded plate, where sparks are then produced.

In practice, the discharging and grounding electrode is usually a wire and a plate, respectively. Ionized molecules are formed from the corona glow caused by the high electrical gradient around the discharging wire, and the space outside the corona is filled with a dense cloud of negative ions. The dust particles collide with ions, giving them a negative charge. These charged particles are driven by the electric field toward the plate where they are collected. Understanding the distribution of the electric field is necessary because a particle's charge is closely related to electric field force and intensity.

An ESP's electrical properties are determined by the precipitator's configuration, gas state, and gas composition. The electrical field required to start a continuous ionization process depends on the ionization energies of the gas species present and the mean free path between collisions, which is related to the gas state. The corona initiation field strength is thus a function of gas density, whereas the curvature of the wire determines the homogeneity of the electrical field. Due to the complex nature of the gas mixtures, a purely theoretical description has not been validated; however, empirical and semiempirical models have been proposed. For example, the surface field strength of the electrode has been expressed as a function of the relative gas density and the radius of the discharge wire in the semiempirical relationship proposed by Peek [33]:

$$E_s = 3.1 \times 10^6 m\delta\left(1 + \frac{0.03}{\sqrt{\delta r}}\right), \tag{3.3}$$

$$\delta = \frac{T_0}{T}\frac{P}{P_0}, \tag{3.4}$$

where E_s, m, and r represent the surface field strength, V/m, roughness coefficient and radius of the wire electrode, m, respectively.

The discharge current increases with an increase in the strength of the electric field in the discharge gap. At a certain point, the corona current drastically increases, causing a gas breakdown; this point is known as the breakdown voltage (V_{sp}) as can be calculated as follows:

$$V_{sp} = \frac{B \cdot D \cdot P}{\ln\left(\frac{A \cdot D \cdot P}{\ln(1/\gamma)}\right)}, \tag{3.5}$$

where A and B are constants related to the gas properties and D is the discharge gap, m.

In a wire–plate ESP, the charge density of the electric field and ion can be described by the Poisson and current continuity equations as follows:

$$\frac{\partial^2 \varphi}{\partial x_k^2} = -\frac{\rho_{\text{ion}} + \rho_{\text{pc}}}{\varepsilon_0}, \tag{3.6}$$

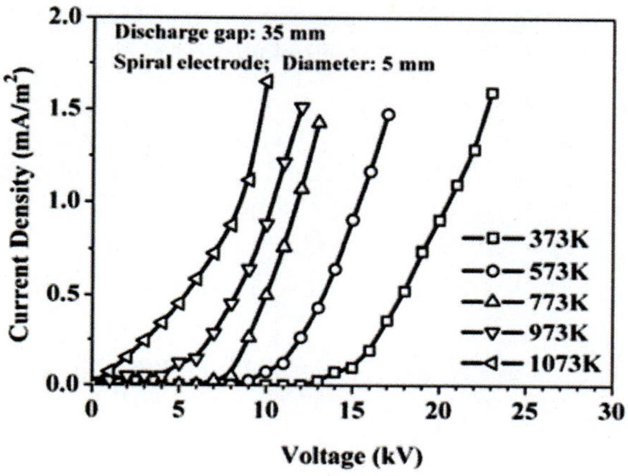

Fig. 3.9 Voltage–current (*V–I*) characteristics under different temperature [32]

$$\frac{\partial}{\partial x_k} = \left[\rho_{\text{ion}}(k_{\text{ion}}E_k + u_k) - De\frac{\partial \rho_{\text{ion}}}{\partial x_k}\right] = 0, \tag{3.7}$$

$$k_{\text{ion}} = \frac{2.1 \times 10^{-4}}{\delta}, \tag{3.8}$$

where φ represents the electric potential, V; ρ_{pc} represents the particle charge density, C/m^3; ρ_{ion} represents the charge density of ions, C/m^3; ε_0 represents the vacuum dielectric constant, F/m; k_{ion} represents the ionic mobility, m^2/(V s); E_k represents the component of the field strength in the x_k direction, V/m; D_e represents the effective diffusion coefficient of ions, m^2/s; and δ represents the flue gas relative density compared with the standard case ($T_0 = 273.15$ K, $p_0 = 101{,}325$ Pa).

The voltage–current (*V–I*) characteristics of an example wire-plate ESP under different temperatures and electrode diameters are shown in Fig. 3.9. As the relative density of the background gas decreases at high temperatures, the mean free path of the electrons and gas molecules increases. The free electrons produced in the initial ionization process are accelerated in the imposed electric field, causing the energy of the electrons to increase. As a result, the ionization coefficient increases at high temperatures. Inelastic collisions between the electrons and gas molecules produce more electron–positive ion pairs in a self-sustaining process at high temperatures, causing the discharge current to increase with increase in temperature.

3.2.2.2 Particle Charging

During particle charging, free ions are captured by particles in an electric field. The charging process can be expressed by two mechanisms, assigned by particle size: field charging mainly for large particles and diffusion charging mainly for small particles. Field charging requires the presence of an electrical field to drive the free movable charge carriers, whereas diffusion is based on the random movement of gas ions caused by temperature and described by kinetic gas theory, i.e., Brownian motion. In an ESP, particles of all sizes experience both situations.

Field charging: Based on a comprehensive physical overview, the theoretical field charging model detailed in the following equations simplifies a charged particle to a conductive sphere suspended in an electric field. Gaseous ions are assumed to be transported to the particle's surface along the field lines.

$$q_f(t) = q_\infty \frac{t/\tau}{1 + t/\tau}, \tag{3.9}$$

$$q_\infty = \pi\varepsilon_0 \frac{3\varepsilon_S}{\varepsilon_S + 2} d_p^2 E_{\text{space}}, \tag{3.10}$$

$$\tau = 4\frac{\varepsilon_0 E_{\text{space}}}{J}, \tag{3.11}$$

where q_f represents the field charge, C; τ represents the field charging time constant for, s; E_{space} represents the space electric strength, V/m; ε_S represents the relative dielectric particle constant and q_∞ represents the saturation charge, C.

Diffusion charging: Although field charging is generally considered to be the dominant charging mechanism of large particles, it becomes less effective as the particle size decreases. The chaotic motion of gaseous ions thus should not be ignored, especially when in the presence of small particles and a low-intensity electric field. To better describe this charging mechanism of airborne particles, the diffusion charge (q_d) proposed can be described as

$$q_d(t) = q^* \ln\left(1 + \frac{t}{\tau^*}\right), \tag{3.12}$$

$$\tau^* = \frac{8\varepsilon_0 kT}{d_p C_i N_i e^2}, \tag{3.13}$$

$$q^* = \frac{2\pi\varepsilon_0 d_p kT}{e}, \tag{3.14}$$

where $\tau*$ represents the time constant, s; k represents the Boltzmann constant, J/K; T represents the temperature, K; $q*$ represents the charge constant, C; e represents the electronic charge, C; C_j represents the ion thermal velocity, m/s; N_j represents the ion density, m^{-3}; and t represents the charging time, s.

Given that particle charge depends on both charging mechanisms, a simple addition of the charge resulting from both mechanisms is sometimes made; however, this simple model cannot accurately predict particles' charge. A variety of combined models have thus been proposed to take both charge mechanisms into account. Murphy [34] made a qualitative analysis trial. Liu and Yeh [35] divided the charging process into two steps, where a particle is subject to both charging mechanisms before reaching the saturation charge limit, after which point only diffusion charging is accounted for. Similarly, Smith and McDonald [36] proposed a complex method involving the sum of the charging rate of three identified regions of a particle's surface. Lawless [37] presented a combination charging model, detailed in Eqs. (3.15)–(3.19), that considers both charging mechanisms and has since been used extensively in the numerical ESP simulations [38–40].

$$\frac{dv}{dt_q} = \begin{cases} f(w)\frac{v-3w}{\exp(v-3w)-1} & v > 3w, \\ \frac{3w}{4}\left(1-\frac{v}{3w}\right)^2 + f(w) & -3w \le v \le 3w, \\ -v + f(w)\frac{-v-3w}{\exp(-v-3w)-1} & v \le -3w, \end{cases} \tag{3.15}$$

$$f(w) = \begin{cases} \frac{1}{(w+0.475)^{0.575}} & w \ge 0.525, \\ & 1w < 0.525, \end{cases} \tag{3.16}$$

$$v = \frac{q_p e}{2\pi\varepsilon_0 d_p kT}, \tag{3.17}$$

$$w = \left(\frac{k_p}{k_p+2}\right)\frac{Eq_p e}{2kT}, \tag{3.18}$$

$$t_q = \frac{N_i Zet}{\varepsilon_0}, \tag{3.19}$$

where w represents the dimensionless electric field strength, v represents the dimensionless charge of the particle, t_q is the dimensionless charging time, q_p represents the particle charge, C, and d_p represents the particle diameter, m.

Some typical charging models are shown in Table 3.3.

There are fundamental limits on the maximum charge that a particle of a given size can acquire [41]. The maximum charge of a spherical negatively charged solid particle is reached when the self-generated electric field on the particle's surface reaches the value required for spontaneous emission of electrons from the surface; this maximum charge number (n_L) is proportional to the square of the particle diameter and can be calculated as follows:

$$n_L = \frac{d_p^2 E_L}{4K_E e}, \tag{3.20}$$

where E_L represents the surface field strength required for spontaneous emission of electrons, 9.0×10^8 V/m. When this value is exceeded, electrons on the particle's surface are ejected from the particle by a mutual repulsion force K_E is the Boltzmann constant, J/K.

Particle charge is affected by many factors that must be considered to enhance charging efficiency, including the temperature, humidity, current density, and dielectric constant, as discussed in detail below.

Table 3.3 Some typical particle charging models

Model	Authors	Interacting forces
Field charging	Pauthenier and Moreau-Hanot (1932)	Coulomb force
Diffusion charging	Fuchs (1947) and Bricard (1949)	Coulomb force
Combined charging	Murphy et al. (1959), Liu and Yeh (1968), Marlow and Brock (1975) and Lawless (1996)	Coulomb force
Limiting sphere	Natanson (1960), Fuchs (1963) and Marquard (2007)	Coulomb and image force
BGK model	Gentry and Brock (1967), Marlow and Brock (1975), Huang et al. (1990) and Lushnikov (2005)	Coulomb, image and Van der Waals force
Charge distribution	Boisdron and Brock (1970), Domat et al. (2014a)	Coulomb and image force
Nonspherical model	Laframboise and Chang (1977), Chang (1981)	Coulomb force
Nonspherical model 2	Gopalakrishnan et al. (2013)	Coulomb and image force
Combined charging 2	Tsai et al. (2012)	Coulomb force
Combined charging 3	Lawless and Sparks (1980), Guo et al. (2014)	Coulomb force

Effect of temperature: As detailed in Fig. 3.10, the particle charge number has been demonstrated to increase with increasing flue gas temperature under the same electric field strength (-4.4 kV cm^{-1}) [42]. Higher temperatures cause an increase in the mean free path of the surrounding gas molecules, which increases the energy gained by an electron when it is accelerated in the electric field before colliding with a background molecule. Furthermore, the electron energy and probability of the gas

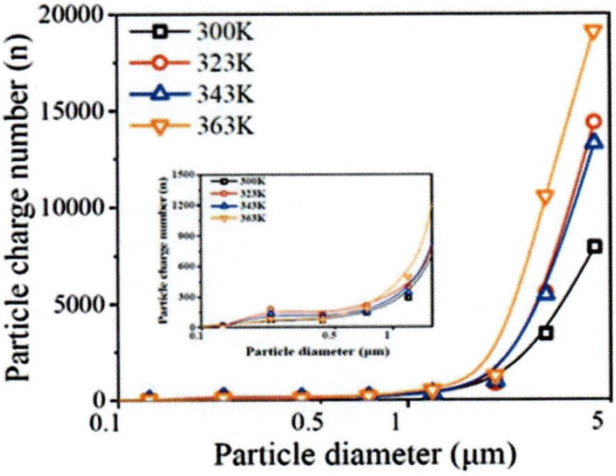

Fig. 3.10 Effect of temperature (300, 323, 343 and 363 K) on particle charging [42]

molecule to be ionized increase with increasing mean free path such that the number of ions and the discharge current increase.

Effect of humidity: Similarly, the particle charge number has been demonstrated to increase with an increased relative humidity of the flue gas, as shown in Fig. 3.11 [42]. The ionization coefficient increases with humidity in the air, causing a more drastic ionization of neutral molecules in corona discharge and thus a higher ion density, which causes more negative ions in the ESP. This causes the probability of a particle's collision with an ion to increase, thereby increasing the current. Furthermore, particles can absorb water vapor in a stable high-humidity environment. Vapor condensed on the surface of a particle increases the water content of the particle. As the relative dielectric constant of water (80) is higher than that of the fly ash from the power plant, water increases the polarization capability of particles in the electric field, thereby allowing more electric field line to reach the surface of the particles and increasing the number of collisions between particles and ions.

Effect of current density and dielectric constant: For particles of the same size, the current density has been demonstrated to correlate positively with the current density, as shown in Fig. 3.12a, because an increased current density increases the ion density and thus the collision frequency between particles and ions, thereby increasing the particle charge number [42]. Further, particles with a high dielectric constant have a stronger polarization capability, increasing the particle–ion collisions and thus increasing the particle charge number, as shown in Fig. 3.12b.

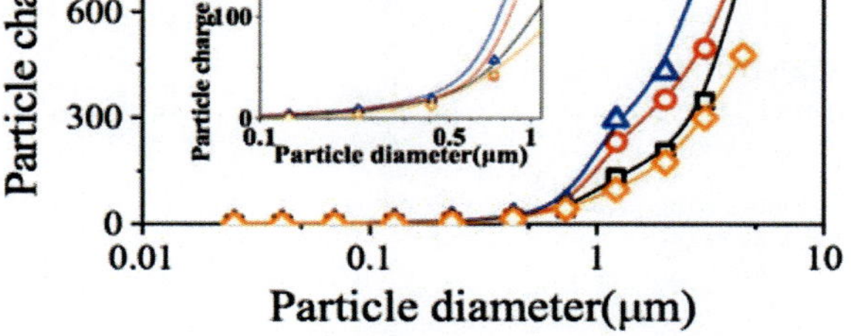

Fig. 3.11 Effect of humidity (10, 30, 60 and 80%) on particle charging [42]

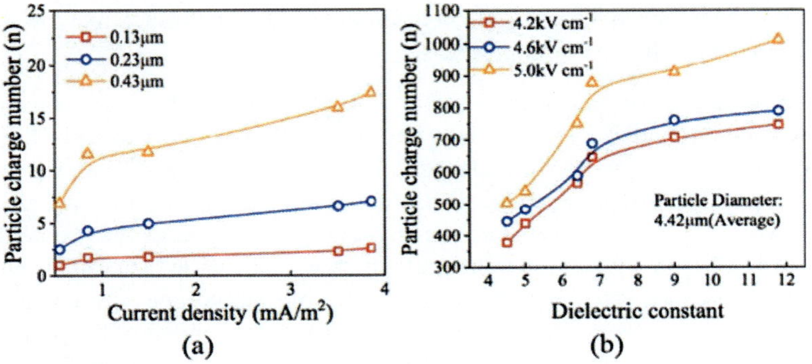

Fig. 3.12 Effect of **a** current density and **b** dielectric constant on particle charge number [42]

3.2.2.3 Particle Migration

Under an external electric field, a wide variety of factors are known to affect a particle's movement as it is deflected from the flue gas and migrates to the collection plates. As summarized in Table 3.4 and Fig. 3.13 and detailed below, these forces include Coulombic, drag, gravitational, buoyancy, Brownian, Saffman lift, thermophoretic, pressure gradient, and virtual mass forces.

As particles pass through an electric field, they are mainly affected by the airflow and electric field, i.e., drag and electrostatic forces. Owing to the low particle proportion present in the flue gas, particle–particle, particle–wall, and aggregation forces can be assumed negligible. In the ESP, the gas density was significantly lower than particle density, furthermore, gas–solid two-phase flow studies have demonstrated that when the density of the fluid (gas or liquid) is much lower than that of the particle, the buoyancy, virtual mass, and Saffman lift forces are several orders of magnitude smaller than the inertia of the particle itself; thus, these forces can be assumed negligible when considering the force of the fluid on the particle. The thermophoretic

Table 3.4 Forces acting on particle transport in an electric field

Force	Equation expression
Gravitational force	$F_G = m_p g = \rho_p v_p g = \frac{\pi}{6}\rho_p d_p^3 g$
Capillary force	$F_c = 4\pi\sigma\beta N_c$
Centrifugal force	$F_{cf} = m_p \frac{v_f^2}{r}$
Van der Waals Force	$F_{s-p} = \frac{-A_a d_p}{12 d_{s-p}^2}$
Drag force	$F_{drag} = \frac{3\pi\eta V_p d_p}{C_c}$
Electrostatic force (Coulombic force)	$F_E = QE + \frac{Q^2}{16\pi\varepsilon_0 z^2} + \frac{QEd^3}{16z^3} + \frac{3\pi\varepsilon_0}{128z^4}$

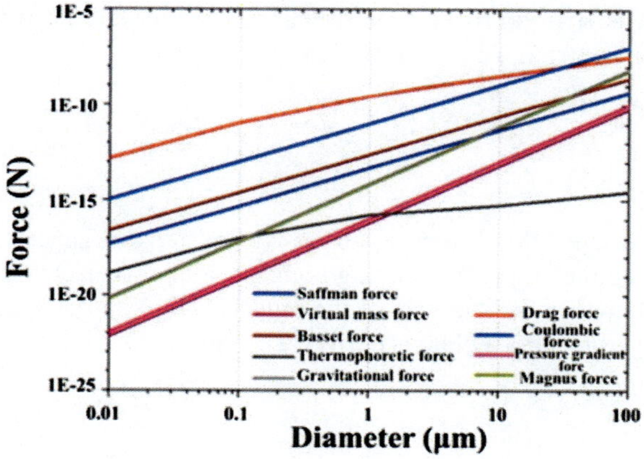

Fig. 3.13 Magnitude analysis of various forces

force, which is related to temperature difference, can also be assumed negligible in an ESP because the temperature difference is low in commercial ESPs. At high fluid pressures and low particle densities (approaching fluid density), both the virtual mass and pressure gradient forces should be considered, but in the ESP, these forces can be reasonably ignored because of the high particle densities.

Gravitational force: The gravitational force, F_G, is proportional to the particle mass as follows:

$$F_G = m_p g = \rho_p v_p g = \frac{\pi}{6} \rho_p d_p^3 g, \quad (3.21)$$

Capillary force: When there are droplets in an electric field, particle behavior is influenced by a force referred to as the liquid bridge force, which is commonly considered as the external surface tension of the liquid or the external performance of the capillary force. The capillary force of coarse particles can be calculated as follows:

$$F_c = 4\pi\sigma\beta N_c, \quad (3.22)$$

where σ is the surface tension of water (0.0735 N/m at room temperature).

If the liquid meniscus is formed around the contact surface, the flattening effect of the protrusion is ignored. If the liquid forms a film around the granules and immerses the bulges, the capillary force of the spherical particles of diameter d is expressed as follows:

$$f_c = 2\pi\sigma d_p, \quad (3.23)$$

Centrifugal force: Centrifugal force is an inertial force that acts to push particles away from the center of rotation:

$$F_{cf} = m_p \frac{v_f^2}{r}, \tag{3.24}$$

Van der Waals force: Van der Waals forces are generated when dipoles in the surface regions of two interacting bodies respond to electromagnetic radiation propagating between the surfaces of the objects. The Van der Waals force between a spherical particle and a plate is derived as follows:

$$F_{s-p} = \frac{-A_a d_p}{12 d_{s-p}^2}, \tag{3.25}$$

Similarly, the Van der Waals force between unit areas of two opposing parallel plates and two spherical particles of radii R_1 and R_2 can be represented as follows:

$$F_{p-p} = \frac{-A_a}{6\pi d^3}, \tag{3.26}$$

and

$$F_{s-s} = \frac{-A_a}{6d^2}\left(\frac{R_1 R_2}{R_1 + R_2}\right), \tag{3.27}$$

respectively.

Drag force: When an object moves relative to a fluid, the fluid resists the motion of that object; this resistance is called drag, as schematically described in Fig. 3.14. An object is subjected to drag whenever there is a difference in velocity between the object and fluid. The drag force can be calculated using the Navier–Stokes equations for a particle moving in air as follows:

$$F_{drag} = 3\pi \eta V_p d_p, \tag{3.28}$$

where V_p represents the particle velocity with respect to gas, m/s.

The continuum regime ($Kn \ll 1$) drag force F_{drag} is derived as Equation above. While in the slip regime ($Kn \gg 1$), the particle may pass through a gap between molecules before colliding with other molecules or objects (Kn means the the ratio of the mean free path of the ion to the particle radius). This results in an increase in the particle speed and a lower drag force than predicted by Eq. 3.28. Small-diameter particles can slip between the gas molecules. In 1910, English mathematician Ebenezer Cunningham introduced a correction factor, known as the Cunningham slip correction factor (C_c) to reduce the drag coefficient for small particles, calculated as follows [44, 45]:

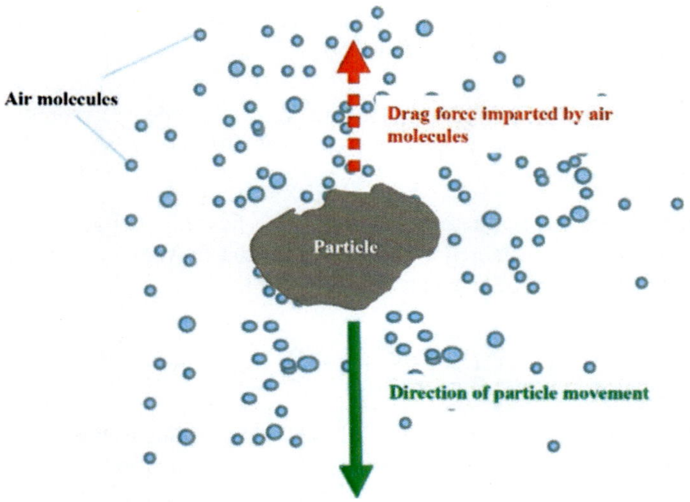

Fig. 3.14 Drag force of air imparted onto a particle moving [43]

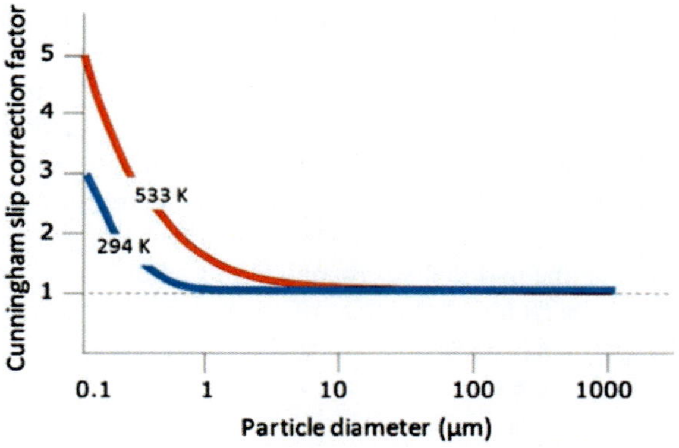

Fig. 3.15 Cunningham slip correction factor as a function of the temperature and particle diameter [43]

$$C_c = 1 + 1.246 \cdot \frac{2\lambda}{d_p} + 0.42 \cdot \frac{2\lambda}{d_p} \cdot \exp\left(-0.87 \cdot \frac{2\lambda}{d_p}\right). \tag{3.29}$$

The modified drag force on a particle of any size in the slip regime can be described as follows:

$$F_{drag} = \frac{3\pi\eta V_p d_p}{C_C}. \tag{3.30}$$

The drag force F_{drag} acting on a spherical particle that is in contact with a smooth surface is expressed as follows:

$$F_{drag} = \frac{C_D \pi f p_g d^2 V_p^2}{C_c}, \tag{3.31}$$

where $f = 1.7009$ is the correction factor for the wall effect, p_g represents the air density at normal conditions, V_p is the mean air velocity at the centroid of the sphere, and C_D is the drag coefficient.

Electrostatic force: Electrostatic force affects the flow and deposition of particles and must always be considered when analyzing the movement and migration of particles. The electrostatic force can be calculated as

$$F_E = QE + \frac{Q^2}{16\pi\varepsilon_0 z^2} + \frac{QEd^3}{16z^3} + \frac{3\pi\varepsilon_0}{128z^4}, \tag{3.32}$$

The first term addresses the Coulomb force, which is due to the extra charge on the plates or particles that produces a graded Coulomb attraction.

When there is a negatively charged $(-q)$ particle in the space distance of z outside of the collecting plate, then the charge induces a positive charge on the surface of the plate. The positive and negative charges will attract each other. The force generated by the attraction between these positive and negative charges is accounted for in the second term, representing the image force or Coulomb force generated by this induction of this charge.

The third term represents the dielectrophoretic force, which is the force induced by a spatially nonuniform electric field on a polarized but uncharged particle. A dipole moment is generated due to polarization, causing the particles to move due to uneven forces at opposite ends of the particle. If the electric field is uniform, the force at both ends of the particle is equal and opposite in direction, so the resultant force in the particle is zero, and the original motion state remains unchanged.

To characterize the state of particle motion, a balance of all forces acting upon the particle is needed. For a particle in an ESP, these forces include the total external force $\boldsymbol{F}_T$, the electrical force $\boldsymbol{F}_E$, and the drag force $\boldsymbol{F}_{drag}$.

$$\boldsymbol{F}_T = -m\boldsymbol{a}, \tag{3.33}$$

The sum of all acting forces has to be zero.

$$\boldsymbol{F}_T + \boldsymbol{F}_{drag} + \boldsymbol{F}_E = 0, \tag{3.34}$$

Therefore, the motion of a charged sphere in an electric field can be calculated as follows:

$$\frac{\mathrm{d}w}{\mathrm{d}t} + \frac{3\pi\eta d_p}{m_p(\cdot C_c)} w = \frac{Q_p^\infty}{m_p} E, \tag{3.35}$$

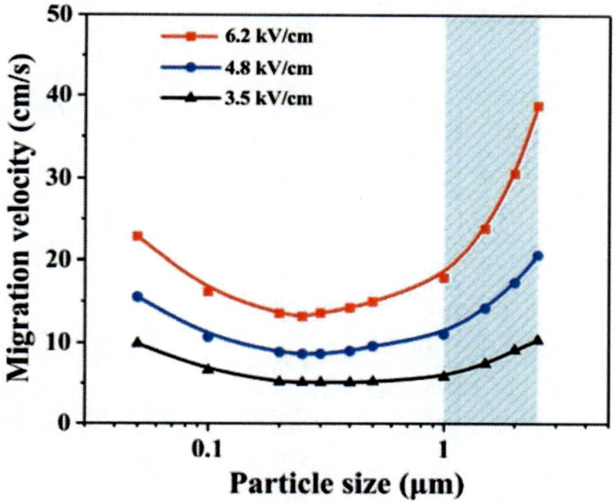

Fig. 3.16 Particle migration velocity under different applied voltages at 323 K with a charging time of 1.75 [46]

Taking ($t = 0$) as the initial condition, Eq. (3.36) can be used to describe the time dependence of the particle migration velocity (the velocity of particles moving to the collecting plate), where the relaxation time τ_p characterizes the dynamic behavior of the particle. The following equation presents the migration velocity, which is the steady-state velocity of the particle.

$$w(t) = w_{th}\left(1 - \exp\left(-\frac{t}{\tau_p}\right)\right), \tag{3.36}$$

$$\tau_p = \frac{m_p(\cdot C_c)}{3\pi\eta d_p} = \frac{\rho_p d_p^2}{18\eta}(\cdot C_c), \tag{3.37}$$

$$w_{th} = \frac{Q_p^\infty E}{3\pi\eta d_p}(\cdot C_c), \tag{3.38}$$

The particle migration velocity under different applied voltages in an ESP is shown in Fig. 3.16. Particle migration velocity shows a U-shape trend with respect to particle size and there is a critical particle size corresponding to the bottommost point.

3.2.2.4 Particle Deposition

Inside an ESP, charged particles are removed from the flue gas and deposited on collection plates to form a dust layer with increasing thickness, as shown in Fig. 3.17. These deposited particles can inhibit particle migration in the space discharge gap. Further, the resistivity of the dust plays an important role in ESP operation because when the drifting, electrically charged particles land on the collecting electrode, the

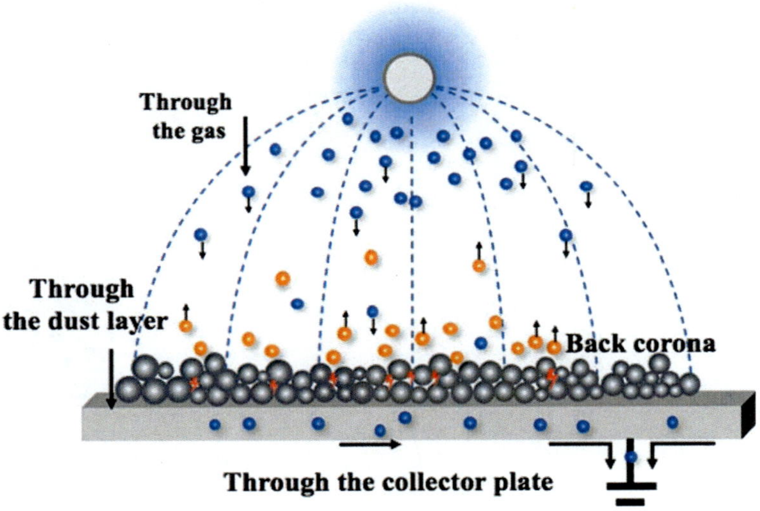

Fig. 3.17 During the operation of an electrostatic precipitator (ESP), the electrical charges have to pass through the dust layer

electric charge has to flow from the particles, through the dust layer, and then into the ground electrode.

If the dust layer has a high resistivity and thickness, it can significantly reduce particle charging, ion density, and electric field strength, further reducing particle migration velocity. This effect is particularly apparent at high temperatures, as shown in Fig. 3.18.

As the charge emitted by the discharge electrode accumulates on the surface of the dielectric layer and produces a high-strength electric field, the increasing electric strength can cause the gap of the dust layer to break down. The sudden increase in the discharge current can be several times higher than the previous state once the

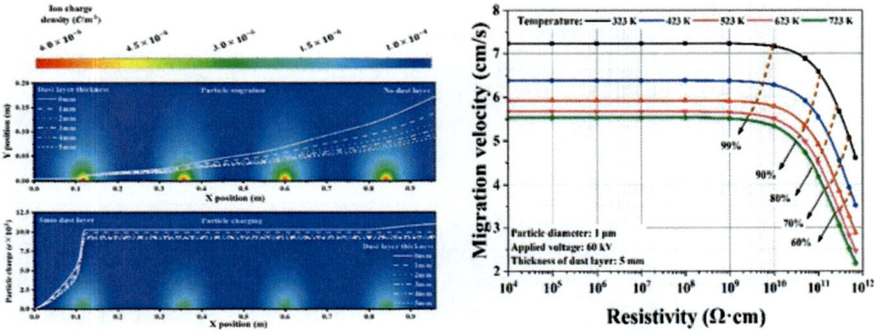

Fig. 3.18 Relationship between dust layer and particle migration at different temperatures [47]

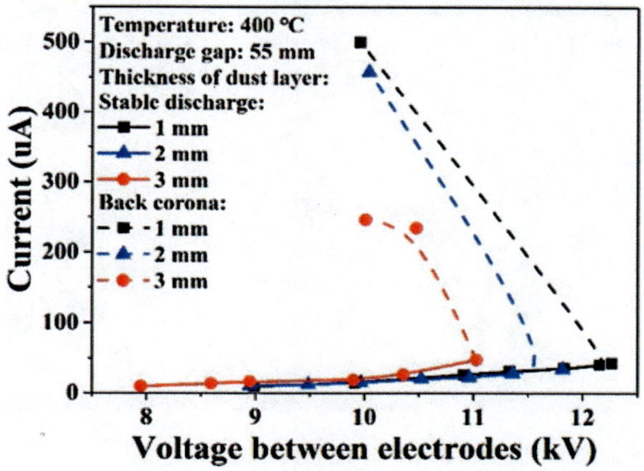

Fig. 3.19 *V–I* characteristics of corona discharge at different dust layer thicknesses at 623 K with a discharge gap of 55 mm [48]

voltage between electrodes increases above the onset voltage, severely inhibiting particle migration. This phenomenon, shown in Fig. 3.19, is called back corona.

The low resistivity caused by the dust layer causes low cohesive forces and thus allows some particles to undergo re-entrainment owing. Once re-entrained, these particles can return to the discharge gap, which increases the particle emission concentration. Future research efforts are required to clarify the critical points leading to back corona and re-entrainment effects and reduce the negative influence of the deposited dust layer on particle migration.

3.3 PM Control Enhancement and Applications

ESPs employ electrical energy to ionize flue gases, electrically charge dust particles, and precipitate said particles on the collection electrode, as summarized in Fig. 3.20. However, the fractional collection efficiency of $PM_{2.5}$ particles is relatively low because of the insufficient particle charge; as was shown in Fig. 3.16, the relationship between particle size and migration velocity (and therefore collection efficiency) reaches a minimum around 200–500 nm. To increase the collection efficiency of $PM_{2.5}$ particles, researchers have investigated the mechanisms of key subprocesses, including particle condensation and growth, charging, and transport. Some proposed industry-applied $PM_{2.5}$ removal enhancement methods employing electrostatic forces and phenomena, such as increasing the particle size, conditioning particle properties, the use of a precharger, and electrode optimization, are discussed in the following subsections.

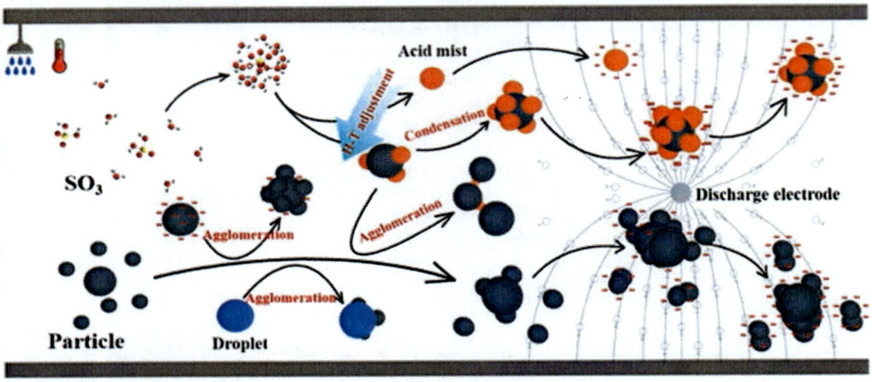

Fig. 3.20 Typical PM removal process in an ESP

3.3.1 Particle Growth

A high concentration of fine particles hampers the performance of the discharge of an ESP, making it difficult to meet lower PM emission targets (e.g., < 5 mg/m^3). The efficient removal of fine particles is key to reducing overall PM emissions because particle removal is influenced by particle size regardless of the collection method used [49]. Increasing particle size can thus improve the removal efficiency of fine particles in the electrostatic field.

Recent attempts to increase particle size have included gas humidification and cooling. For example, adding water vapor in a limestone–gypsum desulfurization tower was recently demonstrated to improve particle removal efficiency from 30 to 40% to above 60% [50]. By combining a heat exchanger with a novel wet ESP (WESP), Yang et al. [51] reduced particle concentration emissions to < 1 mg/m^3. These studies have demonstrated that particle growth is vital in improving particle removal efficiency. In this section, experimental studies detailing the process of particle growth and its influencing factors are introduced, and methods to promote particle growth and enhance particle removal are proposed.

3.3.1.1 Particle Coagulation

Particle growth of CPM can be realized by promoting particle coagulation. To explore the particle size distribution evolution of a flue gas under different parameter conditions, Zheng et al. [52] designed a particle coagulation experimental system comprising a flue gas generator, a sprayer, a growth tube, and a measurement system, as shown schematically in Fig. 3.21.

Understanding the general evolution of particle size distribution along a growth tube and classification of the governed aerosol growth dynamics are important before investigating the specific activation methods. To obtain high-humidity flue

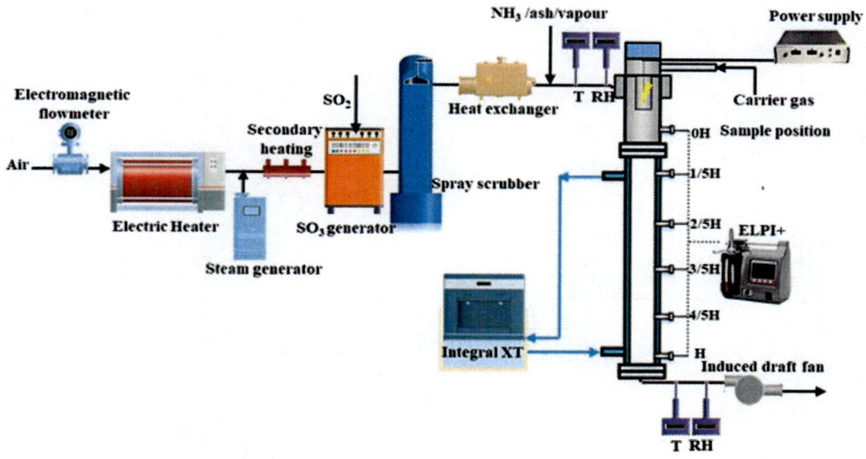

Fig. 3.21 Particle coagulation experimental system designed by Zheng et al. [52]

gas containing SO_3 to cover all possible conditions in real power plants, researchers have produced simulated flue gas by an induced air fan with a flow rate of 60 L/min, heated it to 250 °C via a double-stage electric heater, and passed it through a water scrubber, similar in function to a wet desulfurization tower, to ensure saturated conditions. Gaseous SO_3 is assumed to be completely converted into sulfuric acid aerosol after the gas was scrubbed to a saturated condition [53]. The temperature of the flue gas was kept below 50 °C before entering the growth tube.

Using the particle coagulation system developed by Zheng et al. [52], Yang et al. [54] found a particle number concentration as high as 2.64×10^7 cm^{-3} at the initial position ($h = 0$), as shown in Fig. 3.22a. Coagulation is a fundamental interparticle phenomenon, particularly when the collision possibility is noticeably high under high concentration conditions. Although this interaction leads to complicated variations of fractional size distribution, it causes the overall number concentration to decrease continuously; the overall particle number concentration decreased from 2.64×10^7 cm^{-3} to 1.94×10^7 cm^{-3} as the flue gas flowed from $h = 0$ to $h =$ H. If coagulation occurs alone in an aerosol system, the mass concentration remains unchanged. However, the mass concentration increased as the number concentration decreased, as shown in Fig. 3.22b, suggesting the presence of condensation (i.e., mass transfer from gas to aerosol phase under supersaturated gas conditions) [55].

Furthermore, the coupled effect of coagulation and condensation can allow for increased particle size and accordingly varied fractional size distribution, as shown in Fig. 3.22c. Here, the variation in particle size distribution from positions $h = 0$ to $h =$ H was caused by the coupled effect of coagulation and condensation. As clarifying the contribution of coagulation and condensation toward particle growth through experimental approaches alone is difficult, dynamics modeling can be used to supplement experimental characterizations. The final relative concentration using the dynamic model considering only coagulation, only condensation, and the coupled case is

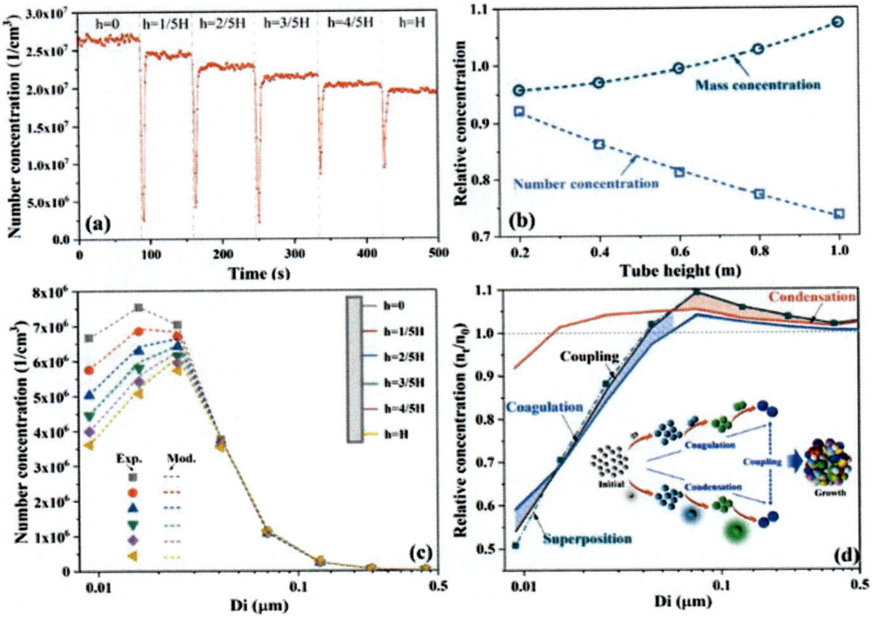

Fig. 3.22 **a** Overall particle number concentration during continuous sampling; **b** relative number and mass concentration under different sampling positions; and **c–d** fractional particle size distribution at positions from $h = 0$ to $h = H$ [54]

shown Fig. 3.22d. For small particles (0.01–0.05 μm), the coupled case is consistent with coagulation, whereas it is more consistent with condensation for larger particles (>0.1 μm). Therefore, the primary contribution to aerosol evolution varies with particle size. The decrease in number concentration of particles smaller than 0.05 μm in diameter was caused by the collision of these aerosols with aerosols in any size range. In contrast, condensation occurs more easily on large particles according to the Kelvin effect, whereas the growth rate of small particles is slowed according to the Knudsen effect. Overall, aerosol dynamics are dominated by coagulation and condensation mechanisms for small and large aerosols, respectively.

The dotted green line in Fig. 3.22d represents the summation of the only coagulation and only condensation cases. This superposition almost exactly matched the coupled case for particles larger than 0.1 μm, indicating that the coagulation and condensation behavior operate independently in this size range. A slight deviation existed between the summed and coupled case for particles > 0.05 μm in diameter, indicating that the coagulation and condensation behavior are influenced by each other in this size range.

Effects of gas temperature: Heterogeneous condensation is generally driven by the saturation ratio. Gas temperature is one of the most important activation factors to change this ratio. In actual power plants, flue gas is quickly cooled to < 50 °C

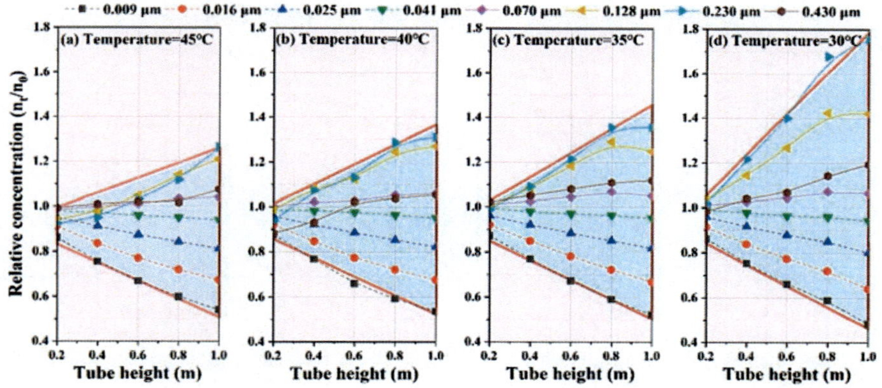

Fig. 3.23 Fractional concentration under different sampling positions corresponding to **a** $T = 45$ °C, **b** $T = 40$ °C, **c** $T = 35$ °C, and **d** $T = 30$ °C, respectively [54]

through the wet desulfurization tower, wherein the overall number concentration decreases while the overall mass concentration increases with tube height at different gas temperatures. As shown in Fig. 3.23, which details the fractional particle evolution concentration under different temperatures experimentally obtained by Yang et al. [54], a low temperature can cause large differences between the number and mass concentrations. Here, the increase in the relative concentration of particles 0.05–0.5 μm in size was triggered by the decreasing gas temperature. Consequently, a higher fraction of particles approximately 0.230 μm in diameter was present; at a gas temperature of 30 °C, a relative concentration of particles of this size reached 1.76. Dynamics analysis indicated that the aerosol dynamics were dominated by condensation for large aerosols. This result confirmed that condensation growth can be effectively enhanced by reducing gas temperature.

Effects of water content: The particle size distribution has been demonstrated to vary significantly with water content, as shown in Fig. 3.24a; as the water content was increased from 0 to 160 g m^{-3}, the overall number concentration increased from 1.85×10^7 to 2.42×10^7 cm^{-3}. Adding vapor promoted particle formation, thereby causing an increase in particle concentration, indicating that the accumulative size distribution is primarily dominated by growth rather than formation. Nevertheless, subsequent coagulation is substantially influenced due to concentration variations.

In their work, Yang et al. [54] sampled data from the center of the growth tube to minimize the effects of wall deposition; therefore, only coagulation and condensation were considered in their particle dynamics model. As such, particle dynamics considering only coagulation and condensation can be governed by a general dynamic equation (GDE) in terms of particle volume (V_p) at time (t) as

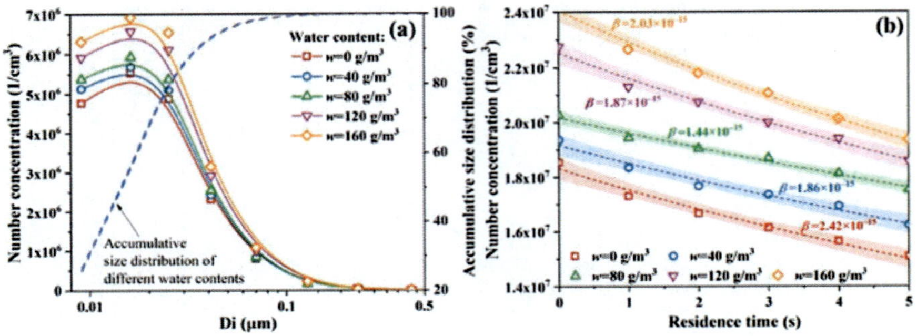

Fig. 3.24 **a** initial particle size distribution and **b** overall particle reduction along the growth tube [54]

$$\frac{\partial}{\partial t}[n(V_p, t)] = \frac{1}{2}\int_{V_{p*}}^{V_p} \beta(V_p - V'_p, V'_p)n(V_p - V'_p, t)n(V'_p, t)\mathrm{d}V'_p$$
$$- \int_{V_{p*}}^{\infty} \beta(V_p, V'_p)n(V_p, t)n(V'_p, t)\mathrm{d}V'_p + \frac{\partial G(V_p, t)n(V_p, t)}{\partial V_p}, \quad (3.39)$$

Here, the first two terms on the right-hand side of the equation represent the increase and decrease in aerosol concentration brought about by coagulation, respectively, i.e., the formation rate of aerosols with diameter d_p from smaller ones and the removal rate of aerosols with diameter d_p colliding with aerosols of any size. The third term represents the gain or loss of aerosols by condensation growth. The expression $\beta(V_p, V_p')$ is the coagulation coefficient or kernel for coagulation between two aerosols with volumes of V_p and V_p'. Brownian coagulation is the dominant mechanism among interparticle interactions [56]; the Brownian coagulation coefficient can be calculated as follows [57]:

$$\beta(V_p, V'_p) = \beta_0 \frac{\left(d_{V_p} + d_{V'_p}\right)^2}{d_{V_p} d_{V'_p}}, \quad (3.40)$$

where β_0 is 1×10^{-17} m³/s, and d_{Vp}, $d_{Vp'}$ represents the particle diameter in volume of V_p, V_p'.

Accordingly, the particle number concentration as a function of residence time varies significantly with water contents, as shown in Fig. 3.24b. The slope of the curve represents d_n/d_t, which is a function of the coagulation coefficient and aerosol number concentration. Rather than reducing particle emissions, adding vapor increased particle emissions by favoring a high terminal concentration and reducing the coagulation coefficient.

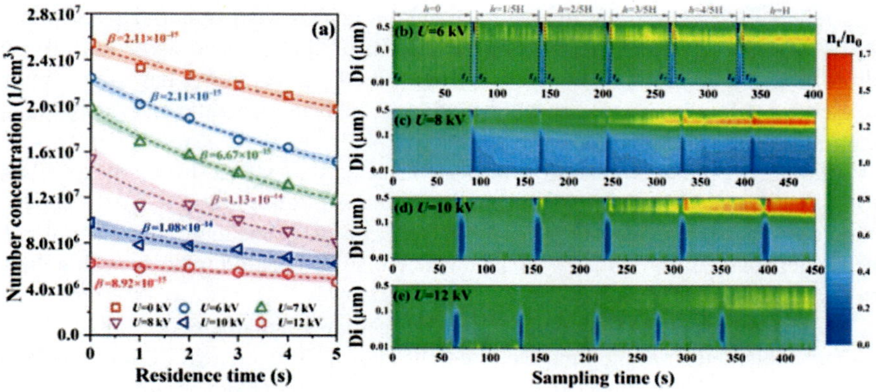

Fig. 3.25 **a** overall number evolution of particle along with the tube, and particle size evolution under. **b** $U = 6$ kV, **c** $U = 8$ kV, **d** $U = 10$ kV, **e** $U = 12$ kV [54]

Effects of corona discharge: As flue gases pass through a WESP, entrained particles are exposed to and charged by high-density negative ions and an electric field. When compared with neutral particles, these charged particles have an increased relative motion. Varying the applied electric field in a WESP demonstrated a clear increase in coagulation coefficient with increasing voltage until reaching a maximum at 8 kV, as shown in Fig. 3.25, indicating that the collision frequency is positively influenced by the number concentration and external force [57]. The particle number concentration decreased with increasing voltage because high voltages make electrostatic scavenging of initial particles more efficient in the mixing region of the growth tube. Conversely, increased charging and a higher electrical force can be realized because the corona current increased with the increasing voltage. These simultaneous electric field effects on electrostatic scavenging and charging led to the variation in aerosol behavior and increased the coagulation coefficient.

The observed particle size evolution under various discharge voltages is further detailed in Fig. 3.25b–e. At a voltage of 8 kV, particles < 0.1 μm in diameter decreased rapidly, owing to interparticle coagulation effects. Furthermore, the induced charge of a small aerosol sufficiently reduced the Gibbs free energy for nucleation and condensation [58, 59], thereby accelerating the aerosol growth. Coupling the coagulation and condensation effects, particles grew rapidly at $h = 3/5H$. Overall, imposing an electrical field significantly influences the coagulation and condensation mechanisms and thus the particle size evolution within a stack. The size evolution consumed more electric power when increasing operation voltage. The decrease in number concentration and growth of particle size became less distinct because the overall number decreased under high voltages.

Effects of flue gas composition: In most CFPP, selective catalytic reduction (SCR) and selective noncatalytic reduction (SNCR) employing NH_3 as a reducing agent are used for NO_x removal. In this process, some NH_3 escapes from the system and reacts with SO_3 in the flue to form sulfate aerosols, such as $(NH_4)_2SO_4$ and NH_4HSO_4 [60, 61]; this leads to the formation of sulfuric acid aerosols and salt droplets in the high-humidity flue gas after WFGD. Researchers have thus aimed to clarify agglomeration and particle growth methods in this complex system under varying flue gas compositions.

Adding 10 ppm of NH_3 shifted the peak of the particle size distribution from 0.04 to 0.06 μm and led to a decrease in sulfate aerosol concentration at 0.04 μm, as shown in Fig. 3.26. The sulfate aerosols can act as condensation nuclei and cause heterogeneous nucleation [62]. Thus, unreacted sulfuric acid molecules can adhere to and enlarge sulfate aerosols.

By introducing various amounts of fly ash (median ash diameter of approximately 10 μm) obtained from the final electric field of an ESP from a CFPP into the front end of the growth tube under the same conditions, Zheng et al. [52] investigated the variation in particle number and mass concentration in the growth tube under more realistic conditions; their results are summarized in Fig. 3.27a. Increasing the ash concentration decreased the particle number concentration in the growth tube that varied with the position in the growth tube. Near the inlet of the tube, the decrease in particle concentration was more evident. Adding 20 mg/m^3 of ash led to a 16.7, 16.5, and 9% decrease in particle number in the inlet ($h = 0$), middle ($h = 3/5H$), and outlet ($h = H$) of the tube, respectively. Here, the fly ash and particles combined to form ash–salt droplets comprising a core of fly ash with particles adhering to the surface. These wet, viscous particles could then capture more smaller particles during the flow process. Furthermore, these ash–salt droplets collided and agglomerated to form large multi-component complex particles. The removal efficiency of larger particles increased considerably compared with small-size particles. Overall, the particle concentration of all sized particles decreased with initial ash addition as measured by an electrical low pressure impactor (ELPI$^+$) in the growth tube (Fig. 3.27b).

With further increases to the ash content, the particle concentration slightly reduced and then stabilized; particles could not adhere to ash without restriction

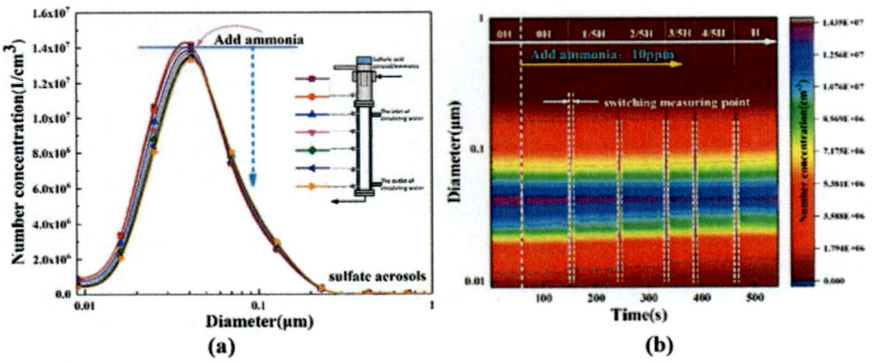

Fig. 3.26 Concentration variation of sulfate acid aerosol [52]

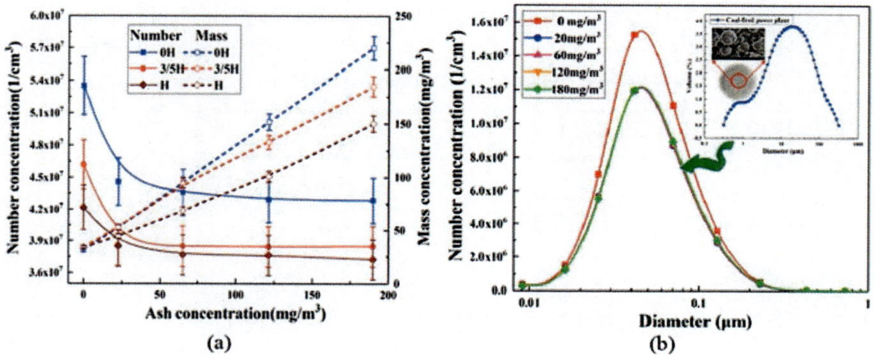

Fig. 3.27 **a** Variation in particle concentration and size distribution after adding ash and **b** particle size distribution of fly ash in a CFPP [52]

or further combine given the limited residence time in the growth tube. Interactions between the ash–salt droplets and unadhered particles formed a steady-state system in the growth tube, involving a balance of collision, agglomeration, repulsion, and breakage of various types of particles. Thus, further increases in the fly ash concentration caused no substantial decrease in the particle concentration.

3.3.1.2 Particle Agglomeration

Particle agglomeration methods can be used to increase mean particle size and thus increase the collection efficiency of ESPs in CFPPs, where the particle charge and migration velocity are low for submicron particles [63, 64]. Electric agglomeration methods employ an external electric field or mixers to charge particles, generate electrostatic attractive forces, and thereby increase particle collision [65]. In this section, recent research efforts to strengthen the adhesive force and surface tension between positively and negatively charged particles by introducing water droplets into the agglomeration process are discussed. This is close to the real case in industrial flue gases with high humidity, such as in a WESP. A laboratory bipolar agglomerator employing water droplet spray was designed for fine particle agglomeration enhancement, and an ESP was placed downstream to collect pre-charged and agglomerated particles (shown in Fig. 3.28).

Adhesive forces between charged particles: During and after particle collisions, a wide variety of adhesive forces act between particles and promote agglomeration, including dispersive adhesion due to Van der Waals attraction, electrostatic adhesion due to electrical attraction between charged particles, and liquid bridge force due to the water layer between particles [67, 68]. Each of these forces can be calculated based on theoretical and/or semiempirical formulas and to explain the promoted agglomeration. However, the overall adhesive force cannot be calculated by summing these

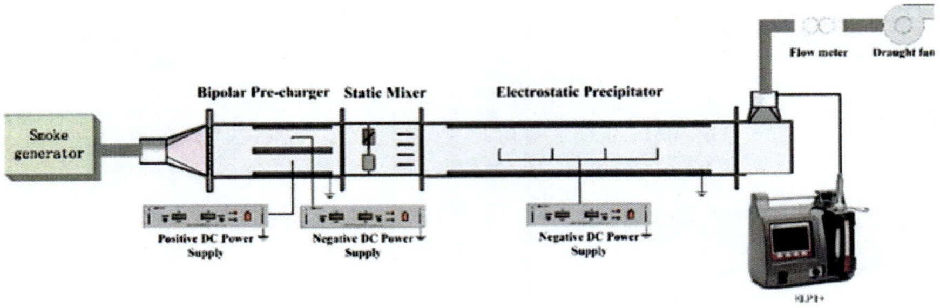

Fig. 3.28 Schematic of bipolar agglomeration experimental system [66]

forces as London–van der Waals forces are affected by water. When two particles are at a short distance apart relative to their diameter, London–van der Waals forces act as an attractive force, are responsible for adhesion in most systems, and can be expressed as follows:

$$F_{vw} = \frac{A_a}{12Z^2} \times \left(\frac{d_1 d_2}{d_1 + d_2}\right), \tag{3.41}$$

where A represents the material-related Hamaker constant, d_1 and d_2 represent the diameters of each particle, and Z represents the distance separating the particles.

The electrostatic attractive force between particles with opposite charges is derived as follows:

$$F_c = \frac{k \times q_1 \times q_2}{Z^2}, \tag{3.42}$$

where $k = 9.0 \times 10^9$ N m^2/C^2 is the electrostatic constant and q_1 and q_2 represent the charge of each particle.

When a water droplet attaches onto a particle's surface and fills in the capillary space between particles, the particle's adhesive force is increased by surface tension. This so-called liquid bridge force, or capillary force, can be calculated as follows:

$$F_{\text{cap}} = 2\pi\gamma R\cos\theta\left[1 - \left(1 + \frac{2V_L}{\pi Z^2 R}\right)^{-\frac{1}{2}}\right], \tag{3.43}$$

where γ represents the surface tension of the liquid; R represents the effective particle radius, $1/R = 1/r_i + 1/r_j$, where r_i and r_j are the radii of each particle; V_L represents the liquid bridge volume; and h represents the contact angle.

When the particles vary by an order of magnitude, the liquid bridge force can be simplified as the force between a particle and a smooth surface; the maximum liquid bridge force is thus expressed as follows [69]:

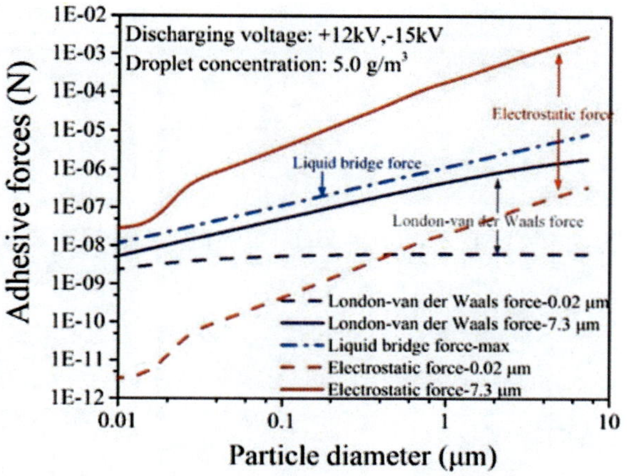

Fig. 3.29 Adhesive force analysis between charged particles [66]

$$F_{\text{cap}} = 2\pi\gamma R. \tag{3.44}$$

Calculated electrostatic forces, London–van der Waals forces, and liquid bridge forces between particles with different particle diameters and under different applied voltages are shown in Fig. 3.29. The particle diameters ranged from 6 to 10 mm, and the distance between particles was 0.4 nm when they were about to collide, as determined from relevant research [67, 70].

Numerical results indicate that for a given diameter of a negatively charged particle, the electrostatic force increases with an increasing positively charged particle diameter. Similarly, the London–Van der Waals and liquid bridge forces increase with an increase in particle diameter. The liquid bridge force is always five or more times higher than the London–Van der Waals force. The Van der Waals forces dominate for sufficiently small particles (e.g., < 0.5 μm in diameter) with the aid of the liquid bridge force, whereas the electrostatic force dominates for sufficiently large particles (e.g., > 1.0 μm in diameter). Therefore, adding water droplets increases the adhesive forces and promotes the agglomeration of submicron particles onto large particles.

Observation of agglomeration characteristics: Before agglomeration, coal-fired ash particles comprise smooth spheres with few attached fine particles. When water droplets are added, some submicron particles attach to the surface of large particles. Passing the flue gas through an agglomerator causes submicron and micron particles to attach to each other. Furthermore, the combined effect of an agglomerator and adding water droplets, i.e., of agglomeration and liquid bridge force, promoted submicron–micron agglomeration. To quantify this effect, researchers counted the number of particles attached to a single large particle (15–25 μm in diameter). Scanning electron microscope (SEM) images of sample agglomerated particles are shown in Fig. 3.30, and average results are presented in Fig. 3.31. Overall, the number of attached particles of all sizes increased when using the agglomerator and adding

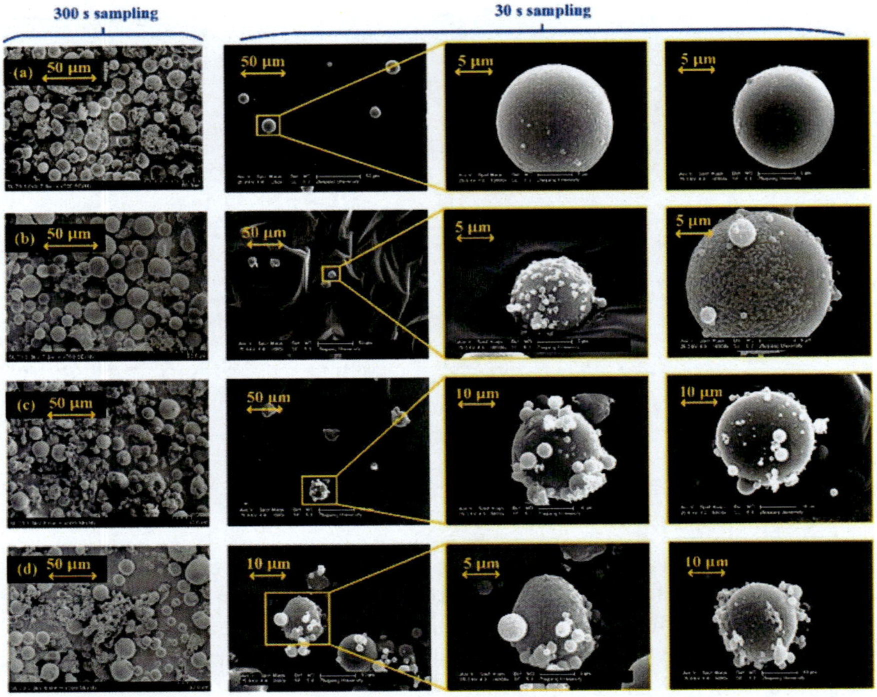

Fig. 3.30 SEM images of particles agglomerated on large particles (15–25 μm in diameter), **a** condition 1: before agglomerator, without water droplets; **b** condition 2: before agglomerator, with water droplet concentration of 5.0 g/m^3; **c** condition 3: after agglomerator, without water droplets; and **d** condition 4: after agglomerator, with water droplet concentration of 5.0 g/m^3 [66]

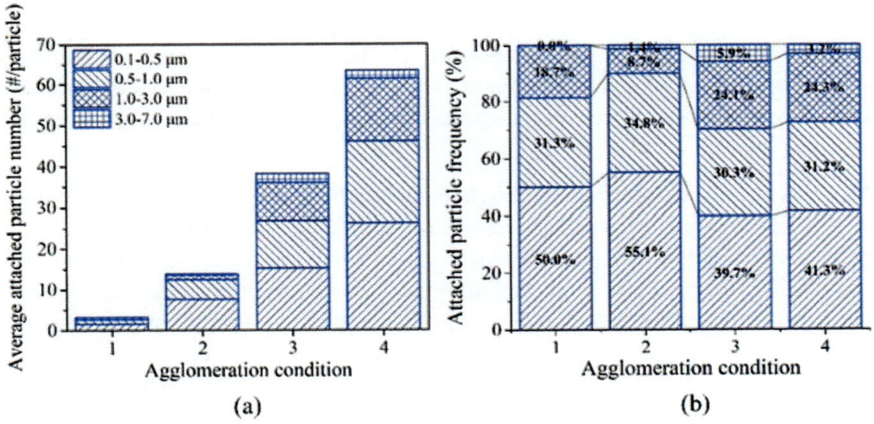

Fig. 3.31 **a** Average attached particle number and **b** attached particle frequency on large particles [66]

water. In particular, the number of attached submicron particles increased under electric agglomeration. Adding water droplets increased particle adhesion via the liquid bridge force. The attached particle percentages of submicron particles increased from 81.3 to 89.9% and from 70.0 to 72.5% before and after electric agglomeration, respectively.

Proposed agglomeration models: Based on these experimental results and adhesive force analysis, Chang et al. [66] proposed particle electric agglomeration modes, as summarized in Fig. 3.32. According to their model, small (submicron in diameter) and large (microns in diameter) particles agglomerate at steady-state because the adhesive force is large enough for fine particles to be attached, whereas large particles' high inertia prevents the formation of agglomerates between large particles. The addition of water droplets allows fine particles to be absorbed by water droplets and attached

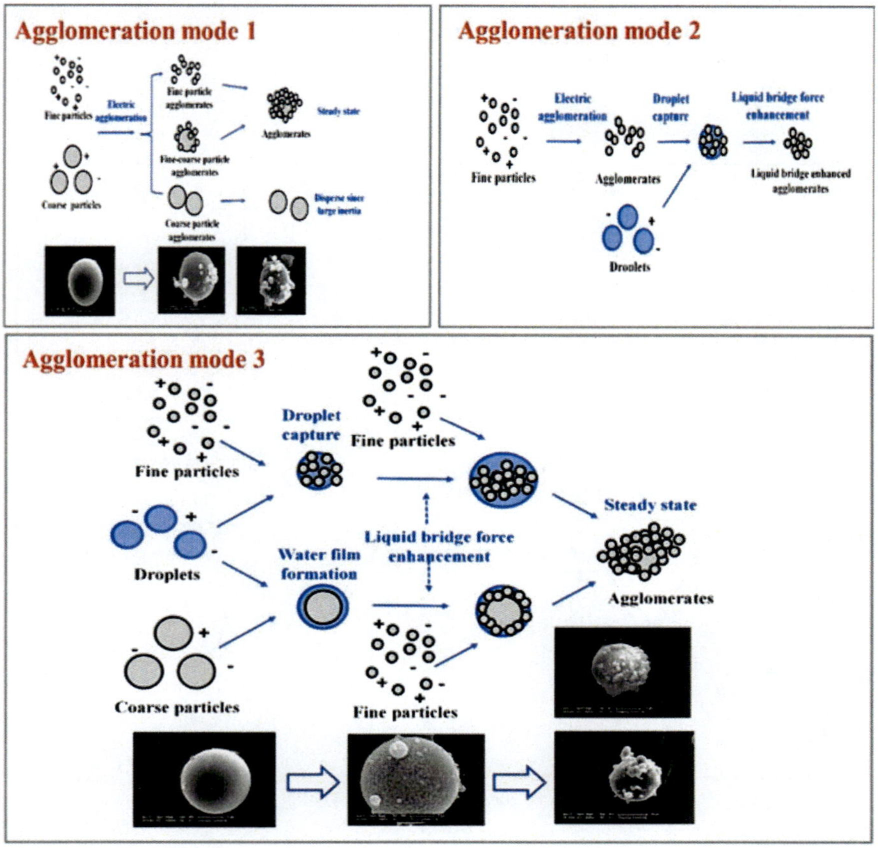

Fig. 3.32 Particle electric agglomeration modes under different conditions [66]

to large particles, further enhancing the generation of agglomerates between fine and large particles.

Overall, the proposed agglomeration modes, observed phenomena, and experimental results presented here indicate that three main factors must be considered to improve particle agglomeration. First, the discharge voltage and ion density of both channels in the agglomerator must be sufficiently high to improve and strengthen the particle charges and electrostatic forces, respectively. Second, attraction forces allow small particles can effectively agglomerate to large particles. Third, additives such as water droplets or other viscous liquids should be used to help generate liquid bridge and increase adhesive forces between particles, are suggested to be utilized.

Effects of applied voltage: Particle percentage distributions under an applied negative voltage of − 15 kV on the negative channel and positive voltage ranging from 9 to 15 kV are shown in Fig. 3.33. The decrease of particle percentages in the first and second stages and the increase of larger particles of other stages demonstrate agglomeration, i.e., smaller particles agglomerated into larger particles. The agglomeration enhancement ratio (R_{ae}) demonstrates the increase of median particle diameter as follows:

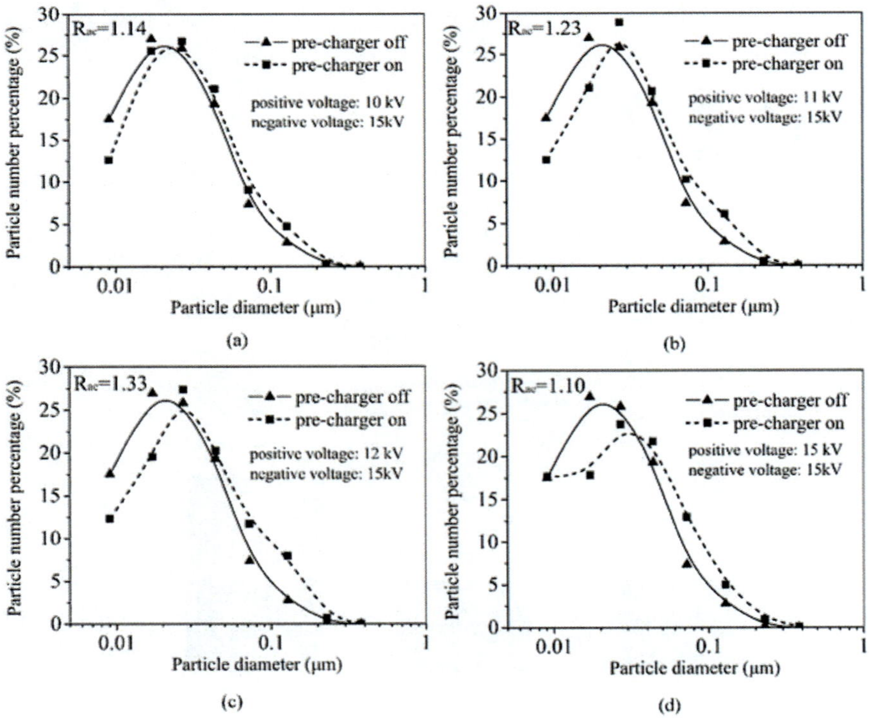

Fig. 3.33 Particle percent distributions under different voltage [71]

$$R_{ae} = \frac{D_{agg}}{D_0}, \tag{3.45}$$

where D_0 and D_{agg} represent the median particle diameter before and after agglomeration in μm. At each voltage studied, $R_{ae} > 1$, indicating an increased average particle diameter and enhanced agglomeration. Increasing the positive discharge voltage from 9 to 12 kV led to an increase in R_{ae}, suggesting that particles with a high mean positive charge are produced with high discharging voltages, which can then mix and agglomerate further with negatively charged particles under the effect of Coulomb force. Further increasing the positive discharging voltage, however, caused R_{ae} to decrease from 1.33 to 1.10. This finding can be presumably explained through submicron particles in the positive channel, wherein multiple neutral particles can agglomerate on charged particles because of polarization. However, when overcharged, an excess of positive ions is produced, and those parts of agglomeration are inhibited.

Effects of perforated plate: The effect of particle plate porosity on particle percentage distribution as demonstrated by Chang et al. [71] is summarized in Fig. 3.34, where the voltage of the positive and negative channels were 12 and

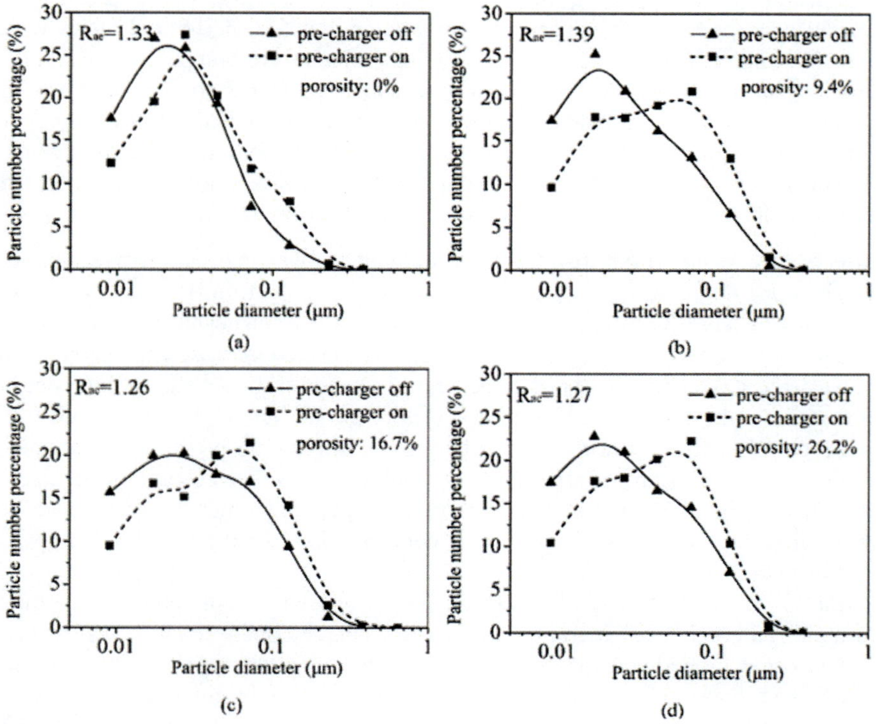

Fig. 3.34 Particle percent distributions under different perforated plate porosities (precharging: + 12 kV, – 15 kV) [71]

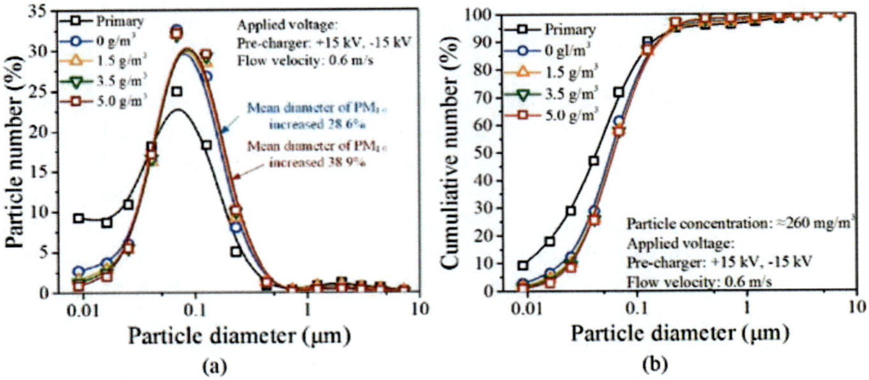

Fig. 3.35 **a** Particle number distributions and **b** cumulative number distribution under different droplet concentrations using a precharger with a positive and negative voltage of + 15 and − 15 kV, respectively [66]

− 15 kV, respectively. At all porosities that have been studied, $R_{ae} > 1$, further demonstrating the increase in median particle diameter caused by the precharger. Considering the porosities under varying voltage matches, perforated plates with porosities under 16.7% achieved better agglomeration results compared with regular smooth plates, whereas the maximum porosity under nearly every voltage match obtained the lowest agglomeration enhancement ratio. Perforated plates with proper porosities can promote flow exchange and particle collision in discharging regions. However, large holes in plates with high porosity can cause inhomogeneities in the electric field and disturbance and excessive neutralization with ions.

Effects of adding water droplets: The particle number and cumulative number distributions under water droplet concentrations ranging from 0 to 5.0 g/m^3 are summarized in Fig. 3.35, where the voltage of the positive and negative channels were 15 and − 15 kV, respectively. As the droplet concentration increased, the percentage of particles < 0.05 μm in diameter decreased, demonstrating clear agglomeration of particles < 1 μm. As water droplets attach to the surface of particles, results are interpreted as adding adhesive forces between particles and promoting agglomeration. The count mean particle diameter (arithmetic number average of all sizes) of $PM_{1.0}$ increased by approximately 28.6% using the agglomerator without added water droplets; by increasing the droplet concentration to 5.0 g/m^3, this value was increased to 38.9%.

Based on these efforts, Chang et al. [66] proposed a comprehensive method combining particle agglomeration and particle charge enhancement. Accordingly, they selected applied voltages of the agglomerator (positive and negative) and ESP of 12, − 15, and − 25 kV, respectively, to flue gas with an original particle concentration and flow velocity of approximately 260 mg/m^3 and 0.6 m/s, respectively. The resulting particle migration velocity (ω) under different droplet concentrations, summarized in Fig. 3.36, increased from 3.5–8.8 to 3.6–13.6 cm/s by adding the

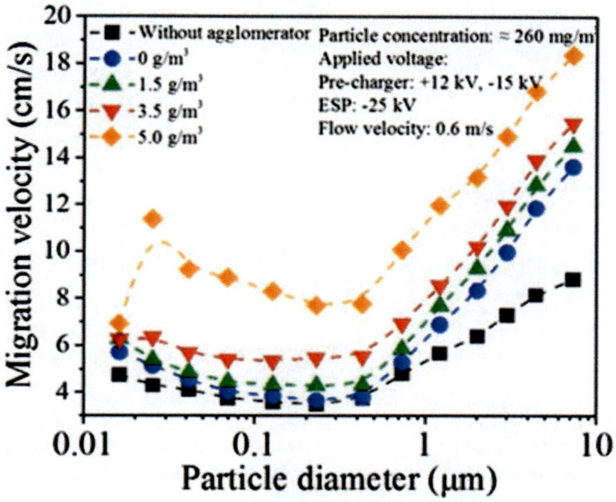

Fig. 3.36 Migration velocity under different droplet concentrations with a positive and negative voltage of 12 and − 15 kV, respectively [66]

agglomerator, and then further increased to 6.9–18.4 cm/s by also adding water droplets. Therefore, the combined effect of the agglomerator and water droplet injection can improve particle migration velocity and allow a higher collection efficiency.

3.3.1.3 Applications

Enhancing particle coagulation: To promote the coagulation and agglomeration of particles in an ESP and meet ultra-low emission standards, an integrated temperature management and electrostatic removal method was applied to a full-scale functional CFPP, as shown in Fig. 3.37; the particle size variation at the inlet and outlet of the

Fig. 3.37 Application of enhancing particle coagulation in a full-scale CFPP

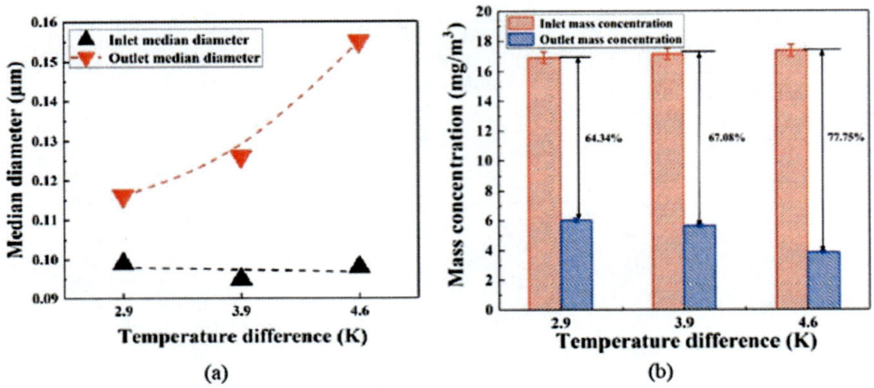

Fig. 3.38 Change in particle **a** median diameter and **b** mass concentration and removal efficiency in the heat exchanger under different temperature differences

condensing heat exchanger and the system emission performance was then examined. Increasing the flue gas temperature difference led to an increase in median particle diameter at the outlet of the heat exchanger from 0.116 to 0.155 μm, as shown in Fig. 3.38a, and a decrease in particle mass concentration at the outlet, as shown in Fig. 3.38b, corresponding to an increase in removal efficiency from 67.89 to 77.75%. Furthermore, PM and SO_3 emissions of < 1.5 and < 0.5 mg/m^3, respectively, were maintained, as shown in Fig. 3.39a. The stacks plume of the final emission from the demonstration system is shown in Fig. 3.39b. Overall, increasing the flue gas temperature drop effectively improved particle removal efficiency in a full-scale CFPP, confirming lab-scale data trends.

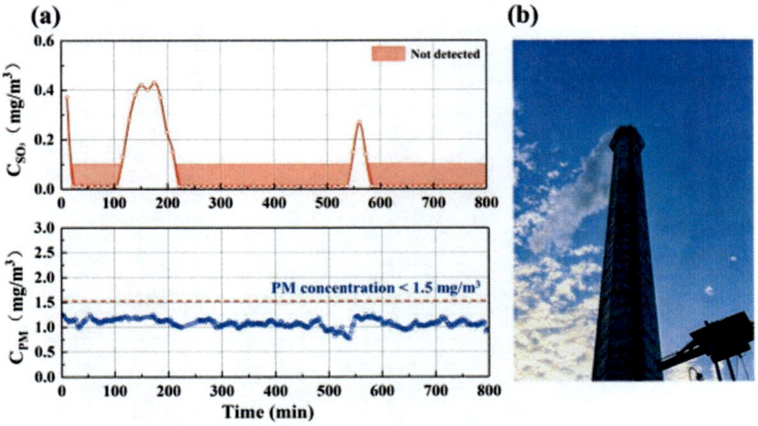

Fig. 3.39 **a** Emission concentration of SO_3 and PM and **b** stacks plume of the CFPP equipped with an integrated temperature management and electrostatic removal system

Fig. 3.40 Electrostatic agglomeration device and ESP Layout of power plant

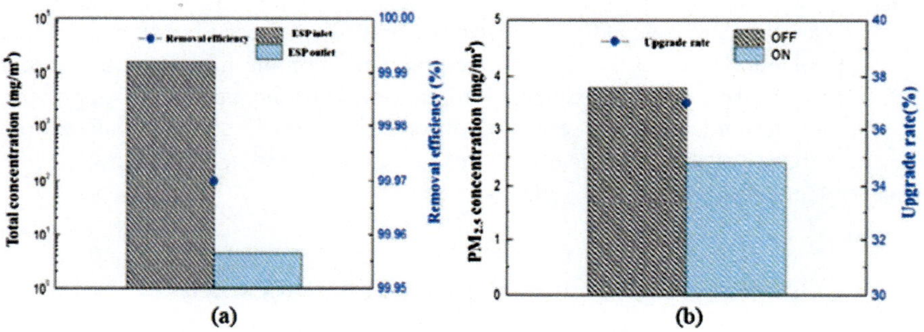

Fig. 3.41 Improvement in total, **a** PM and **b** $PM_{2.5}$ concentration achieved with the bipolar agglomeration system

Enhancing particle agglomeration: Due to their low inherent electrical mobility, particles 0.1–1.0 μm in diameter can easily escape from the ESP [72]. Electric agglomeration has been demonstrated at the lab scale to improve fly-ash particle collection efficiency in traditional ESP systems by increasing particle diameter. Based on particle electric agglomeration technology, the bipolar agglomeration system, shown in Fig. 3.40, was thus designed and applied to a 1000 MW CFPP. By combining the optimized agglomeration system with an ESP, the total PM and $PM_{2.5}$ concentrations were reduced by 20.3 and 30.1%, respectively, as shown in Fig. 3.41. Furthermore, when coupled with a rotating electrode and a low-temperature ESP, the $PM_{2.5}$ emissions were reduced by 37% from the original value.

3.3.2 *Particle Conditioning*

The chemical composition, adhesion, specific resistivity, size distribution, morphological structure, and other particle parameters directly affect particles' electrical conductivity and agglomeration tendencies, and therefore the performance of an

ESP [73, 74]. In some industries, flue gases have high temperatures and complex particle properties. For example, flue gases from steel sintering and waste incineration facilities contain a high moisture content and heavy metal pollutants, whereas those generated by glass furnaces contain particles with a high alkali metal concentration, which can reduce the lifespan of SCR catalysts and affect the operation of SCR systems. Fly ash also contains several hazardous heavy metal elements that can poison catalysts and lead to catalyst deactivation.

A wide variety of particle conditioning properties have thus been proposed to manage particles with high resistivity, viscosity, and other characteristics that are not suitable for the operation of ESPs. Some examples include adding chemical additives such as water, sulfur trioxide, or ammonia to the flue gas to change its characteristics, water conditioning to increase its relative humidity, reduce surface resistivity, and enhance fly ash agglomeration, and adding sulfur trioxide (sulfuric acid) and ammonia to reduce fly ash resistivity and adjust the acid dew point of flue gas [75]. Shanthakumar et al. demonstrated that an ammonia flue gas conditioning method increased particle viscosity; agglomeration was further enhanced by applying an SO_3 and ammonia double flue gas conditioning method [76, 77]. In this section, therefore, the use of solid chemical additives to condition particles with a low specific resistivity and strong adhesion are introduced.

3.3.2.1 Granular Component Conditioning by Mixed Particles

Particle conditioning aims to:

(1) alter the chemical composition and surface conductivity of particles to adjust particle specific resistivity;
(2) increase the space charge ion concentration; and
(3) adjust particle viscosity to improve the effectiveness of the rapping system [78].

In this section, an enhanced removal system of fly ash from a glass furnace with low resistivity and high viscosity employing particle conditioning is introduced. The composition of the fly ash, shown in Table 3.5, shows that the main elements of the fly ash included Na (15.3%), Ca (14.5%), and S (50.6%); heavy metals were also present, such as V (1.5%) and Sb (4.72%). As shown in Fig. 3.42, the maximum particle resistivity was approximately 3.68×10^9 Ω cm at 363 K; above 363 K, the resistivity decreased with the increasing temperature throughout the entire temperature range studied, reaching 2.77×10^4 Ω cm at 873 K. Comparatively, the maximum resistivity of calcium carbonate particles was 6.48×10^{13} and 3.59×10^9 Ω cm at 393 and 873 K, respectively. Furthermore, this fly ash contained many low-resistivity sodium compounds.

The flue gas particles contained many sodium compounds and exhibited strong adhesion, as evidenced in Fig. 3.42b, where the particles resembled an irregular block or flocculent structure with parts of small particles adhered to each other. To adjust the particle characteristics of the fly ash and improve the collection efficiency, fine

Table 3.5 Elemental composition of fly ash from a glass furnace

Composition	Content (%)	Composition	Content (%)
C	2.64	Ni	0.99
Na	15.3	Cu	0.0088
Mg	2.04	Zn	0.102
Al	1.93	Ga	0.0045
Si	2.59	As	0.0373
P	0.223	Sr	0.226
S	50.6	Y	0.0297
K	0.166	Mo	0.0195
Ca	14.5	Sn	0.0353
Ti	0.0952	Sb	4.72
V	1.5	I	0.0241
Cr	0.22	Ba	0.0605
Mn	0.0455	Tl	0.0028
Fe	1.8	Pb	0.0154
Co	0.0538		

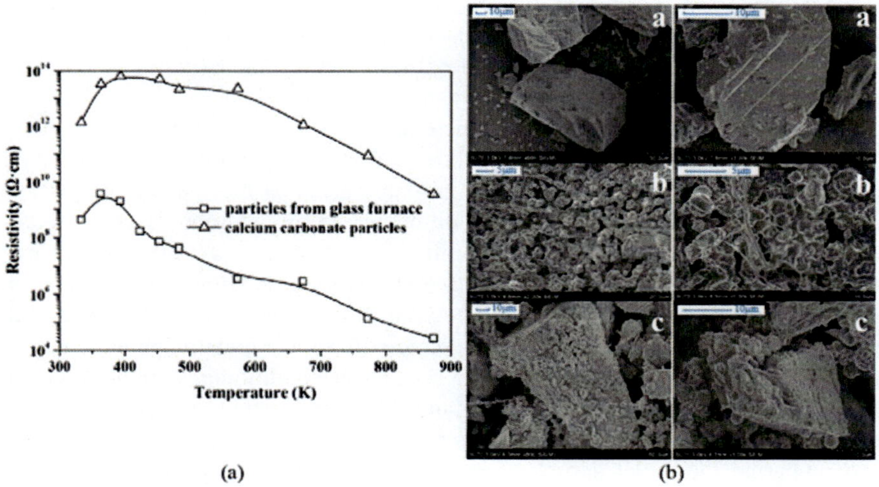

Fig. 3.42 **a** Resistivity of two kinds of particles at different temperatures. **b** Microstructure of the particles with conditioning [79]

calcium carbonate powders (median particle diameter of 15.57 μm) were delivered in a 1:1 mass ratio of fly ash particles to the inlet gas stream of the experimental system using a separate feeder. The inlet particle concentration with and without particle conditioning was 400.98 and 376.81 mg/Nm^3, respectively, the flue gas temperature

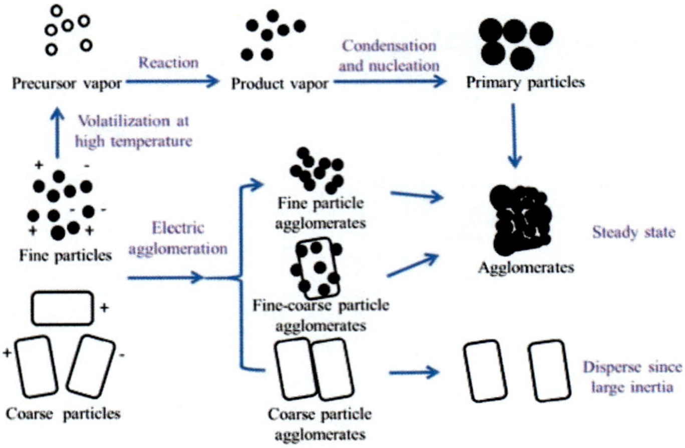

Fig. 3.43 Proposed particle removal enhancement mechanism [79]

was 623 K, and the gas flow velocity was about 0.466 m/s in the ESP. Their findings demonstrated that each calcium carbonate particle appeared as a coarse, blocky structure that allowed many irregular fine particles to adhere to it, and that the adhesive flue gas particles mixed well with the conditioning particles, possibly owing to the high viscosity of the flue gas particles, thus allowing fine particles to condense or attach to coarse particles subsequent to a series of collisions.

Adhesive forces between particles, including electrical (Coulomb) and London–Van der Waals forces, can be promoted through particle conditioning, thereby aiding the removal of fine particles [80]. Two possible mechanisms causing the particle removal enhancement seen during conditioning via the addition of coarse additives are summarized in Fig. 3.43. In the first, introduced fine and fine-coarse particles agglomerate with coarse particles to a steady-state. Agglomerates increase in size, whereas the coarse ones disperse because of their large inertia. In the second, volatilizable components in the fly ash are volatilized at a high temperature before condensing and nucleating to a solid state. As a result, conditioning particles improve the resistivity of the original particles and adjust the particle adhesion and size distribution.

Comparing the outlet particle concentration and particle collection efficiency with/without particle conditioning shown in Fig. 3.44 indicates a clear improvement in particle collection efficiency when particle conditioning is employed; at 24.8 kV, the collection efficiency increased from 65.8 to 87.6%. This result may be preliminarily attributed to two reasons. First, because calcium carbonate particles have high resistivity, their addition may have increased the particle resistivity and weakened the particle re-entrainment. Second, mixing the strong adhesion particles with conditioning particles increased the particle concentration, thereby increasing the chances of collisions among particles and causing particle coagulation.

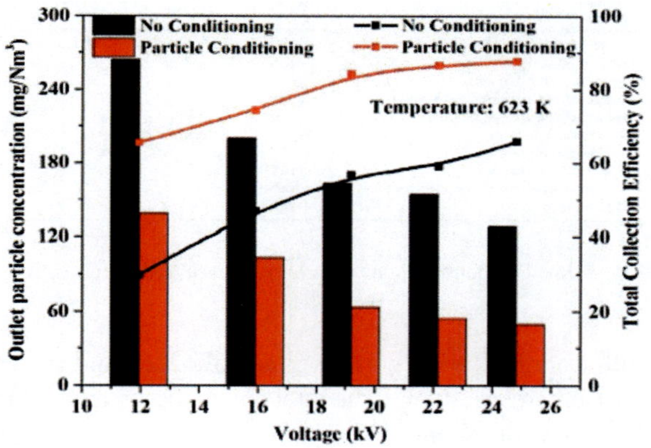

Fig. 3.44 Outlet particle concentration and particle collection efficiency with/without particle conditioning [79]

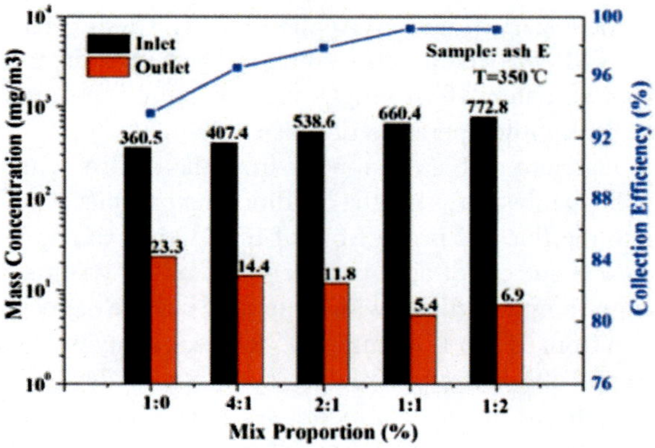

Fig. 3.45 Inlet/outlet concentration and particle removal efficiency under various mix proportion of particle conditioning

This particle condition method has also been applied to another high-viscosity fly ash sample (coal pyrolysis fly ash) at 673 K; the resulting particle inlet and outlet concentrations and the removal efficiency under different mix proportions are shown in Fig. 3.45. Particle conditioning increased the efficiency of particle removal. Increasing the particle mix proportion further increased the particle removal efficiency, although this increase was diminished after a mix proportion of 1:1, at which point the removal efficiency had increased 5.7%.

3.3.2.2 Applications

A particle property conditioning method was applied to enhance the particle removal efficiency of a high-temperature ESP in a glass furnace; the design parameters and an image of the ESP are shown in Fig. 3.46. Here, there were two electric fields in the

Name	Unit	Parameter
Gas flow rate	Nm^3/h	30,000
Operating temperature	K	573
Number of electric field		2
Length of the each electric field	mm	4000
Height of the electric field	mm	7055
Number of channels		16
Total collecting area	m^2	1568
Specific collection area	$m^2/(m^3/s)$	87.5
Space between collection electrode	mm	400
Gas velocity	m/s	0.4

Fig. 3.46 The panorama of the glass furnace and specification of the high-temperature ESP

full-scale ESP, each with a voltage of 42 kV and current range of 107–117 and 174–182 mA, respectively. The total collection area was 1568 m^2, and the specific collection area was about 87.5 $m^2/(m^3/s)$. The flue gas was sampled by a sampling probe, an electric heater, and a vacuum pump at both the inlet and the outlet of the high-temperature ESP, and the particle concentration was measured using an automatic particle tester (Laoshan Applied 3012H, China). The inlet particle concentration of the ESP was about 600–700 mg/m^3, whereas the outlet particle concentration was approximately 103.23 mg/Nm^3, indicating an original particle collection efficiency of the high-temperature ESP of < 90%.

Calcium carbonate powder from the facility with a median particle diameter of 175.5 μm was used as the conditioning particles. Conditioning particles were added into the flue gas in the front of ESP. Using an applied voltage of 42 kV induced a corona current of approximately 120 mA. As summarized in Fig. 3.47, adding the conditioning particles with no applied voltage increased the outlet particle concentration from 612 to 1140 mg/Nm^3, whereas applying voltage with particle conditioning caused the outlet particle concentration to decrease from 103.23 to 56 mg/Nm^3. Despite the increase in fly ash sedimentation at the bottom of the high-temperature ESP, the particle collection efficiency improved from 83.2 to 95.1%. By testing particle dosages of 30, 75, and 150 kg/h, as shown in Fig. 3.48, the particle emissions of the ESP were decreased from 103.23 to 39.25, 49.67, and 56.26 mg/Nm^3,

Fig. 3.47 Effect of particle conditioning on high-temperature ESP performance [79]

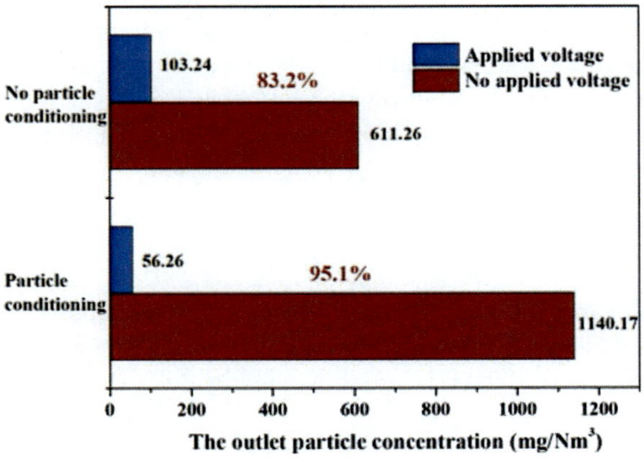

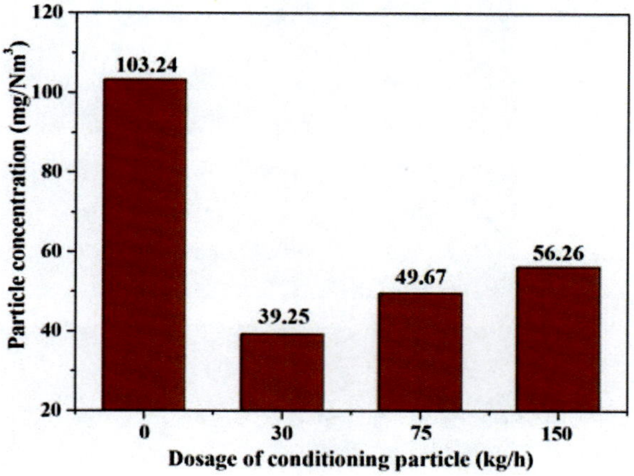

Fig. 3.48 Outlet particle concentration with different dosages of conditioning particles [79]

respectively. Thus, at an original: conditioning particle mass ratio of 1:1.5, an optimal ESP collection efficiency of 39.25 mg/Nm^3 was achieved, representing a decrease of about 64 mg/Nm^3. However, further increases to the conditioning particle dosages from 30 to 150 kg/h further increased the total inlet mass concentration, causing a higher outlet mass concentration. Particle conditioning was thus demonstrated to effectively promote the attachment of small particles to large particles, enhance the removal of submicron particles, and improve the total particle collection efficiency. Overall, particle conditioning can enhance particle removal efficiency of high-temperature ESP and subsequent SCR to satisfy increased emission standards for exhaust pollutants in the glass industry.

3.3.3 Particle Precharging

3.3.3.1 Development of a Novel Precharger

The presence of ultrafine particles (approximately 0.1 μm) in flue gas passed through an ESP can cause a severe current reduction, also called corona suppression or corona quenching, and inhibit fine particle removal. Visualizations of ion density of different $PM_{0.1}$ loadings within an electric field, shown in Fig. 3.49, demonstrate a clear influence of $PM_{0.1}$ loading on the ion density. As the particle space charge shows a competitive influence on ion density distribution, increasing the $PM_{0.1}$ loading suppresses the ion density, particularly at the middle electrodes; by a $PM_{0.1}$ loading of $\geq$ 40 mg/m^3, electron emission from all electrodes except for the first electrode was negligible.

As a result, particle migration velocity decreases in an ESP with increased $PM_{0.1}$ concentration, as shown in Fig. 3.50, leading to a decrease in overall particle collection efficiency. Here, the reduced relative collection efficiency, indicated by a red

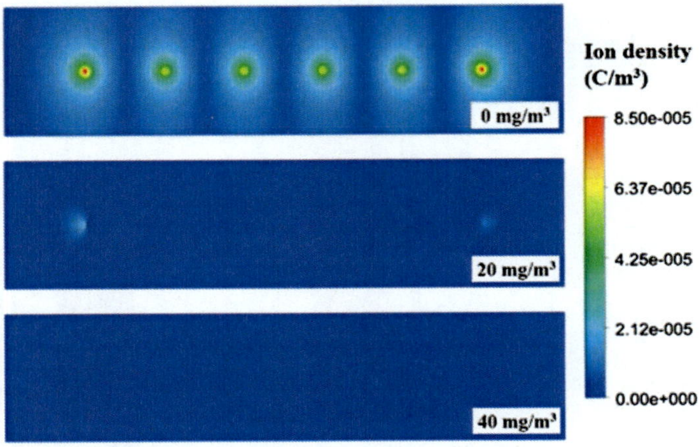

Fig. 3.49 Visualizations of the electric field and ion density under different $PM_{0.1}$ loadings [81]

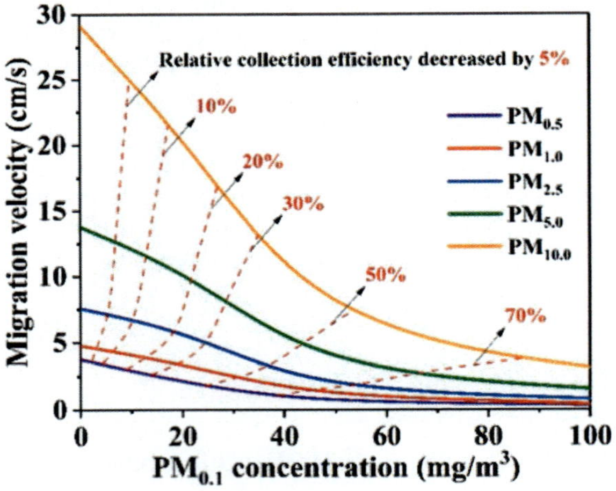

Fig. 3.50 Migration velocity of particles with diameters of 0.5–10 μm under different $PM_{0.1}$ loadings (applied voltage = 50 kV) [81]

dashed curve, represents the degree of reduction in collection efficiency compared with a $PM_{0.1}$ loading of 0 mg/m^3. Overall, the particle collection efficiency of small particles was more significantly influenced by the $PM_{0.1}$ space-charge effect. By increasing the $PM_{0.1}$ concentration from 0 to 100 mg/m^3, the particle migration velocity of particles 10 μm in diameter decreased from 29.1 to 3.1 cm/s; the migration velocity of particles < 2.5 μm diameter dropped to near-zero values when the $PM_{0.1}$ concentration exceeded 80 mg/m^3. The relative collection efficiency of $PM_{0.5}$ decreased by more than 20% when the $PM_{0.1}$ concentration was only 10 mg/m^3.

As the $PM_{0.1}$ concentration increases, so does their probability of collision with electrons, leading to more electrons being captured by these ultrafine particles. As a result, large particles are insufficiently charged and do not move toward the collecting

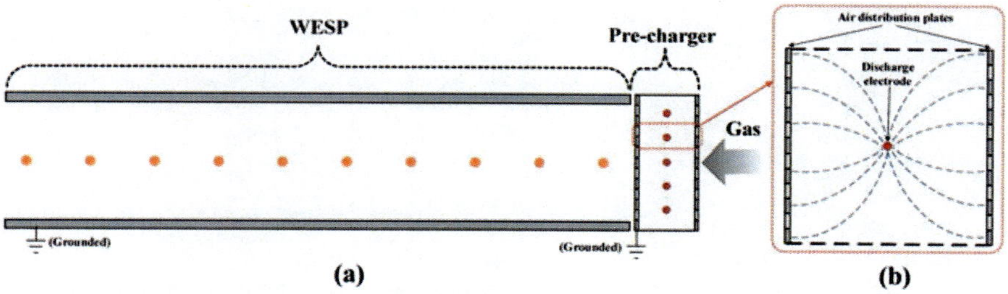

Fig. 3.51 Schematic of **a** combined wet ESP (WESP) and **b** perforated precharger [51]

plates. Moreover, the electric field around the discharge electrode is weakened due to the reverse electric field formed by particle space charges, thereby suppressing the electron avalanche and reducing the electron number.

A recently developed novel combined WESP with a perforated precharger, shown in Fig. 3.51, can be used to reduce the effects of ultrafine particle space charges. Implementing this perforated precharger improved the ion density and efficiently charged many more ultrafine particles. In this system, the discharge electrode is connected with a high-frequency power supply to generate ions for aerosol charging, and the air distribution plates are used as grounding plates. Several benefits have been demonstrated, including (a) no extra installation space, (b) lower specific surface area (the ratio of the collection-plates area to the gas flow, SCA) increase, and (c) higher turbulence intensity.

Electrically neutral particles 5 μm in diameter were uniformly injected into the entrance of the WESP. The particle trajectories, with their colors scaled to particle charges under different electrical conditions, are shown in Fig. 3.52. The installed precharger exerted significant effects on particle charging and transport through two possible mechanisms. In the first, particles being transported toward the collection stage by drag and electric forces are disturbed by the perforated plates. The presence of the precharger thus increases the turbulence of the flue gas and captures some particles on the perforated plates, thereby supporting the subsequent particle collection. In the second, the particles, particularly those near the electrodes, are quickly charged in the precharger and are subjected to deflection in the collection stage.

In a WESP, gas molecules are ionized by a high-strength electric field to produce ions around the discharge electrode. Particles are charged when passing through the WESP, and a certain amount of charges are carried by an escaping particle. The charging characteristics of an escaping particle under different electrical conditions are shown in Fig. 3.53. Here, the elementary charge number decreases drastically with decreasing particle size; the presence of a partial charging regime, where the charge number is < 1, indicates that some of the escaping particles were not charged. As the applied voltage was increased from 16 to 32 kV, the particle size above the partial charging regime shifted from 0.430 to 0.025 μm. Particle charging is a size-dependent process that comprises field and diffusion charging. As field charging is positively influenced by an external electric field around a spherical particle and diffusion

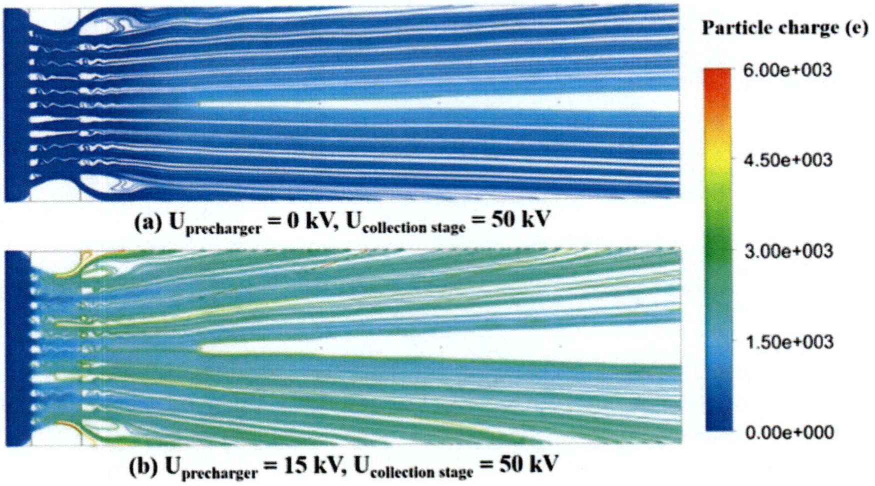

Fig. 3.52 Particle trajectories with colors scaled to particle charge at an inlet velocity of 1 m/s and a particle diameter of 5 μm in a combined WESP ($PM_{0.1}$ concentration = 40 mg/m^3) [81]

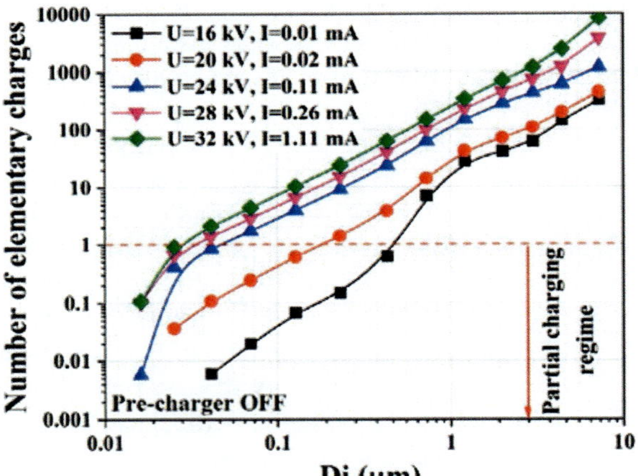

Fig. 3.53 Charge of an escaping particle amount under different electrical conditions (without precharging) [82]

charging is affected by the ion concentration, and ion concentration increases with increasing applied voltage because the corona current increases quadratically as the voltage increases, particle charging can thus be improved by increasing electrical parameters.

The influences of precharging and applied voltage on particle migration velocity are compared for three typical fine and ultrafine particle size categories in Fig. 3.54a. The migration velocity across the entire voltage range was improved using the precharger, with more drastic improvements for smaller particles; the migration velocity increased by 69.9, 65.7, and 34.2% for $PM_{0.1}$, $PM_{1.0}$, and $PM_{2.5}$, respectively.

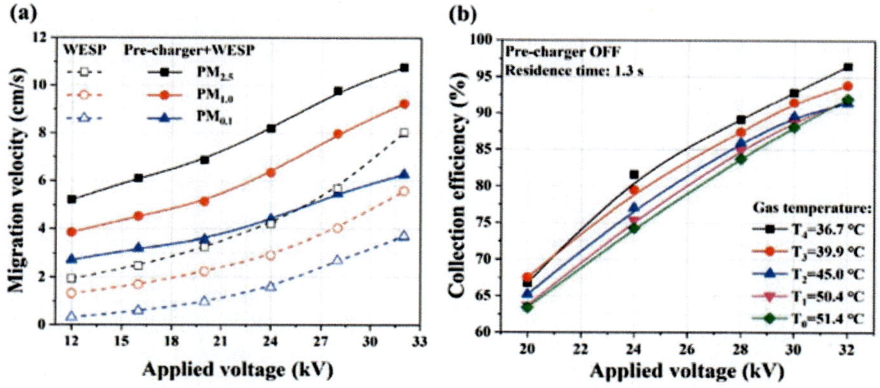

Fig. 3.54 **a** Particle migration velocity improved by the precharger and **b** overall collection efficiency under different gas temperatures [82]

In real power plants, WESPs are usually operated under saturated gas conditions with a gas temperature below 50 °C. Under such conditions, heterogeneous vapor condensation on a particle's surface can occur. The influence of gas temperature on collection efficiency was investigated in the range of 36.7–51.4 °C; as shown in Fig. 3.54b, particle collection efficiency increased with decreasing gas temperature.

The collection efficiency of a WESP and combined WESP (WESP + precharger) are compared in Table 3.6. The overall surface collection area (SCA) can be reduced by using the precharger; to obtain a collection efficiency of 95.4%, the required SCA was reduced from 28.80 to 22.27 $m^2/(m^3/s)$ by using a precharger with a voltage of 8 kV.

The impact of including a precharger and heat exchanger to WESP operation on the removal efficiency and emission concentration of sulfuric acid aerosol is demonstrated in Fig. 3.55. Here, adding a precharger to WESP operation increased the removal efficiency of sulfuric acid aerosol from 90.3 to 95.8% (representing an emission concentration of < 2 mg/m^3). Including a heat exchanger downstream of WFGD further increased the removal efficiency to 97.8% (for an emission concentration of < 1 mg/m^3). In real applications, technologies can be determined based on the emission requirements of sulfuric acid aerosol.

Table 3.6 Comparison of WESP with and without a precharger under different SCAs [51]

	SCA ($m^2/(m^3/s)$)	U_{WESP} (kV)	U_{pre} (kV)	η (%)	Data
WESP	21.60	32	0	90.3	Exp
	24.30	32	0	92.7	Predicted[a]
	28.80	32	0	95.5	
Combined WESP	22.27	32	8	95.4	Exp

[a]Calculated with the Deutsch equation

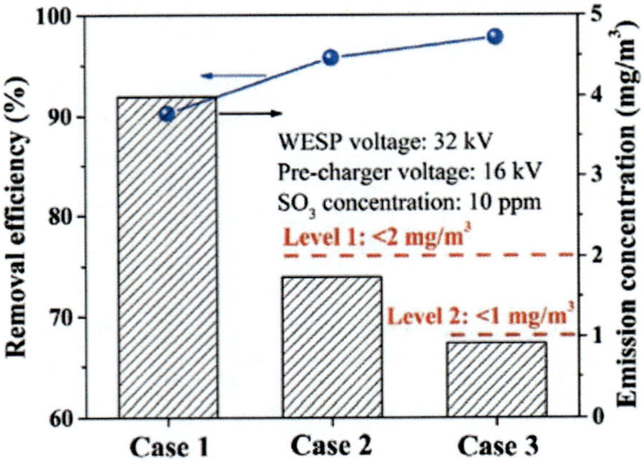

Fig. 3.55 Removal of sulfuric acid aerosol in different cases (case 1: WESP, case 2: WESP + precharger, and case 3: WESP + precharger + heat exchanger) [51]

3.3.3.2 Application

A combined WESP was designed and implemented in a 1000 MW CFPP, as shown in Fig. 3.56. Here, because the main flue gas is divided into two streams before entering the stack, two WESPs are used. Each WESP contains a single electric field with an electric field length of 4.19 m, 28 channels, and a designed gas velocity of 3.62 m/s. A pin-type discharge electrode has been demonstrated to provide high-intensity corona discharge during lab-scale testing (see Sect. 3.3.4.1) and was used in the WESP. A perforated precharger using the existing gas distributor as the grounding electrode was designed and installed at the entrance horn of the WESP. The minute-averaged PM emission concentration collected by a continuous emission monitoring system

Fig. 3.56 Application of the combined WESP in a real CFPP [82]

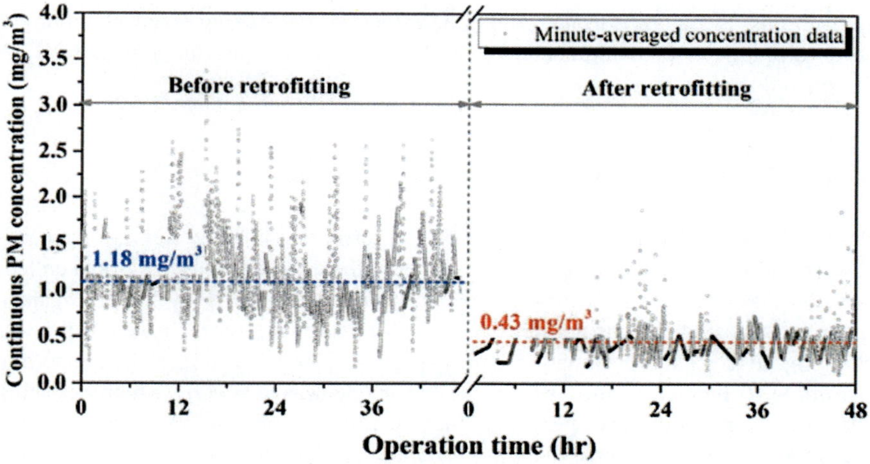

Fig. 3.57 PM emission concentration before and after installing the precharger in a 1000 MW CFPP [82]

before and after installing the precharger is compared in Fig. 3.57. Before retrofitting with the precharger, the average PM emission level was 1.18 mg/m^3, with emission concentrations as high as $\geq$ 3 mg/m^3 when rapping of dry ESP caused an increased concentration at the WESP inlet; after installing the precharger, the average PM emission level was reduced to 0.43 mg/m^3. Further, the increase in PM concentration due to rapping was suppressed and generally maintained below 1.0 mg/m^3. Further, the additional power consumption of the precharger accounted for < 3% of the power consumption of the combined WESP. Therefore, using the precharger effectively increased reliability and cost-effectively reduced PM emissions.

3.3.4 Electrode Optimization

The discharge current density and distribution in an ESP significantly influence operating performance. The distribution of an electric field and the electron/ion concentration bears a solid relationship with an electrode. A high discharge current density leads to effective particle charging, thereby increasing migration velocity and removal efficiency. The removal efficiency can also be improved by increasing the uniformity of the discharge current density distribution. The presence of aerosols or droplets in humid flue gas affects the corona discharge process, and gas molecules are more easily ionized in high-temperature flue gas. As these two extreme conditions are quite different from conventional electrostatic precipitation conditions, careful consideration is needed in electrode optimization. In this section, particle removal in high-humidity or high-temperature flue gas, representing a bottleneck of practical applications, is analyzed to determine the effects of optimizing electrode configuration on particle removal enhancement. The conclusions provide a basis for ESP design and application.

3.3.4.1 Electrode Optimization in Humid Flue Gas

WESPs are typically installed after WFGD to purify wet flue gas and remove particles, including dust and gypsum slurry. Washing the deposited dust off of the collection plates can significantly improve the performance of a WESP by preventing back corona, increasing the current density, and, thus, allowing more sufficient particle charging. In CFPPs, especially those using high-sulfur coal, a sulfuric acid mist (of a size around 0.1 μm) is formed in the saturated wet flue gas of WFGD, which will bring challenges for subsequent operations. Similar problems also arise in metallurgical industries. Researchers have thus proposed and implemented several techniques to improve removal efficiency and WESP operation in high-humidity flue gas, such as optimizing the electrical parameters, transporting charged particles with ionic-wind propulsion, and improving the uniformity of the electric field.

As shown in Fig. 3.58a, b, the type of discharge electrode used significantly influences the corona current and collection efficiency, respectively. Under the same applied voltage, the highest corona current and collection efficiency were achieved in all cases using a pin electrode, followed by, in order, a sawtooth and rod electrode. This is attributable to the nonuniformity of different electrodes because corona current can only be produced when the regional electric field strength is sufficiently high to ionize gas molecules, and different electrodes can generate different electric field strengths under the same applied voltage. As a pin-type electrode was the sharpest among the three electrodes, corona discharge occurred under the lowest

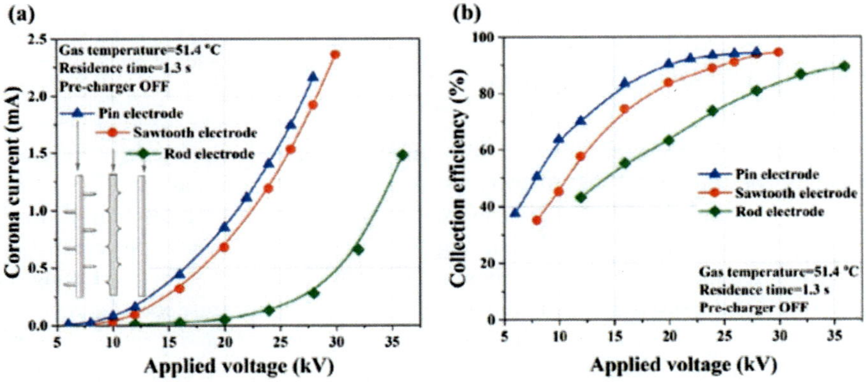

Fig. 3.58 a Corona discharge characteristics and b particle collection efficiency using different electrode types [82]

voltage. Different electrodes also led to different spark-over voltages. The nonuniform pin-type electrode thus led to a lower spark-over voltage, a lower corona onset, and a higher collection efficiency.

This relationship among collection efficiency, operating voltage, and corona current was further analyzed experimentally; the resulting collection efficiency and corona current are plotted as a function of operating voltage in Fig. 3.59. At lower voltages, the rate of increase of collection efficiency was higher than that of the corona current, whereas the rate of increase of collection efficiency was lower than that of the corona current at higher voltages. The boundary between these two regions, where the slopes of the two curves are equal (i.e., $k_1 = k_2$), is indicated by a dashed line. Thus, the collection efficiency is restrained by corona current in Region I and operating voltage in Region II. Particle migration in the electrostatic field is primarily driven by electrostatic force, which is a product of electric field strength and particle charge number. Particles are poorly charged because of the small space-ion density when corona current is extremely low. Under such conditions, a slight increase in corona current can significantly improve particle charging and subsequent migration. With increasing voltage, the corona current rises dramatically and particle charging occurs more easily. The insufficient particle driving force can be then ascribed to the low increase rate of electric field strength. To operate ESPs at a high efficiency in real applications, where the electrical parameters can vary within wide ranges, the discharge electrode thus should be designed with low corona onset voltage to guarantee high corona current for particle charging and with a high spark-over voltage to provide enough electric field strength.

Gaseous particles near the tip of needles are ionized and produce a corona as long as the applied voltage reaches the corona onset voltage. During negative corona discharge, ionized electrons in the cathode corona region are subjected to strong acceleration by an electric field, causing them to collide with other atoms and release more electrons. The moving electrons and ions collide with neutral gas molecules,

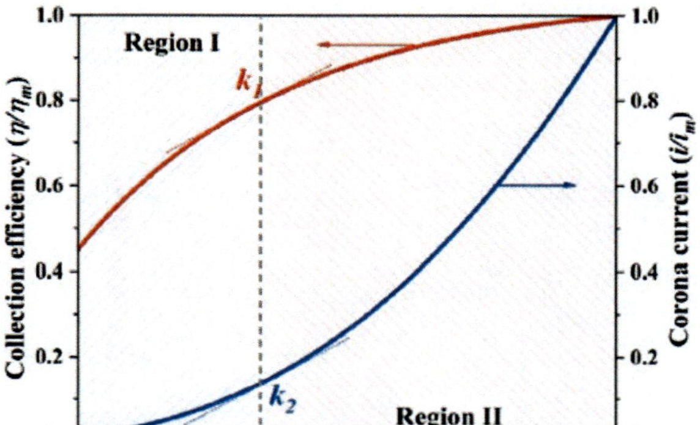

Fig. 3.59 Region division of electrical parameters for particle collection performance [83]

dragging them toward the collection electrode, and thus creating an ionic wind comprising ionized and neutral molecules.

The flue gas flow field and ionic wind direction also affect particle deposition in an ESP, as shown in Fig. 3.60a, which details the flow field (including magnitude and direction of velocity) in a latitudinal cross-section. In the ionic-wind-affected region, the velocity of flue gas was increased, owing to the acceleration effect of high-velocity ions, and the electric field was stronger. Particles flowing through this region move flow faster and thus incur higher drag and electric forces. Two passive vortices in the vicinity of the collection electrode were indirectly induced by the high-velocity ionic wind. A backflow region was consequently formed between the two vortices and pointed to the center of the ESP. The drag force created by the flow field in the backflow region is in the opposite direction as the electric force, which is detrimental to the collection of particles and correspondingly reduced the overall particle migration velocity.

The induced ionic wind also affected the flow of the flue gas in the longitudinal direction, as shown in the cross-section in Fig. 3.60b. In the ionic-wind-affected region, the flue gas was accelerated, enhancing the velocity in the z-direction. A counter-clockwise vortex was formed under the electrode pin and near the collection electrode. The local velocity near the collection electrode was much lower than that in the main stream; this decrease could allow for easier particle deposition. In addition, the generated vortex caused particles to accumulate in this region. Upstream flue gas coming bypassed this vortex; flue gas was deflected and the flue gas particles were moved away from the collection electrode owing to the drag force. This backflow region is detrimental to particle precipitation.

When electrically neutral spherical particles were injected into the entrance of the ESP, the charged particles were transported by drag and electric forces, leading to nonhomogeneous particle deposition distribution because of ionic wind. More

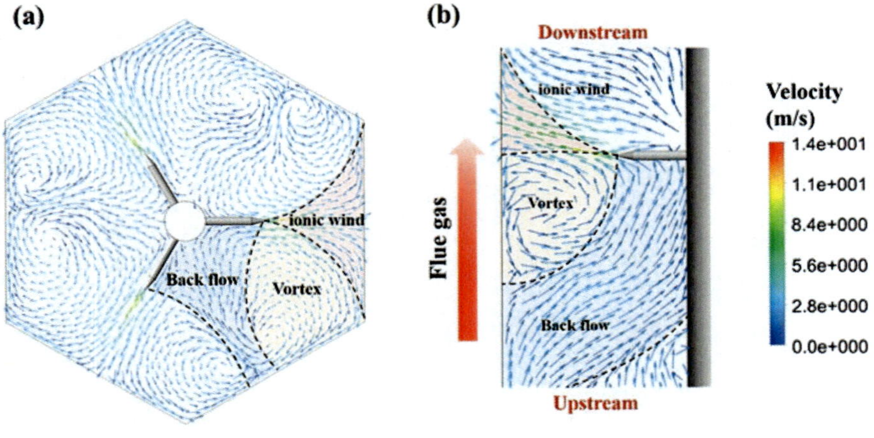

Fig. 3.60 Effect of ionic wind on flue gas: **a** latitudinal and **b** longitudinal cross-sections [46]

intense vortices were formed with an increase in applied voltage, thereby promoting the particle precipitation process, as shown in Fig. 3.61. Upstream vortices caused deposited particles to cluster in the front of needles; these aggregated regions increased with applied voltage. As most particles near the collection electrode were dragged upstream by the vortices, fewer particles were deposited downstream behind the needles, although the length of the affected region decreased with applied voltage. Overall, a higher applied voltage generated a stronger ionic wind and achieved higher collection efficiency by enhancing the upstream aggregation of particles and reducing the dead zone of particles deposited downstream.

The migration velocity of fine particles was further improved by varying the rotation angle of the discharge electrode to increase the uniformity of the discharge current density distribution; the resulting current density distributions on the collection electrodes are shown in Fig. 3.62, where the dashed line represents the position of the spikes. In each case, a triple-spike electrode was used as the discharge electrode, and the discharge current was 3 mA. As the rotation of the discharge electrode was changed from 0° to 30° in the WESP, the ion diffusion region changed; the shortest distance between the sharp point and the collection electrode increased and the peak current density decreased. The location of the peak current density also changed with the change in rotation angle. However, both regions on the collection electrode where ions could reach and the current density of both sides of the spike increased with increasing rotation angle. The current density distribution on the collection electrode gradually expanded. The average discharge current remained unchanged at the same voltage under two opposite effects on the current density, whereas the current density distribution changed significantly with the slight rotation of the triple-spike electrode. Thus, the accurate installation of the electrode is vital in practical applications.

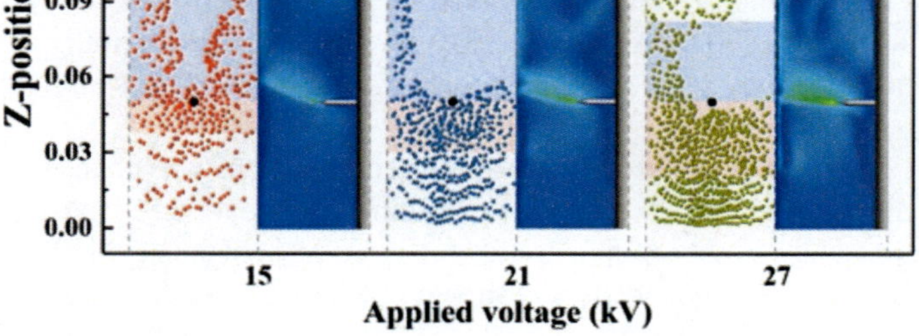

Fig. 3.61 Correlation between ionic wind and deposited particles under different applied voltages [46]

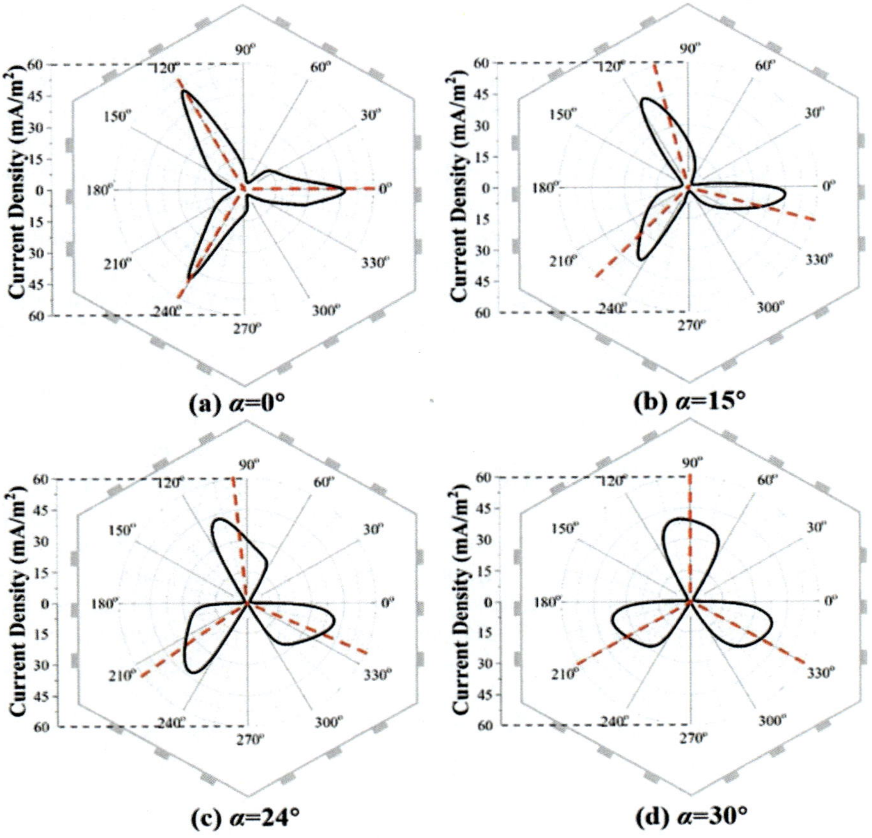

Fig. 3.62 Current density distribution on the collection electrode of different rotation angles [84]

The effect of the rotation angle of the discharge electrode on the effective migration velocity of sulfuric acid aerosol under an applied voltage of 20.08 kV is shown in Fig. 3.63. Here, the effective migration velocities all exceeded 0.088 m/s considering the high discharge intensity of the electrode. Increasing the rotation of the discharge electrode from 0° to 30° led to an increase in the effective migration velocity of sulfuric acid aerosol by 17.1% for a removal efficiency of 99.2% at an SCA of 23.09 $m^2/(m^3/s)$.

A pilot-scale, single field WESP experimental system was then designed and operated under high-humidity flue gas conditions to evaluate the resulting SO_3 mass concentration and number concentration of sulfuric acid mist; the resulting SO_3 removal efficiency for all test cases is shown in Fig. 3.64. At the same gas velocity, the SO_3 removal efficiencies of the three electrodes ranked in the order of electrode C > electrode B > electrode A. Although electrode type clearly influenced the SO_3 removal efficiency, it remained below 75% for all tested electrodes at a gas velocity

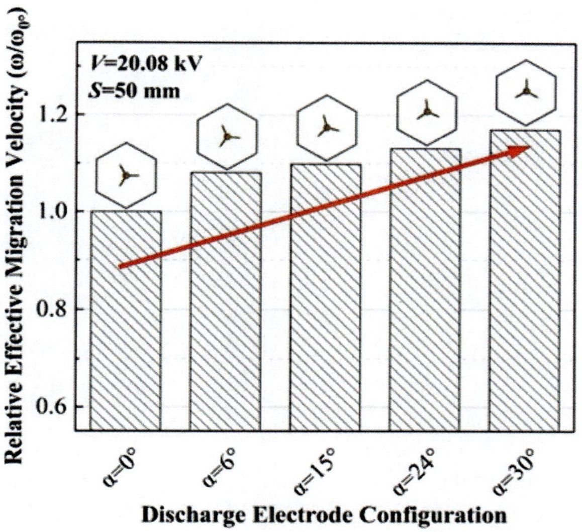

Fig. 3.63 Effect of discharge electrode configuration on the effective migration velocity of sulfuric acid aerosol [84]

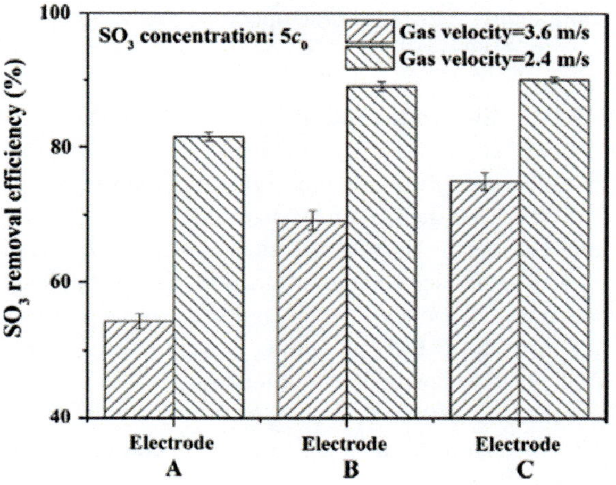

Fig. 3.64 SO_3 removal efficiencies for three types of discharge electrodes [85]

of 3.6 m/s. The removal efficiency of electrode C was further increased to 90.1% by reducing the gas velocity to 2.4 m/s, demonstrating the importance of selecting an appropriate discharge electrode and an appropriate gas velocity when aiming to obtain a high removal efficiency and low emissions of sulfuric acid mist. In this pilot-scale WESP, the collection tube was 2.4 m long, which is much shorter than full-scale WESPs with a similar field configuration (typically 4–6 m). Thus, a full-scale WESP under the same configuration and operating parameters would achieve a higher removal efficiency under the same SO_3 concentration; the SO_3 removal efficiency is predicted to be as high as 97.9–99.7% when the collection tube is 4–6 m long,

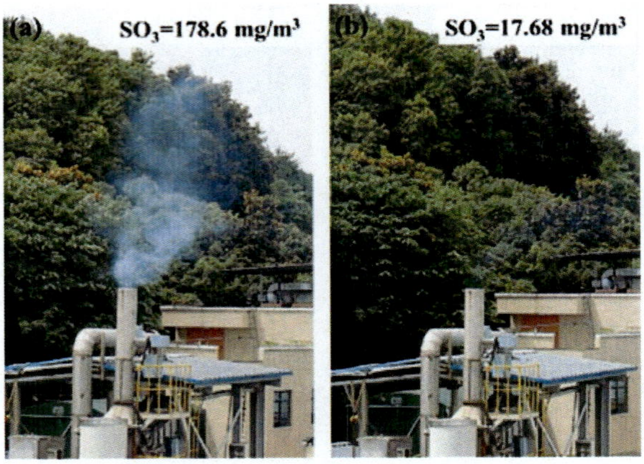

Fig. 3.65 Stack plume: **a** WESP off and **b** WESP on [85]

and the efficiency can be even higher for double- or multiple-field WESPs. A typical stack plume with a WESP inlet SO_3 concentration of approximately 178.6 mg/m^3, as shown in Fig. 3.65, was retrofitted with a WESP using electrode C and a gas velocity of 2.4 m/s. The designed WESP removed 90.1% of SO_3 from the flue gas, which significantly reduced plume opacity.

3.3.4.2 Electrode Optimization in High-Temperature Flue Gas

Particles in high-temperature gas may cause problems in downstream equipment and affect the quality of by-products [86]. Improving particle removal from high-temperature gas is thus vital for reducing particle emissions.

Although electrode optimization has been demonstrated to enhance particle collection efficiency [73], design strategies used to optimize commercial ESPs are not suitable for ESPs operating at high temperatures [87]. Corona onset and spark voltages have been demonstrated numerically and experimentally to decrease as the temperature increases, as shown in Fig. 3.66 [64, 88]. Here, the corona onset voltage decreased from 15 to 6.7 kV as the temperature increased from 293 to 873 K, and the spark voltage decreased from 29 to 13.5 kV. Thus, the operational voltage range for corona discharge on the wire–plate reactor decreased from 14 to 6.8 kV as the temperature increased from 293 to 873 K; this tendency mitigated after 873 K. The Townsend ionization coefficient increases with increasing temperature, thus facilitating the easy transition from nonself-sustained discharge to self-sustained discharge, which is the onset of corona [89]. Sparks occur more easily because of the decrease in gas density at high temperatures. Thus, continuous sparking and a leap in the discharge current occurs. Owing to the small operational voltage range, it becomes difficult to maintain a stable corona discharge and develop discharge structures suitable for corona discharge at high temperatures [90, 91].

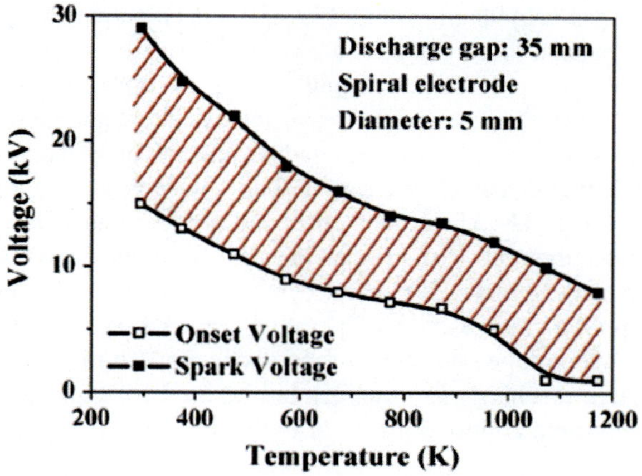

Fig. 3.66 Effect of temperature on corona onset and spark voltages [32]

Effect of electrode discharge gap: The electrode discharge gap affects the intensity of the electric field in the ESP, which is a key factor of electrode optimization. The *V–I* characteristics of different discharge gaps at 673 K are shown in Fig. 3.67a; here, while the applied voltages on the electrode were maintained at 13 kV, the current density decreased from 1.119 to 0.071 mA/m^2 as the discharge gap increased from 30 to 70 mm. The effect of the discharge gap on the current density became more significant as the discharge gap decreased. The external electrostatic field of the short discharge gaps was greater than that of the long discharge gaps under the same applied voltage, thus indicating a greater ionization coefficient for short discharge gaps than long discharge gaps. A large number of negative ions were produced and increased the local electric field. The current density of the short discharge gaps was higher than that of long discharge gaps, since the negative corona current is mainly caused by the mobility of negative ions at 673 K. The observed corona onset and spark voltages at

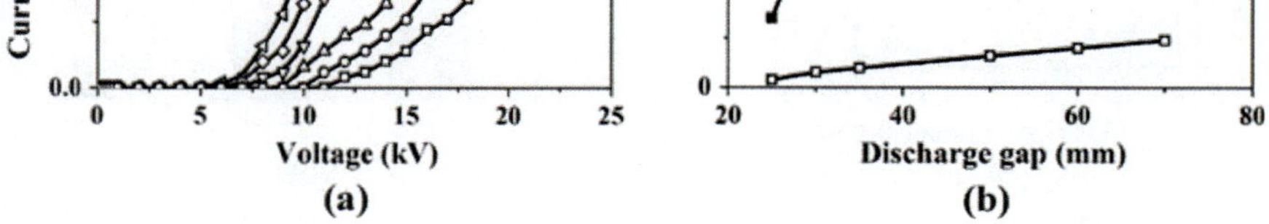

Fig. 3.67 **a** Effect of discharge gap on *V–I* characteristics. **b** Effect of discharge gap on the corona onset and spark voltages [32]

1073 K as a function of the discharge gap is shown in Fig. 3.67b. Here, both the corona onset and spark voltages increased as the discharge gap increased, although the spark voltage increased more rapidly. Thus, the operating voltage range between the spark and onset voltages increases as the discharge gap increases. With a discharge gap of 30 mm, a corona discharge could only be obtained with an applied voltage of 0.8–9 kV; with a discharge gap of 70 mm, however, the applied voltage could range from 2.4 to 16 kV. The corona onset and spark voltages thus likely coincide when a small discharge gap is used, causing the corona discharge to disappear and the electron avalanche to develop directly into a spark. Increasing the discharge gap increases the applied voltage and the operating voltage range for high-temperature ESP, allowing a wide range of stable corona discharge. Although the corona current decreases, a large discharge gap thus should be used for high-temperature ESP because of the sufficient current density for highly efficient precipitation at high temperatures.

Effect of electrode geometry: The shape and diameter of the electrode can also affect the discharge current density and operating voltage range. The *V–I* characteristics of wire, spiral, and ribbon electrodes at temperatures of 673 and 873 K are shown in Fig. 3.68. Here, the current density of the ribbon electrode was significantly higher than that of the others, whereas the wire and spiral electrodes showed better discharge stabilities than the ribbon electrodes at high temperatures. Compared with the wire electrodes, the cusps of the ribbon electrodes produced a higher electric field around the electrode's surface, likely leading to a stronger electron avalanche. The corona onset and spark voltages of the three electrodes at 973 K are shown in Fig. 3.68b. The corona onset voltages of the ribbon electrode were much lower than those of the others. Sparks were more easily produced when using the ribbon electrode within the studied temperature range because of its unstable corona discharge that leads to localized breakdowns. In general, increasing the surface roughness of electrodes can increase the current density and decrease the spark voltage and corona stability. The current density is always high enough for dust removal in a high-temperature ESP.

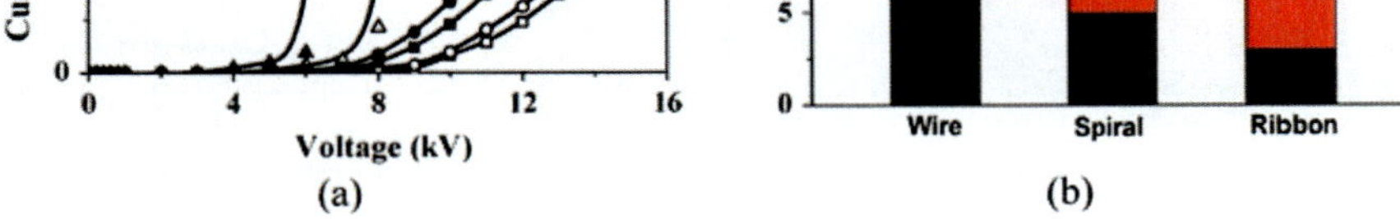

Fig. 3.68 Effect of **a** discharge electrode geometry on *V–I* characteristics, and **b** discharge gap on corona onset and spark voltages [32]

Thus, wire electrodes with low surface roughness are preferred over spiral or ribbon electrodes for increasing the spark voltage and corona stability.

The *V–I* characteristics of wire electrodes with diameters ranging from 1 to 6 mm at 673 K are shown in Fig. 3.69; under the same applied voltage, the current density decreased as the wire diameter increased. The electric field strength on the surface of the thin wire electrode was higher than that of on thick wire electrode, causing the ionization rate to increase. The secondary electrons were considerably energetic when emitted from the surface of the thin wire. Thus, the diameter of the wire electrode affects the corona current and onset voltage but has little influence on the spark voltage. The corona onset voltage of the wire–plate discharge configuration is a function of the wire diameter, wire surface roughness, and gas density, whereas the spark voltage depends on the discharge gap and gas density. Generally, thin wire electrodes improve the discharge current and reduce the corona onset voltage, although the mechanical strength of wire electrodes should be considered in practice.

Considering these factors, four electrodes (shown in Fig. 3.70a) were designed using three electrode shapes (barbed, sawtooth, and rod) and applied to a typical high-temperature low-resistivity particle removal system (high-carbon fly ash) [92]. The resulting particle mass concentration with the increase of residence time under each electrode is shown in Fig. 3.70b; in each case, the particle concentration decreased to 30 mg/m^3 with a residence time of approximately 50 s. The fluctuation range under the DE1, DE2, and DE3 electrodes after 200 s were 8–15, 10–16, 6–10 mg/m^3, respectively, whereas that of the rod electrode DE4 was 24–30 mg/m^3. When DE1 was under a multiple electric field, the particle concentration was reduced under 25 mg/m^3 within 10 s. To achieve the same concentration, DE2 needed more than 50 s, and DE3 needed approximately 100 s. Hence, DE1 allowed for a large and rapid decrease in PM concentration, indicating that DE1 is conducive for the rapid and stable operation of ESPs.

The fractional particle removal efficiency of high-carbon fly ash under each of these four electrodes is shown in Fig. 3.71. Similarly, the barbed DE1 provided

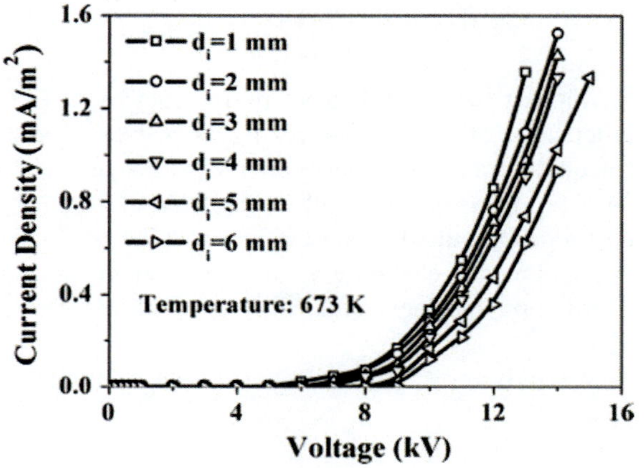

Fig. 3.69 Effect of wire diameter on *V–I* characteristics [32]

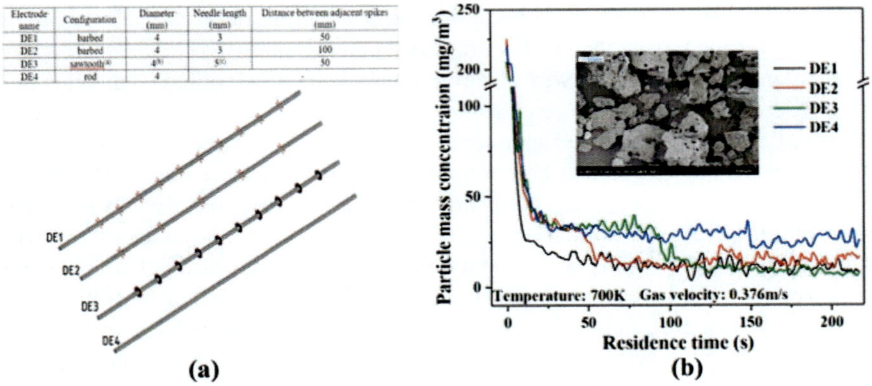

Fig. 3.70 **a** Schematic of various discharge electrodes and their effect on **b** particle mass concentration variation with residence time [92]

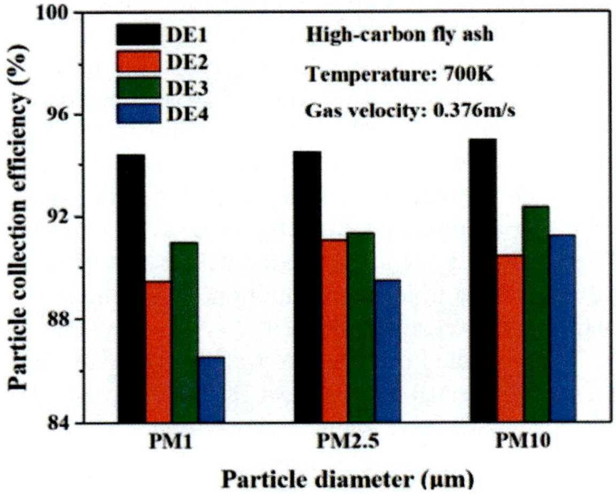

Fig. 3.71 Fractional particle removal efficiency of high-carbon fly ash under various electrodes [92]

the highest overall PM removal efficiency, with PM_{10}, $PM_{2.5}$, and $PM_{1.0}$ removal efficiencies of > 94%. Comparing the standard rod electrode (DE4) with DE1, the fractional removal efficiency of $PM_{1.0}$, $PM_{2.5}$, and PM_{10}, increased from 86.48 to 94.39%, from 89.48 to 94.48%, and 91.20 to 94.96%, respectively. The barbed electrode DE1 thus can most efficiently facilitate high-carbon particle removal under 700 K. Overall, electrode optimization can enhance particle collection at each size distribution to achieve efficient removal of fine PM.

Fig. 3.72 Typical applications of PM removal in treating humid flue gas

3.3.4.3 Applications

Treating humid flue gas: Particle removal enhancement methods employing optimized WESPs have been successfully applied to CFPPs and thermal power boilers to treating humid flue gas, as shown in Fig. 3.72. Electrode optimization via the methods introduced in this section has allowed long-term, efficient, and stable operation of WESPs at a high PM removal efficiency. For example, applying an optimized electrode to an ESP system on a 660 MW CFPP allowed for PM and SO_3 emissions to be maintained below 5 mg/m^3 when the boiler runs at full load.

Treating high-temperature flue gas: Optimized high-temperature ESPs have been applied to particle removal for high-temperature flue gases, such as biomass boilers and nonferrous metal smelting flash furnaces, as shown in Figs. 3.73a, b. Here, electrode optimization allowed for a final PM concentration below 50 mg/m^3. Optimized electrodes have also been used to meet emission standards and realize the efficient and stable operation of ESPs for high-temperature flue gas purification of thermal power boilers, as shown in Fig. 3.73c.

3.4 Simultaneous Control of Multiple Pollutants

Recent research efforts have demonstrated pollutant conversion by injecting energy into the electrostatic field to generate nonthermal plasma (NTP). The use of NTP has demonstrated rapid chemical reactions and simultaneous multi-pollutant removal (including NO_x, SO_x, PM, VOCs, and Hg) via conversion and absorption with high efficiency [93, 94]. In this section, recent efforts to convert and/or capture multiple

Fig. 3.73 Typical applications of PM removal in treating high-temperature flue gas

pollutants (e.g., PM and SO_x, NO_x, or trace elements) in an electrostatic field are introduced.

3.4.1 Simultaneous Control of SO_x

Although the formation and removal characteristics of sulfuric acid aerosol in a WFGD system have been described in several publications, SO_x conversion also occurs in the electrostatic field. Ions and electrons produced by corona discharge are necessary for aerosol charging in WESP, whereas various other reactive species, such as O, OH, O_3, H_2O_2, and HO_2 radicals, are simultaneously produced, which may induce SO_2 oxidation into SO_3. Furthermore, the saturated gas condition inside a WESP is conducive to homogeneous nucleation and thus results in the formation of sulfuric acid aerosol.

Photographs of corona discharge of a WESP taken at night with a 3-min exposure time from a side view under different corona currents are shown in Fig. 3.74. The corona occurred (> 0.001 mA) when the applied voltage exceeded 16 kV, which was considered as the corona onset voltage. In each case, one or more bright spots appeared on the discharge wire surface. As the applied voltage continued to increase, several bright spots appeared along the discharge wire surface with slight hissing

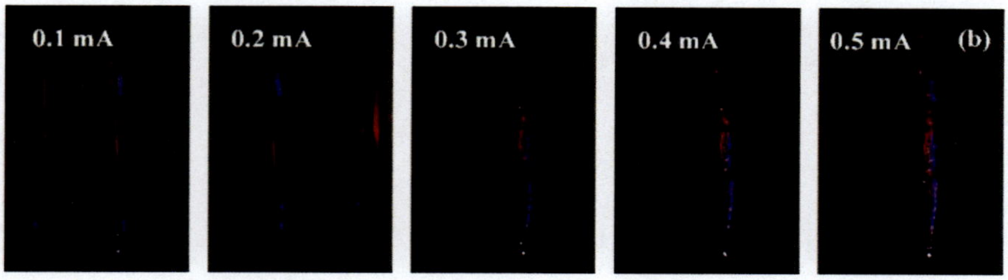

Fig. 3.74 Photographs of corona discharge under different corona currents [95]

sounds. As the applied voltage reached the breakdown voltage, the bright spots spread along the discharge wire but did not become uniform, and spark-over occasionally occurred on some points. Thus, radicals and active species such as O, OH, and O_3 can be produced during corona discharge and may play a significant role in the conversion of SO_2 in a corona discharge reactor. SO_2 can be activated and oxidized by these radicals and active species to form SO_3 and corresponding aerosolized acids (i.e., H_2SO_4) by the following reactions, leading to the formation of large amounts of sulfuric acid aerosols by nucleation.

$$SO_2 + O \rightarrow SO_3 \tag{3.46}$$

$$SO_2 + O_3 \rightarrow SO_3 + O_2 \tag{3.47}$$

$$SO_2 + HO_2 \rightarrow SO_3 + OH \tag{3.48}$$

$$SO_2 + OH \rightarrow HSO_3 \tag{3.49}$$

$$HSO_3 + OH \rightarrow H_2SO_4 \tag{3.50}$$

$$HSO_3 + O_2 \rightarrow HO_2 + SO_3 \tag{3.51}$$

$$SO_3 + H_2O \rightarrow H_2SO_4 \tag{3.52}$$

Introducing SO_2 concentration downstream from the WESP demonstrates a clear increase in aerosol number concentration, as shown in Fig. 3.75. Here, dust particles were added into the flue gas without corona discharge from 40 s, and a relatively stable concentration of 3.5×10^4 cm^{-3} was maintained for 60 s. The voltage was increased to 30 kV 120 s after commencing the experiment, causing the number concentration to drop to approximately 1.3×10^4 cm^{-3}. Beginning at 290 s, SO_2 was introduced into the flue gas in 5 ppm increments every 60 s until reaching 20 ppm.

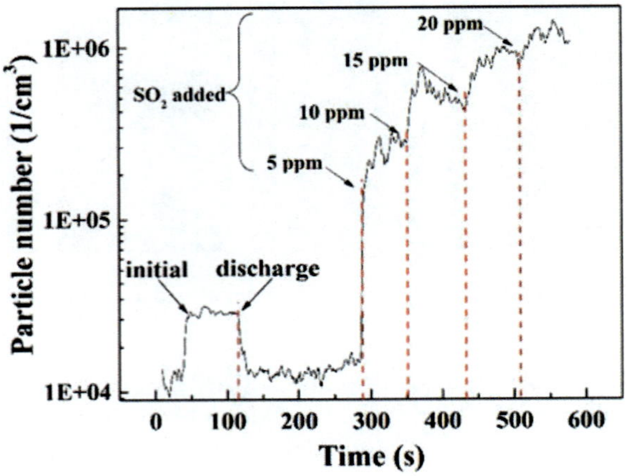

Fig. 3.75 Aerosol number evolution under different SO_2 concentrations. (gas temperature = 28.7 °C; residence time = 1.29 s) [95]

The introduction of 5 ppm SO_2 caused the particle number concentration to jump to more than 2.0×10^5 cm^{-3}, which then increased with increasing SO_2 concentrations.

Although the *V–I* curve has typically been used to describe the corona discharge characteristics of a WESP, the specific energy density (SED) can also be used to characterize the correlation between corona discharge and aerosol formation as follows:

$$\mathrm{SED} = \frac{UI}{Q}, \tag{3.53}$$

where U is the applied voltage (kV), I is the corona current (mA), and Q is the gas flow rate (m^3/s).

The SO_2 conversion rate depends on the intensity of the radicals and active species. The intensities of these radicals and species are closely related to the SED. As shown in Fig. 3.76, both the aerosol number concentration and SED increased exponentially

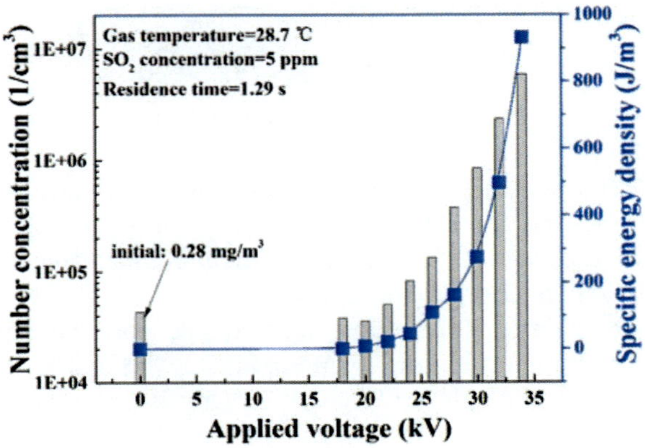

Fig. 3.76 Aerosol number concentration and specific energy density (SED) under different voltages [95]

with applied voltage when it exceeded 20 kV. The fitted relationship between the aerosol number concentration and SED was $N = 2.62 \times 10^6 \exp(\text{SED}/782) - 2.77 \times 10^6$; with $R^2 = 0.9972$, there was a strong dependence of aerosol number concentration on SED.

An aerosol penetration ratio (R_p) can be used to describe the penetration of formed aerosol in a WESP as follows:

$$R_\mathrm{p} = \frac{N_\mathrm{out,on}}{N_\mathrm{out,off}}, \tag{3.54}$$

where $N_\mathrm{out,on}$ and $N_\mathrm{out,off}$ denote the aerosol number concentration (cm^{-3}) with and without corona discharge, respectively. Here, $R_\mathrm{p} > 1$ indicates that the aerosol number with corona discharge is higher than that without corona discharge, i.e., more aerosol is formed than collected.

The penetration ratio under different SEDs with different gas temperatures is shown in Fig. 3.77. Without SO_2, the penetration ratio decreased with an increase in SED. When SO_2 was presented to the flue gas, the penetration ratio decreased slightly to a minimum value and then increased significantly as the SED rose. For example, at a gas temperature of 28.7 °C, the penetration ratio changed from 0.87 to 0.79 when the SED increased from 0.5 to 11.8 J/m^3, whereas the penetration ratio increased from 0.79 to 166.53 when the SED increased from 11.8 to 905.8 J/m^3.

Sulfuric acid aerosol was generated by corona discharge in the WESP, some of which was subsequently collected by charging and migration. Both the formation and collection of sulfuric acid aerosol were positively influenced by the corona discharge intensity. Fewer radical and active species produced by corona discharge were found under a lower SED, indicating that only limited aerosols were formed; as the SED increased, the number of radical and active species produced by corona discharge

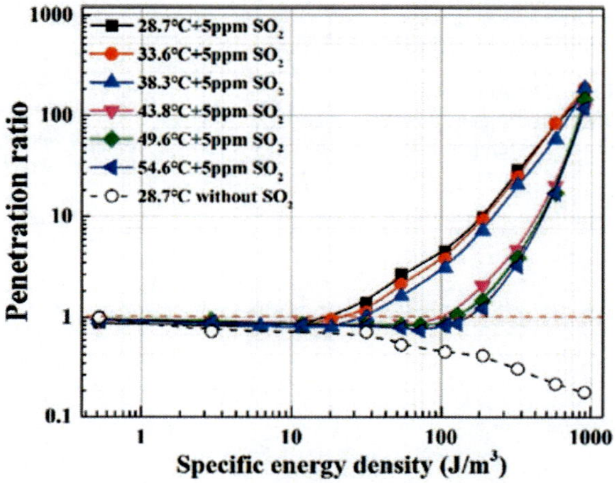

Fig. 3.77 Sulfuric acid aerosol penetration ratio under different SEDs [95]

increased rapidly. Thus, under a low SED, the collection rate dominated, whereas the formation rate increased faster than the collection rate under a higher SED.

The threshold SED, or SED when $R_p = 1$, was thus defined to describe the aerosol formation and collection processes. As is consistent with the variation of the effect of gas temperature on aerosol formation and collection processes, the threshold SED increased as the gas temperature increased, as shown in Table 3.7. As the temperature was increased from 36.0 to 56.0 °C, the aerosol number concentration generated by corona discharge also increased from 4.0×10^6 to 1.0×10^7 cm^{-3}, as shown in Fig. 3.78. Particle collection efficiency increases with an increasing gas temperature as a result of the changes in the corona current and ion thermal mobility. The threshold SED increased with the gas temperature, suggesting that the influence of the gas temperature on the aerosol collection was more obvious than the aerosol formation.

As shown in Fig. 3.79, the WESP's PM collection efficiency of all particulate sizes decreased with the presence of SO_2; the strongly negative collection efficiency of $PM_{0.1}$ indicates that more particles were generated than collected, likely owing to the high number of generated aerosol in this size range. In addition, $PM_{0.1}$ was primarily charged by diffusion charging. The suppression of the corona current as a result of the SO_2 electronegative effects or space charge of charged aerosols may affect the particle charging process, which could also diminish collection efficiency.

Although the fractional collection efficiency of $PM_{0.1}$ decreased much more drastically, the total collection efficiency reduced from 93.0 to 79.5% when the SO_2 concentration increased from 0 to 25 ppm and the corresponding overall particle concentration at the outlet increased from 1.08 to 3.14 mg/m^3. Thus, although the amount of aerosol generated with corona discharge was two orders of magnitude

Table 3.7 Fitted threshold SED at different gas temperatures [95]

Gas temperature (°C)	28.7	33.6	38.3	43.8	49.6	54.6
Threshold SED (J/m^3)	16.7	22.3	30.2	100.7	128.4	154.4

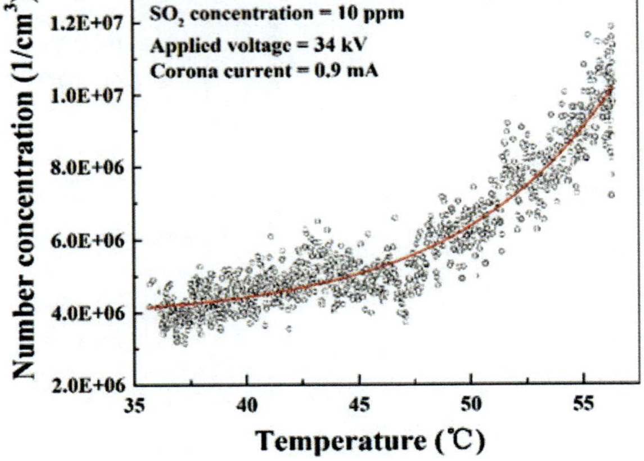

Fig. 3.78 Aerosol number concentration as a function of the gas temperature [95]

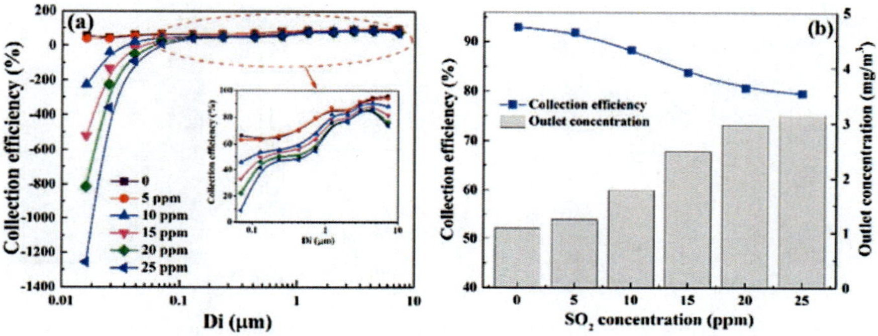

Fig. 3.79 **a** Fractional and **b** total particle collection efficiency under different SO_2 concentrations (gas temperature = 28.7 °C, applied voltage = 30 kV) [95]

higher than that without corona discharge, it only accounted for a small part of the weight of particles downstream of the WESP. However, without strict control of the SED, the total collection efficiency in a CFPP burning high-sulfur coal would see a drastic increase in sulfuric acid aerosol.

3.4.2 *Simultaneous Control of NO_x*

Many physicochemical approaches have been developed to remove NO_x, such as SCR [96] or SNCR [97]. These methods are effective for NO_x removal, but they could be improved. Promoting NO_x removal through NO oxidation to NO_2 or N_2O_5 is also an effective method for NO_x removal.

NTP technology presents a promising alternative for NO_x removal. In a discharge process, radicals such as OH, O, and N are produced by reactions between energetic electrons (5–20 eV) and molecules such as H_2O, O_2, and N_2. More than 95% of the NO_x in flue gas is NO [98], which can either be reduced to N_2 by N radicals or oxidized to NO_2 by OH, O, O_3, etc. [99]. In this section, the recent investigation of a system involving NO_x conversion by plasma discharge by planar laser-induced fluorescence (PLIF), illustrated in Fig. 3.80, is discussed.

3.4.2.1 PLIF Diagnostics in Pulsed Corona Discharge

NO distribution: The two-dimensional NO distribution under laminar gas flow (*Re* = 1379) is illustrated in Fig. 3.81, where the NO-PLIF signal was obtained by subtracting the background signal (without NO injection) from the total fluorescence signal (with NO injection). Five hundred images were accumulated in each case and corrected for laser sheet intensity spatial variation. The collection area of the ICCD

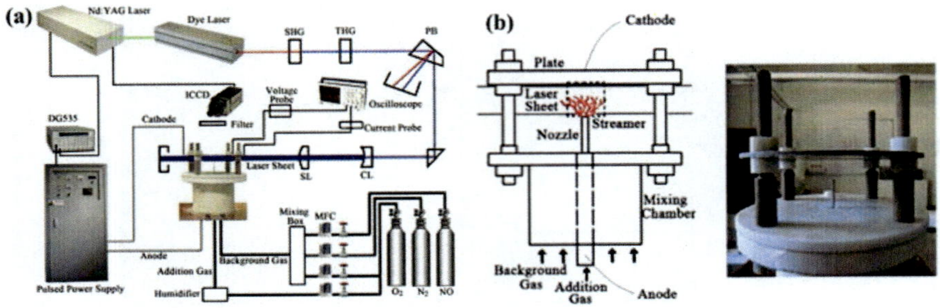

Fig. 3.80 Experimental setup for diagnostics in pulsed corona discharge: **a** planar laser-induced fluorescence (PLIF) system. **b** pulsed corona discharge reactor [100]

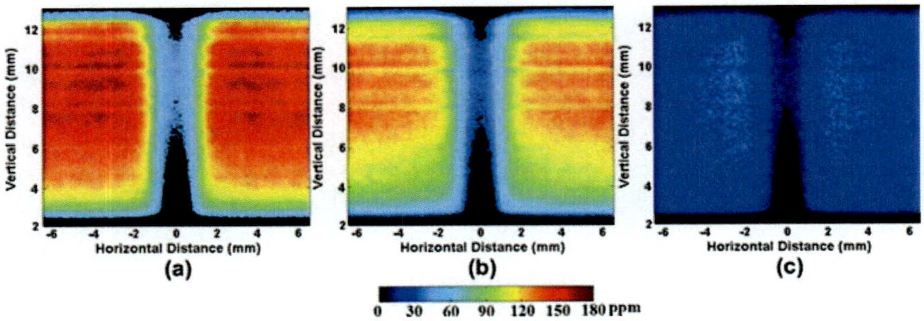

Fig. 3.81 NO-PLIF at *Re* = 1379: **a** without discharge, **b** with discharge, and **c** NO oxidization [100]

camera was − 6.5 mm to 6.5 mm (horizontal) centered by the outlet of the nozzle and 2–13 mm (vertical). The absolute NO concentration was calibrated using the NO fluorescence intensity without discharge under a known NO concentration.

The resulting two-dimensional NO distribution without discharge is shown in Fig. 3.81a. In this case, diffusion is the main mixing process between the added and background gases. Near the outlet of the nozzle, the NO concentration was zero and a sharp edge-like structure was observed. NO in the background gas diffused to the center and mixed with the added gas at vertical distances greater than 7 mm. The two-dimensional NO distribution 6 μs after discharge is shown in Fig. 3.81b. Here, NO was oxidized to NO_2 by radicals (OH, O, O_3, etc.) generated in the discharge region, especially in the mixing region of background and added gas, leading to a decrease in NO concentration. The converted NO concentration distribution, calculated by subtracting the NO concentration after discharge from that before discharge, is shown in Fig. 3.81c. The highest converted NO concentration is in the region from − 3 to 3 mm in the horizontal axis since a high concentration of radical sources (H_2O, O_2) in that region generated many radicals by collision with the energetic electrons.

Mixing between the added and background gases is limited under a laminar flow (*Re* = 1379) of added gas since diffusion is the dominant mixing mechanism; the

OH and O radicals generated in the added gas cannot collide with NO effectively. To enhance NO oxidization, a turbulent added gas (*Re* = 4138) flow was adopted; NO-PLIF measurements at this condition are illustrated in Fig. 3.82. Here, no distinct edge between the added and background gases is visible at the nozzle outlet, and NO is detectable near the nozzle outlet, as shown in Fig. 3.82a, indicating better mixing than was present under laminar flow. Turbulent eddies can rapidly blend radicals (OH, O, and O_3) with NO molecules, likely increasing collisions between radicals and NO. The reaction zone is thicker, and the converted NO concentration, illustrated in Fig. 3.82c, is higher than that of the laminar case.

As mixing between the added and background gases affects the collision probability between radicals and NO, NO oxidization is increased as the *Re* of added gas increases. Single-shot NO-PLIF images recorded for *Re* = 1379 and *Re* = 4138 without discharge are shown in Fig. 3.83 to demonstrate the structure of the added

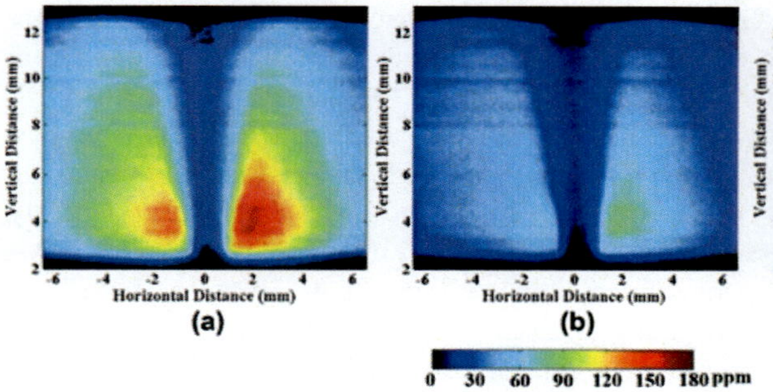

Fig. 3.82 NO-PLIF at *Re* = 4138: **a** without discharge, **b** with discharge, and **c** NO oxidization [100]

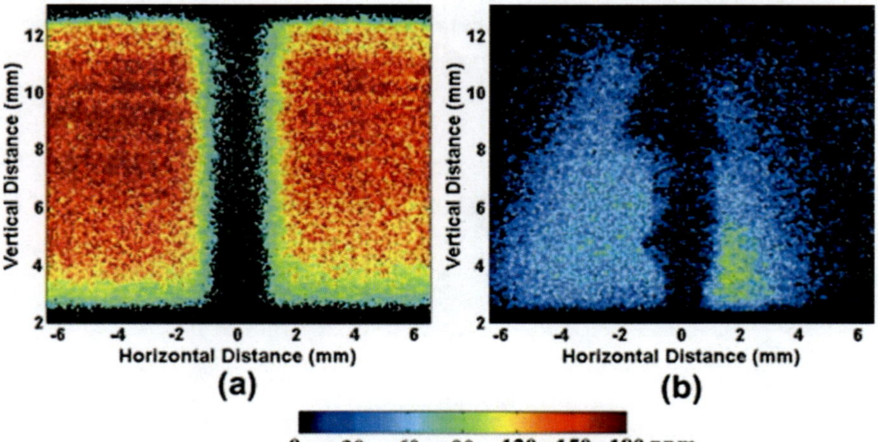

Fig. 3.83 Single-shot NO-PLIF images at **a** *Re* = 1379 and **b** *Re* = 4138 [100]

gas flow and its effect on NO distribution. Here, NO distribution is symmetrical and homogeneous in the laminar case ($Re = 1379$). NO diffuses into the center only at vertical distances greater than 7 mm; radical sources may also diffuse into the background gas. In the turbulent case ($Re = 4138$), the NO distribution is asymmetrical owing to the stretching and squeezing effect of turbulence, and there are many vortices in the mixing region at vertical distances greater than 3 mm. The presence of turbulent eddy structures enhances the interaction between the added and background gases, thus increasing the collision probability between NO and radicals.

OH distribution: Although the NO oxidization process is affected by the combined action of OH, O, and other radicals, OH is detectable in a two-dimensional image using existing laser techniques and thus can be used to illustrate NO oxidization. During discharge, O_2 mainly affects the concentration of active substances such as O, O (^1D), and O_3, H_2O mainly affects the concentration of free radicals such as OH, HO_2, and H, and CO_2 generates free radicals with little effect on OH distribution. As electronegative gases, O_2 and CO_2 reduce the discharge current and thus the NO_x conversion efficiency. OH-PLIF can be performed under the same conditions and image correction processes as discussed above to evaluate the effect of OH radicals on NO oxidization; a simulated OH temporal and spatial evolution process in pulsed corona discharge is shown in Fig. 3.84, with background and added gas conditions of 3% H_2O, 5% O_2, and 15% CO_2 in N_2, and 3% H_2O in N_2, respectively. Here, OH decay much faster before 5 μs than after 5 μs. Primary OH radicals are mainly produced by the electron–molecule interaction of $e + H_2O \rightarrow OH + H + e$, many primary OH radicals form during pulsed discharge before 1 μs; however, the electron–molecule interaction process only has a duration of hundred nanoseconds.

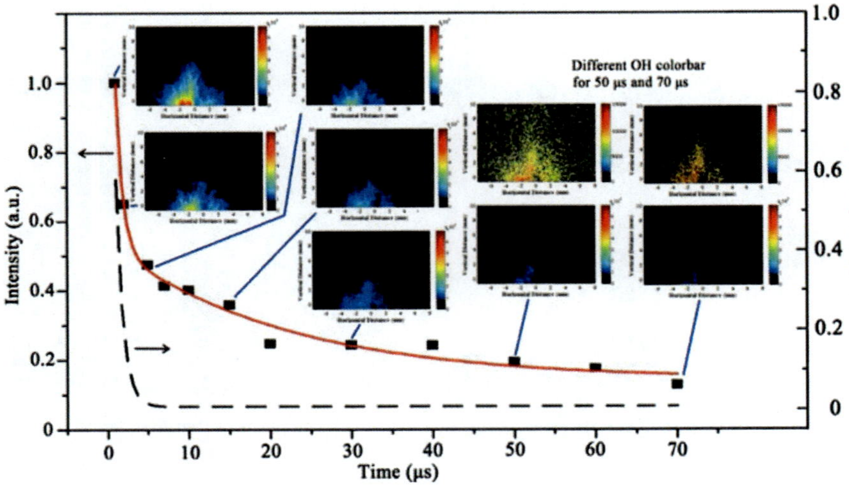

Fig. 3.84 OH temporal evolution during pulsed corona discharge (background gas: 3% H_2O, 5% O_2, and 15% CO_2 in N_2; added gas: 15% CO_2 in N_2) [101]

Table 3.8 Reactions related to the OH radical occurring during pulsed corona discharge

	Reactions	Rate constant[a]
R1	$OH + OH \rightarrow H_2O + O$	2.0×10^{-12}
R2	$OH + OH + M \rightarrow H_2O_2 + M$	6.9×10^{-31}
R3	$OH + O \rightarrow O_2 + H$	3.5×10^{-11}
R4	$OH + O_3 \rightarrow HO_2 + O_2$	6.5×10^{-14}
R5	$OH + N \rightarrow NO + H$	4.9×10^{-11}
R6	$OH + H + M \rightarrow H_2O + M$	4.0×10^{-30}
R7	$OH + HO_2 \rightarrow H_2O + O_2$	1.1×10^{-10}
R8	$O + HO_2 \rightarrow OH + O_2$	5.8×10^{-11}
R9	$N + HO_2 \rightarrow OH + NO$	2.2×10^{-11}
R10	$HO_2 + O_3 \rightarrow OH + O$	2.0×10^{-15}
R11	$H + HO_2 \rightarrow 2OH$	3.0×10^{-12}
R12	$NO + OH \rightarrow HNO_2$	6.6×10^{-12}
R13	$NO_2 + OH \rightarrow HNO_3$	1.0×10^{-11}
R14	$HNO_2 + OH \rightarrow NO_2 + H_2O$	4.9×10^{-12}
R15	$NO_3 + OH \rightarrow NO_2 + HO_2$	2.6×10^{-11}

[a]The unit for two-body reaction is $cm^3\ s^{-1}$ and that for three-body reaction is $cm^6\ s^{-1}$

Secondary OH radicals are generated by recombination reactions after the electron–molecule interaction process; summarized in Table 3.8. Here, OH decay reactions R1–R7 dominated during 1–5 μs owing to the high density of OH radicals, causing a rapid decrease in OH intensity. After 5 μs, the OH decay rate had slowed because OH was also generated by R8–R11.

NO oxidization rate: Another important parameter during pulsed corona discharge, the NO oxidation rate, can be defined as

$$\eta = \frac{[NO]_{off} - [NO]_{on}}{[NO]_{off}} \times 100\%, \tag{3.55}$$

where η is the NO oxidization rate, and $[NO]_{on}$ and $[NO]_{off}$ represent the NO concentration with and without discharge, respectively.

A two-dimensional distribution of the NO oxidation rate under laminar and turbulent conditions is shown in Fig. 3.85. In the turbulent case, the reaction zone is expanded, especially in the region from 2 to 7 mm along the vertical axis, and presents a higher effective oxidization rate and region. The oxidization rate distribution under turbulent conditions is asymmetrical due to the entrainment and squeeze between the added and background gases, whereas it is almost symmetrical under laminar conditions. In the laminar case, the highest oxidation rate was approximately 30% and located between − 4 and 4 mm in the horizontal direction, owing to diffusion; in the turbulent case, a maximum NO oxidization rate of 52% was achieved.

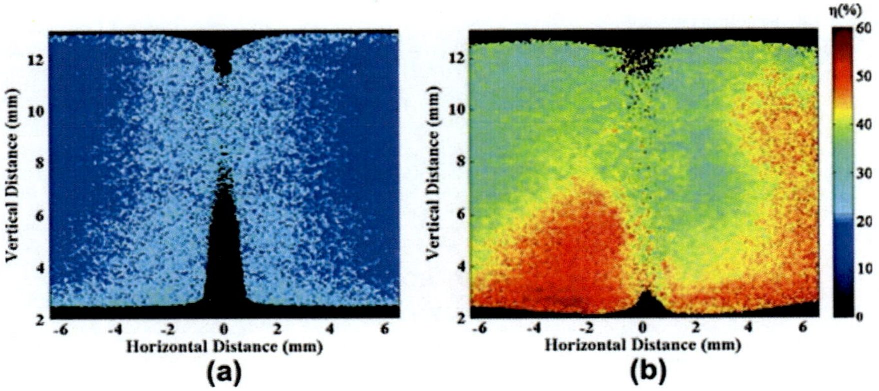

Fig. 3.85 NO oxidization rate distribution when **a** $Re = 1379$ and **b** $Re = 4138$ [100]

The converted NO concentration and oxidization rate along the z-axis for both cases, shown in Fig. 3.86a, demonstrate that more NO is converted in the turbulent case at vertical distances less than 8 mm; afterward, however, more NO is converted in the laminar case, owing to the higher NO initial concentration in that region. Overall, NO oxidization is much more efficient under a turbulent added gas condition, as shown in Fig. 3.86b, which indicates that efficient mixing is important for a high NO oxidization rate. Better mixing conditions increase the collision probability between radicals and NO, leading to more efficient NO oxidization. The average oxidization rate under a turbulent added gas condition was 42%, which was 22% higher than that of the laminar case; the maximum oxidization rate was 47% at 3 mm.

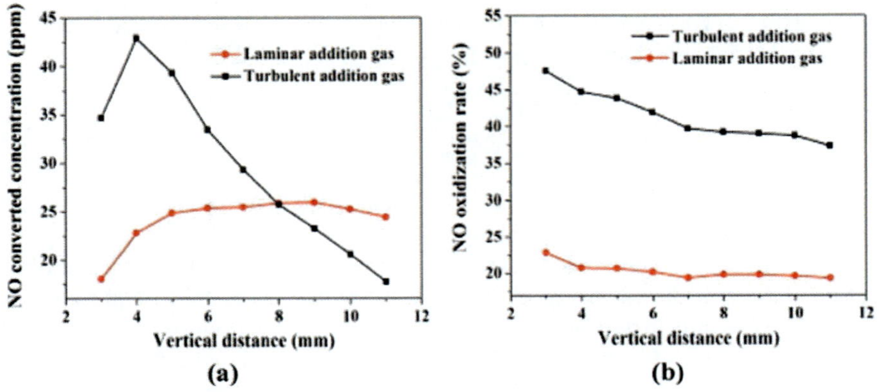

Fig. 3.86 NO **a** converted concentration and **b** oxidization rate along the vertical axis [100]

3.4.2.2 NO_x Conversion Enhancement

A pulsed corona discharge system designed to investigate the characteristics of NO_x conversion, shown in Fig. 3.87, comprised a pulsed power supply, a gas control system, gas analysis units, and a pulsed corona discharge reactor, the latter of which is detailed in Fig. 3.88. This reactor, designed with a wire-to-plate configuration and powered by a pulsed power supply, comprises a quartz glass box 600 mm long, 90 mm wide, and 60 mm high, two stainless steel plates at the inner sides of the reactor as the grounded electrodes, and eight stainless steel discharging wires evenly spaced along the centerline of the top of the reactor as the high-voltage electrodes. The high-voltage electrodes were 2 mm in diameter with an effective discharge length of 40 mm. The distance between the high-voltage electrodes and the grounded electrodes was 40 mm.

Effect of energy density: Using this experimental setup, the NO_2 conversion rate was shown to increase with increasing SED rate for different gas compositions, as

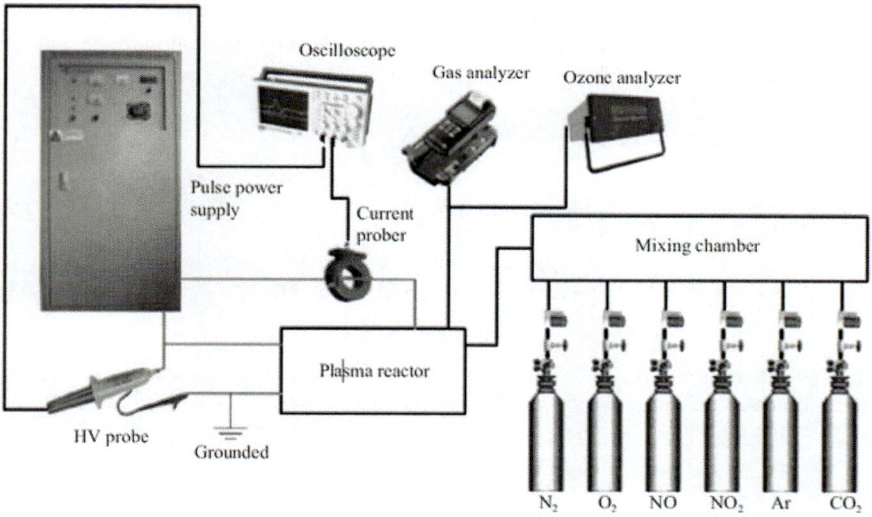

Fig. 3.87 Experimental setup for NO_x conversion by pulsed corona discharge [102]

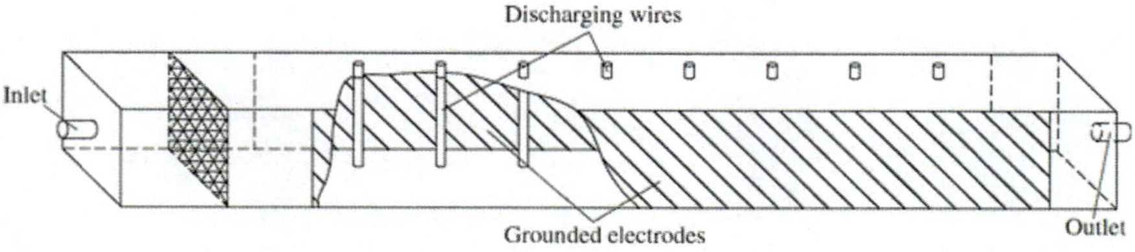

Fig. 3.88 Structure of the pulsed corona discharge reactor [102]

shown in Fig. 3.89. As the SED increased from 1.2 to 38.3 J/L in pure Ar plasma, the NO_2 conversion rate increased from 2.8 to 33.7%. In N_2/Ar plasma, notably, the conversion rate increased from 5.1 to 42.1% as the SED increased from 2.5 to 49.0 J/L; in O_2/Ar plasma, the conversion rate increased from 4.3 to 25.9% as the SED increased from 2.4 to 52.4 J/L. Other researchers have also demonstrated that increasing the plasma energy enhances the electric field, electron density, and gas temperature in the discharge, all of which may contribute to increasing the conversion of NO_2 [103, 104]. Figure 3.89 also shows that when the SED exceeds 31.8 J/L for pure Ar plasma, similar results are observed for N_2/Ar and O_2/Ar when the SED exceeds 43.8 and 50 J/L, respectively.

In pure Ar plasma, due to the low initial concentration of NO_2 molecules compared with background Ar, the probability of NO_2 direct dissociation reactions induced by high-energy electrons is negligible. Metastable $Ar(^3P_2)$ atoms and Ar^+ ions are generated via $Ar + e \rightarrow Ar(^3P_2) + e$ and $Ar + e \rightarrow Ar^+ + 2e$ in the plasma zone. Due to the lower ionization potential of NO_2 than of Ar, Penning ionization and charge transfer mechanisms ($Ar^+ + NO \rightarrow Ar + NO^+ + e$) play key roles in NO_2 conversion. NO^+ then recombines with a low-energy electron via $NO^+ + e \rightarrow NO$ or collides with a high energy electron via $NO^+ + e \rightarrow NO + O$ [105–107].

In the presence of N_2 or O_2, active species, such as N and O radicals and metastable nitrogen, are formed via electron impact reactions as follows:

$$e + \mathrm{N}_2 \rightarrow e + \mathrm{N} + \mathrm{N} \quad k_{1a} = f(e), \tag{3.56}$$

$$e + \mathrm{N} + \mathrm{N} \rightarrow e + \mathrm{N}_2(A) \quad k_{1b} = f(e), \tag{3.57}$$

$$e + \mathrm{O}_2 \rightarrow e + 2\mathrm{O} \quad k_2 = f(e). \tag{3.58}$$

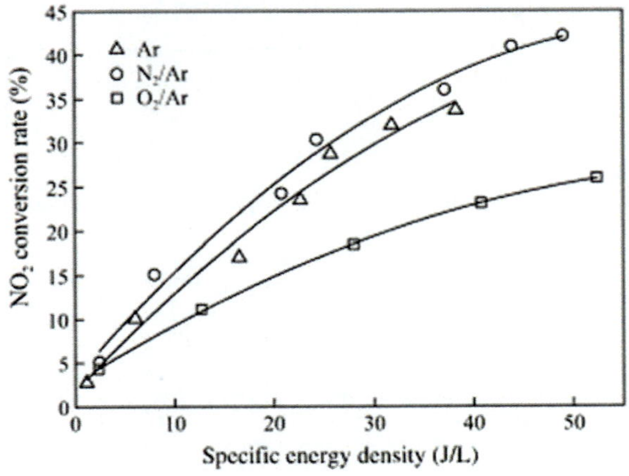

Fig. 3.89 Effect of SED on NO_2 conversion rate during pulsed corona discharge [102]

Many mechanisms for the conversion of NO_2 to NO in plasma once N and O radicals form have been proposed, including [108, 109]

$$M + O_2 + O \rightarrow O_3 + M \quad k_3 = 6.01 \times 10^{-34}[M], \tag{3.59}$$

$$NO_2 + O \rightarrow NO + O_2 \quad k_4 = 1.03 \times 10^{-11}, \tag{3.60}$$

$$N + NO_2 \rightarrow N_2O + O \quad k_5 = 1.21 \times 10^{-11}, \tag{3.61}$$

$$N + NO_2 \rightarrow 2NO + O \quad k_6 = 3.01 \times 10^{-12}, \tag{3.62}$$

$$NO_2 + N_2(A) \rightarrow NO + N_2 + O \quad k_7 = 1.3 \times 10^{-11}, \tag{3.63}$$

$$N + NO \rightarrow N_2 + O \quad k_8 = 2.93 \times 10^{-11}, \tag{3.64}$$

$$M + NO + O \rightarrow NO_2 + M \quad k_9 = 9.02 \times 10^{-32}[M], \tag{3.65}$$

$$NO + O_3 \rightarrow NO_2 + O_2 \quad k_{10} = 1.73 \times 10^{-14}, \tag{3.66}$$

$$N_2O + O \rightarrow N_2 + O_2 \quad k_{11} = 5.00 \times 10^{-3}, \tag{3.67}$$

where M denotes background gas molecules, including N_2, O_2, or Ar. The unit for k of two-body reactions is cm^3/(molecule sec) whereas that for three-body reactions is cm^6/(molecule^2 sec).

Effect of initial NO_2concentration: The effect of initial NO_2 concentration on NO_2 conversion rate in pure Ar at a fixed SED is shown in Fig. 3.90. Here, as the initial NO_2 concentration increased from 63.5 to 271.2 ppm, the NO_2 conversion rate decreases from 39.7 to 26.1%, whereas the converted NO_2 concentration increased nearly linearly from 25.2 to 70.8 ppm. The number of high-energy electrons and reactive species capable of NO_2 conversion produced by pulsed corona discharge remained the same under a given SED. As the initial NO_2 concentration increased, the concentration of electrons and reactive species became relatively low compared with the NO_2 concentration. Thus, the probability of reaction between NO_2 molecules and high-energy electrons or reactive species decreased, leading to lower conversion efficiency. However, increasing the initial NO_2 concentration increased the probability of collisions between NO_2 molecules and high-energy electrons or reactive species, causing an increase in the converted NO_2 concentration.

Effect of gas composition: The effect of varying the O_2 content in the O_2/Ar plasma on NO, NO_2, and O_3 concentration and NO_x conversion is shown in Fig. 3.91, where the gas initial NO_2 concentration was 210 ppm and the SED was the same as the

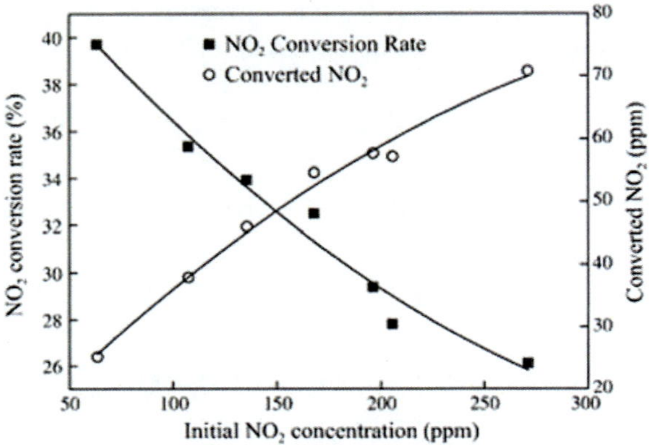

Fig. 3.90 Effect of initial concentration on NO_2 conversion rate during pulsed corona discharge [102]

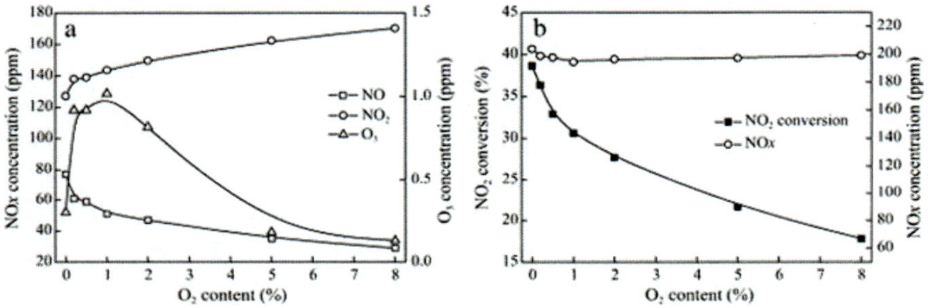

Fig. 3.91 Effect of O_2 content on NO_x conversion during pulsed corona discharge. **a** NO, NO_2, and O_3 concentrations; **b** NO_2 conversion and total NO_x concentration [102]

experiments in N_2/Ar. As the O_2 was increased from 0 to 8%, the concentration of NO_2 increased from 127.1 to 170.3 ppm whereas the NO concentration decreased from 77 to 29 ppm (Fig. 3.91a), indicating that the addition of oxygen suppressed the conversion of NO_2 to NO.

In an O_2/Ar atmosphere, the oxidation pathways, i.e., ozone- and O radical-containing reactions, dominate NO_2 conversion [110]. Electron-induced oxygen dissociation and quenching of metastable Ar by O_2 molecules produces ground-state O (3P) and excited O (^{1}D), which oxidize NO to NO_2. These O radicals can also react with O_2 via a three-body reaction to form O_3. Both O radicals and ozone can contribute to NO oxidation, further inhibiting NO_2 conversion. Although O can also reduce NO_2 to NO, the rate constant of NO oxidation with O radicals in an oxygen-rich environment is 3.00×10^{-11} cm^{-3}/(mol s), which is much larger than that of NO_2 reduction. Thus, with increasing oxygen content, the NO_2 conversion rate decreases, as shown in Fig. 3.91b. Since no N_2O and N_2 were formed in the oxygen-rich plasma, the change in total NO_x conversion can be regarded as negligible.

Separated radical injection discharge: Although nozzle electrodes can increase the yield of free radicals by spraying the electrode gas, the free-radical-source substances carried by the nozzle into the discharge reaction zone are mainly distributed around the outlet of the nozzle, resulting in an insufficient utilization rate. Moreover, a nozzle electrode is more complicated and expensive to produce than a needle tip electrode and thus has not been widely implemented in industry. To further improve the utilization rate of free radicals, a needle-tip electrode using air-jet separation has been developed; the reactor structure is shown in Fig. 3.92.

The concentrations of NO and NO_2 at the outlet of a typical nozzle-plate discharge and separated radical injection discharge are shown in Fig. 3.93a, b respectively. Here, the NO concentration is lower at the outlet of the separated radical injection discharge than that of the nozzle-plate discharge, and the NO_2 concentration is higher. At an injection energy of 50 J/L, the outlet concentration of NO of separated radical injection discharge was 12 ppm lower than that of nozzle-plate discharge. In both cases, the $NO + O \rightarrow NO_2$, $NO + O_3 \rightarrow NO_2 + O_2$, and $NO + OH \rightarrow HNO_2$

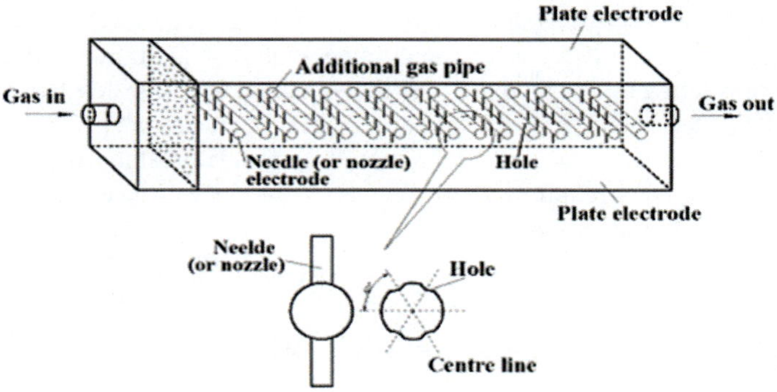

Fig. 3.92 Reactor structure

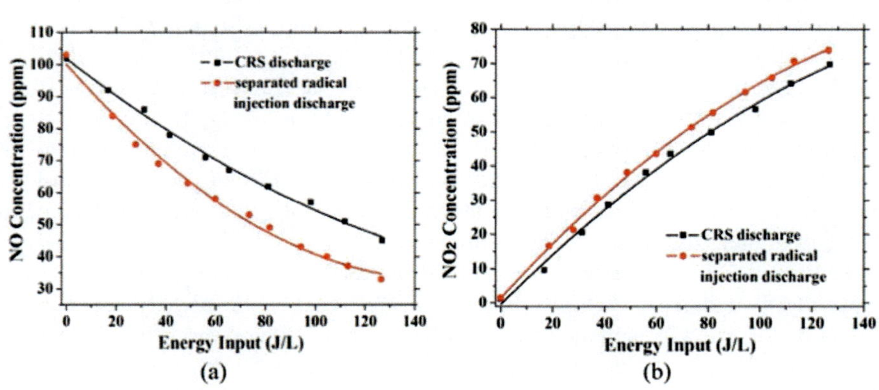

Fig. 3.93 Outlet concentration of **a** NO and **b** NO_2 in two discharge forms

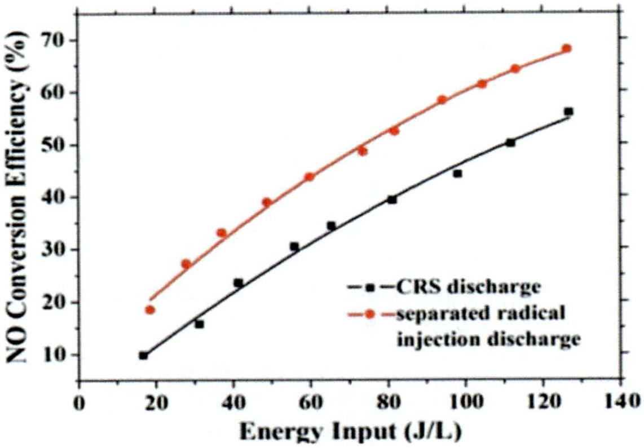

Fig. 3.94 NO conversion efficiency of two discharge forms

reactions lead to a decrease in NO concentration and increase in NO_2 concentration with increasing injection energy. Thus, the NO conversion efficiency increased with increasing injection energy, as shown in Fig. 3.94. Further, separated radical injection discharge allowed for a NO conversion efficiency of 70% at an injection energy of 130 J/L, representing a 15% increase over that obtained with nozzle-plate discharge, owing to the difference in the yield and distribution of free radicals.

Overall, NO_x conversion can be promoted in an electrostatic field. However, the converted products require further treatment to reduce NO_x emissions. This is discussed in detail in other chapters.

3.4.3 Simultaneous Control of Trace Elements

Dust removal devices, such as ESPs and electrostatic fabric filters (EFF), can also be used to reduce emissions of trace elements. During coal combustion, hazardous trace elements such as mercury (Hg), arsenic (As), and selenium (Se) are volatilized and then partitioned into bottom ash, fly ash, and flue gas. Air pollution control devices are the most promising way to control emissions of hazardous trace elements from CFPPs. In this section, the concentrations and distributions of Hg, As, and Se in coal, bottom ash, and fly ash from 15 CFPPs (detailed in Table 3.9) equipped with low–low temperature (LLT)-ESPs, cold-side (CS)-ESPs, and EFFs to achieve ultra-low emission standards are detailed and compared to clarify the trace element abatement capacity of these different dust removal devices.

ESPs and EFFs can effectively remove particulate-bound Hg, As, and Se by capturing fly ash. The removal efficiency of Hg by ESP/EFF heavily depends on the ratio of Hg^p in Hg^T and the oxidation efficiency of Hg^0 and varies from 5 to 91% [111–115]. Physical adsorption plays an important role in Se adsorption [116]; the distribution of Se in fly ash varies from 49 to 98% [117]. As chemical and physical

Table 3.9 CFPP Parameters

Unit	Installed capacity (MW)	Boiler type	Coal type	DRD[a]	T (°C)[b]
#1	1000	PC	Bit[c]	LLT[f]	90
#2	1000	PC	Bit	LLT	87
#3	660	PC	Bit	LLT	92
#4	660	PC	Bit	LLT	94
#5	660	PC	Bit	CS[g]	118
#6	300	PC	Bit	CS	115
#7	660	PC	Bit	CS	125
#8	220	PC	Bit	CS	120
#9	330	PC	Bit	CS	132
#10	330	PC	Bit + lig[d]	CS	128
#11	330	PC	Bit	CS	145
#12	300	PC	Bit	CS	140
#13	220	CFB	Bit	EFF[h]	118
#14	50	CFB	Bit + S[e]	EFF[h]	120
#15	50	CFB	Bit + S	EFF[h]	122

[a]Dust removal devices; [b] Inlet temperature of ESP/EFF (°C); [c] Bituminous coal; [d] Lignite; [e] Sludge; [f] Low–low temperature ESP; [g] Cold-side ESP; [h] electrostatic fabric filter

adsorption determine the enrichment of As in fly ash, the removal efficiency of As in ESP/EFF ranges from 35 to 99% [117, 118].

LLT-ESPs and EFFs are the most commonly used ultra-low emission dust removal devices. LLT-ESP is employed to lower the ESP inlet gas temperature below the acid dew point (typically < 95 °C) using a low-temperature economizer or hot media gas–gas heat exchanger [119]. EFFs operate under the assumption that 70–90% of the dust is collected by the ESP; the rest of the precharged particulates are removed by a fabric filter. Before WFGD, the PM concentration is < 20 mg/m^3. The development of such a dust removal device may have an enormous effect on the control of trace elements. However, little research has focused on capturing trace elements at lower temperatures, i.e., < 95 °C in LLT-ESP. The abatement capacity for trace elements in different dust removal devices was examined.

Pearson correlation and Grey relational analyses can be used to analyze the effects of the eight main factors known to influence the enrichment of fly ash with Hg, As, and Se, including the temperature and pressure, the content of iron and calcium oxides (Fe_2O_3 and CaO, respectively) in the fly ash, the content of Cl and S and the Cl/S ratio in the feed fuel, and the condensation of SO_3 on the fly ash in the LLT-ESP. The relative enrichment factor (REF) is widely used to evaluate the enrichment of trace elements in bottom ash and fly ash [120] and can be calculated as:

$$\text{REF} = C_{i,ash} \times \frac{C_{ash-in-coal}}{C_{i,coal}}, \tag{3.68}$$

where $C_{ash\text{-}in\text{-}coal}$ represents the ash content in the feed coal, $C_{i,ash}$ and $C_{i,coal}$ represent the mean concentrations of trace elements in ash and coal, respectively.

The relationship between the REF of Hg, As, and Se in fly ash as a function of the eight above factors is shown in Fig. 3.95. As EFFs have a different dust removal mechanism than LLT-ESPs and CS-ESPs do, CFPP employing EFFs are not included. The enrichment of Hg, As, and Se in fly ash was negatively correlated with ESP temperature, particle size, and S contents in the feed fuel and positively correlated with Cl, Cl/S in feed fuel, Fe_2O_3, and REF values of S.

The inlet temperature of an ESP has a significant effect on the enrichment of highly volatile elements. The simultaneous removal of SO_3 and trace elements could be achieved in LLT-ESP. The particle size of fly ash also has a great influence on the adsorption of Hg, As, and Se. The inhibitory effect of sulfur content was significant for Hg and Se. An increase in chlorine content and a decrease in sulfur content in the feed coal were correlated with an increase in Hg, As, and Se in fly ash. However, the correlation between the content of oxides (CaO and Fe_2O_3) and trace elements in the fly ash was limited compared with other factors. Overall, optimal conditions for Hg, As and Se removal involve a lower ESP inlet temperature, smaller fly ash particle size, higher chlorine and oxide (CaO and Fe_2O_3) content, and lower content of sulfur in the feed coal.

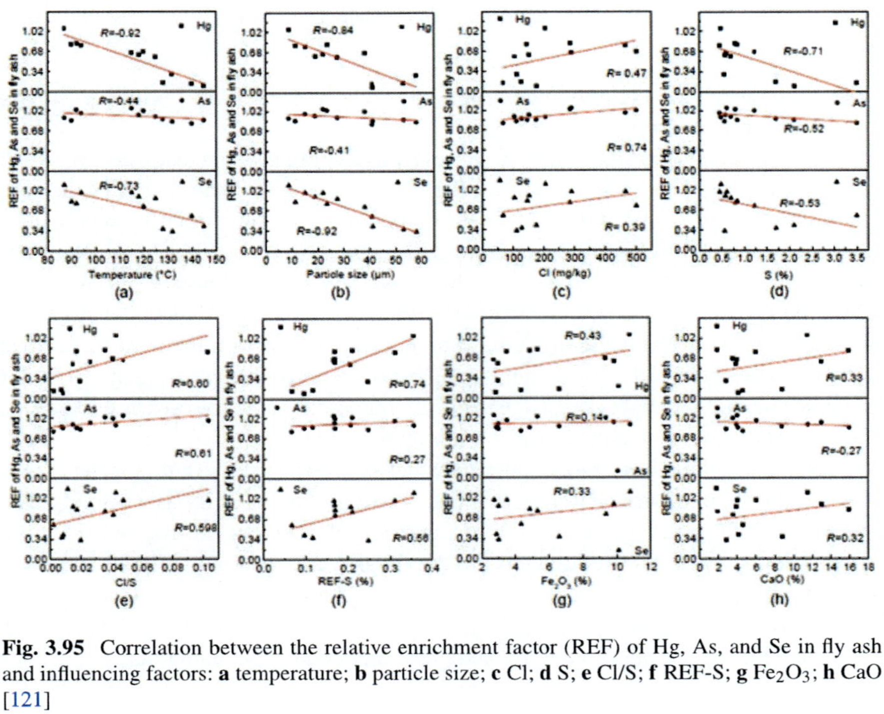

Fig. 3.95 Correlation between the relative enrichment factor (REF) of Hg, As, and Se in fly ash and influencing factors: **a** temperature; **b** particle size; **c** Cl; **d** S; **e** Cl/S; **f** REF-S; **g** Fe_2O_3; **h** CaO [121]

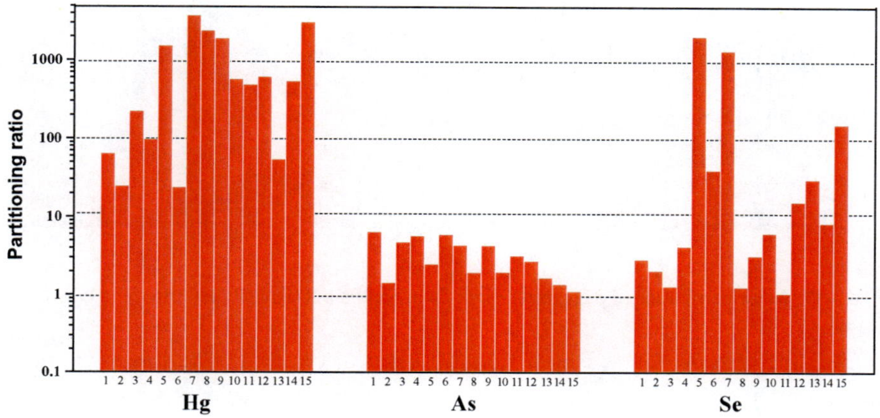

Fig. 3.96 Partition ratio of Hg, As, and Se (i.e., mass concentration ratio between fly ash and bottom ash) [121]

Abatement of Hg, As, and Se in a Dry ESP: The partition ratio, or ratio of an elements' mass concentration in the fly ash and bottom ash, of Hg, As, and Se is shown in Fig. 3.96. In all cases, the partition ratios of each trace element were greater than 1, indicating that the fly ash had a higher concentration of each element than did the bottom ash. Consistent with documented volatility of the trace elements studied as Hg > Se > As, Hg had a higher partition ratio (23.7–3740) than As (1.1–6.5) or Se (1.1–2027) for all tested CFPPs.

To compare the trace element mitigation capability of different dust removal equipment, the mass distribution ratios of Hg, As, and Se after the bottom ash, fly ash, and dust collector were calculated. The mass flow rate of heavy metals in coal, fly ash (FA), and bottom ash (BA) were determined according to the operating parameters during the unit test. The mass distribution was then calculated as follows:

$$P_{i,FA} = \frac{C_{i,FA} \times q_{m,FA}}{C_{i,coal} \times q_{m,coal}} \times 100\%, \tag{3.69}$$

$$P_{i,BA} = \frac{C_{i,BA} \times q_{m,BA}}{C_{i,coal} \times q_{m,coal}} \times 100\%, \tag{3.70}$$

$$P_{ESP-OUT} = (1 - P_{i,FA} - P_{i,BA}) \times 100\%, \tag{3.71}$$

where $P_{i,FA}$, $P_{i,BA}$, $P_{ESP\text{-}OUT}$ represents the mass proportion of heavy meals in FA,BA, ESP outlet (ESP-OUT), and $q_{m,FA}$, $q_{m,BA}$, $q_{m,coal}$ represents the mass flow rate of heavy metals in FA,BA, coal.

As summarized in Fig. 3.97, much less Hg was emitted when using an LLT-ESP or EFF than a CS-ESP (0.8–36.1% versus 42.1–90.6%, respectively). Using an EFF and ESP (inlet temperature < 125 °C), 72.3–99.1% of As and 70.7–100% of Se were

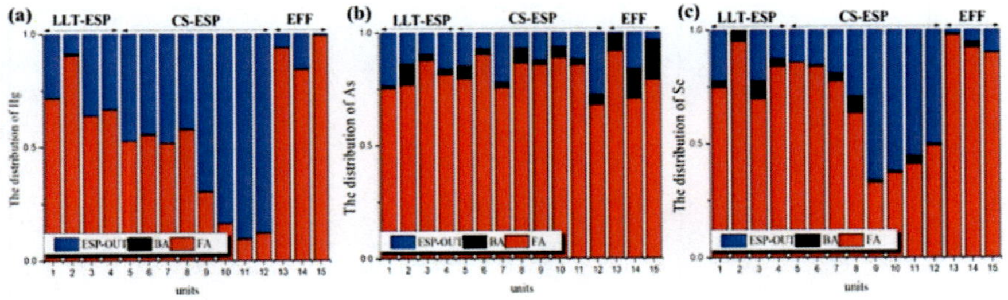

Fig. 3.97 Mass distribution of **a** Hg, **b** As, and **c** Se in bottom ash (BA), fly ash (FA), and ESP-OUT [121]

removed before the WFGD system. Therefore, CFPPs equipped with LLT-ESP can effectively capture dust and hazardous trace elements to achieve ultra-low emissions of multiple pollutants.

Abatement of Hg, As, and Se in WESP: In this section, the partitioning of typical hazardous trace elements (Hg, As, Cd, Se, Cr, and Pb) in solid and liquid streams and the REFs in the fly ash and bottom ash in five typical ultra-low emission CFPPs employing WESPs from the above units (#1, #5, #7, #12, #14) is discussed.

The mean concentrations of hazardous trace elements in the WESP wastewater and the corresponding slag contained are presented in Table 3.10, where WESP wastew-

Table 3.10 Mean concentration of hazardous trace elements in WESPs

	Hg	As	Cd	Se	Cr	Pb
Slag in WESP wastewater (mg/kg)						
#1	8.67	133	4.51	61.3	274	255
#2	10.9	280	2.22	106	1303	74.9
#3	–	–	–	–	–	–
#4	17.5	45.2	9.26	121	1061	136
#5	–	–	–	–	–	–
Wastewater (μg/kg)						
#1	1	1.7	0.9	5	11.8	5.8
#2	1.5	1	1.6	51.3	19.6	< 1
#3	–	–	–	–	–	–
#4	0.9	1.2	0.7	12.2	4.9	< 1
#5	–	–	–	–	–	–
C_{WESP}/C_{FA}						
#1	16.8	3.4	4.0	6.4	2.1	2.0
#2	20.4	7.0	2.4	7.7	17.3	1.0
#4	60.1	2.0	14.0	9.6	27.1	0.8

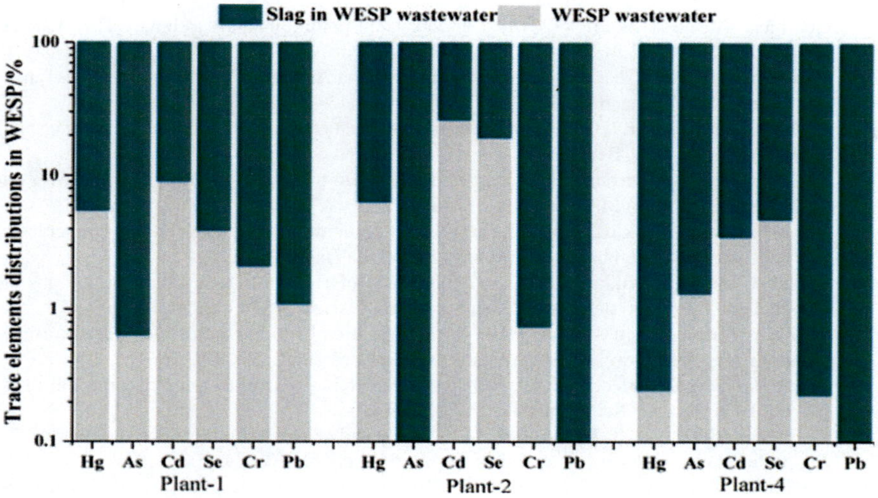

Fig. 3.98 Trace element partitioning in the WESP in three CFPPs [123]

ater slag comprises the submicron particles captured by the WESP. The concentrations of Hg, As, Cd, Se, Cr, and Pb in the WESP wastewater slag ranged from 8.67 to 17.5, 45.2 to 133, 4.51 to 9.26, 61.3 to 121, 274 to 1303, and 74.9 to 255 mg/kg, respectively. The ratio of trace element concentrations in the WESP wastewater slag to that in the fly ash ($C_{\mathrm{WESP}}/C_{\mathrm{FA}}$) demonstrated that the concentration of Hg in the WESP wastewater slag was 16.8 – 60.1 times higher than that in the fly ash. Previous studies have indicated that ≤ 0.1 μm was the dominant sink for Hg in the fly ash [122]. Thus, mercury was enriched in the WESP wastewater slag. The $C_{\mathrm{WESP}}/C_{\mathrm{FA}}$ of As, Cd, Se, and Cr ranged from 2 to 7, 2.4 to 14, 6.4 to 9.6, and 2.1 to 27.1, respectively. No obvious enrichment of Pb was observed in the WESP wastewater slag compared with that in the fly ash. However, the concentrations of hazardous trace elements in the WESP wastewater were extremely low.

The distribution of hazardous trace elements in the WESP wastewater and slag in plants 1, 2, and 4 is shown in Fig. 3.98. The WESP system removed hazardous trace elements mainly by capturing fine particles; only 0.2–26% of the hazardous trace elements were retained in the wastewater. The contaminated submicron particles could deteriorate the health of people if inhaled. Hence, capturing fine particles with high concentrations of hazardous trace elements is of great importance, and WESP plays an essential role.

References

1. Agarwal RA, Agarwal AK, Gupta T, Sharma N (2019) Introduction to pollutants from energy sources: characterization and control. Springer Singapore, Singapore

2. Guarnieri M, Balmes JR (2014) Outdoor air pollution and asthma. The Lancet 383(9928):1581–1592
3. Hamilton LH (1981) Char morphology and behavior of Australian vitrinites of various ranks pyrolyzed at various heating rates. Fuel 60(10):909–913
4. Zereini F, Wiseman CL (2011) Urban airborne particulate matter. Environmental Science and Engineering, Springer, Berlin
5. Kikuchi R (2001) Environmental management of sulfur trioxide emission: impact of SO_3 on human health. Environ Manage 27(6):837–844
6. Menasha J, Dunn-Rankin D, Muzio L et al (2011) Ammonium bisulfate formation temperature in a bench-scale single-channel air preheater. Fuel 90(7):2445–2453
7. Mitsui Y, Imada N, Kikkawa H et al (2011) Study of Hg and SO_3 behavior in flue gas of oxy-fuel combustion system. Int J Greenhouse Gas Control 5:S143–S150
8. Zhang Y, Zheng C, Liu S et al (2019) An investigation of SO_3 control routes in ultra-low emission coal-fired power plants. Aerosol Air Qual Res 19(12):2908–2916
9. Zhang Y, Zheng C, Hu F et al (2020) Field test of SO_3 removal in ultra-low emission coal-fired power plants. Environ Sci Pollut Res 27(5):4746–4755
10. Jaworek A, Sobczyk AT, Krupa A et al (2019) Hybrid electrostatic filtration systems for fly ash particles emission control: a review. Sep Purif Technol 213:283–302
11. Maffei T, Frassoldati A, Cuoci A et al (2013) Predictive one step kinetic model of coal pyrolysis for CFD applications. Proc Combust Inst 34:2401–2410
12. Council NR (2004) Research priorities for airborne particulate matter: IV. Continuing research progress. The National Academies Press, Washington, DC
13. Jaworek A, Marchewicz A, Sobczyk AT et al (2018) Two-stage electrostatic precipitators for the reduction of $PM_{2.5}$ particle emission. Prog Energy Combust Sci 67:206–233
14. Pui DYH, Chen S, Zuo Z (2014) $PM_{2.5}$ in China: Measurements, sources, visibility and health effects, and mitigation. Particuology 13:1–26
15. Zheng C, Wang Y, Liu Y et al (2019) Formation, transformation, measurement, and control of SO_3 in coal-fired power plants. Fuel 241:327–346
16. Raizer YP, Braun C (1991) Gas discharge physics. Appl Opt 31:2400–2401
17. Goodfellow H, Tahti E (2002) Industrial ventilation design guidebook. Chem Health Saf 9(1):38–39
18. Caliskan B, Özengin N, Cindoruk SS (2020) Air quality level, emission sources and control strategies in Bursa/Turkey. Atmos Pollut Res 11(12):2182–2189
19. Jaworek A, Sobczyk AT, Marchewicz A et al. (2020) Particulate matter emission control from small residential boilers after biomass combustion. A review. Renew Sustain Energy Rev 137:110446
20. Tang L, Xue X, Jia M et al (2020) Iron and steel industry emissions and contribution to the air quality in China. Atmos Environ 237:117668
21. Hall HJ (1990) History of pulse energization in electrostatic precipitation. J Electrostat 25(1):1–22
22. Castle G (2001) Industrial applications of electrostatics: the past, present and future. J Electrostat 51:1–7
23. Schnelle KBJ, Dunn RF, Ternes ME (2015) Air pollution control technology handbook, 2nd edn. CRC Press, Boca Raton
24. Zheng C, Liu X, Yan P et al (2018) Measurement and prediction of fly ash resistivity over a wide range of temperature. Fuel 216:673–680
25. Chen Y, Shah N, Huggins FE et al (2005) Transmission electron microscopy investigation of ultrafine coal fly ash particles. Environ Sci Technol 39(4):1144–1151
26. Buhre BJP, Hinkley JT, Gupta RP et al (2006) Fine ash formation during combustion of pulverised coal–coal property impacts. Fuel 85(2):185–193
27. Chen Y, Shah N, Braun A et al (2005) Electron microscopy investigation of carbonaceous particulate matter generated by combustion of fossil fuels. Energy Fuels 19(4):1644–1651
28. Bickelhaupt RE, Altman RF (1984) A method for predicting the effective volume resistivity of a sodium-depleted fly-ash layer. J Air Pollut Control Assoc 34(8):832–838

29. Chandra A (2009) Some investigations on fly ash resistivity generated in Indian power plants. Springer, Berlin Heidelberg, pp 399–405
30. Hughes JF (1998) Applied electrostatic precipitation. J Electroanal Chem 45:209–213
31. Yan P, Zheng C, Zhu W et al (2016) An experimental study on the effects of temperature and pressure on negative corona discharge in high-temperature ESPs. Appl Energy 164(9):28–35
32. Yan P, Zheng C, Xiao G et al (2015) Characteristics of negative DC corona discharge in a wire–plate configuration at high temperatures. Sep Purif Technol 139:5–13
33. Zhang H, Wang Y, Gao W et al (2020) Minimizing the adverse effects of dust layer on the particle migration in electrostatic precipitator under various temperature. Fuel Process Technol 106659
34. Murphy AT, Adler FT, Penney GW (1959) A theoretical analysis of the effects of an electric field on the charging of fine particles. Trans Am Inst Electr Eng Part I: Commun Electron 78(4):318–326
35. Liu B, Yeh H (1968) On theory of charging of aerosol particles in an electric field. J Appl Phys 39(3):1396–1402
36. Smith W, McDonald JR et al (1975) Calculation of the charging rate of fine particles by unipolar ions. J Air Pollut Control Assoc 25(2):168–172
37. Lawless PA (1996) Particle charging bounds, symmetry relations, and an analytic charging rate model for the continuum regime. J Aerosol Sci 27(2):191–215
38. Long Z, Yao Q (2012) Numerical simulation of the flow and the collection mechanism inside a scale hybrid particulate collector. Powder Technol 215–216:26–37
39. Guo B, Yang S, Xing M et al (2013) Toward the development of an integrated multiscale model for electrostatic precipitation. Ind Eng Chem Res 52(33):11282–11293
40. Lancereau Q, Roux JM, Achard JL (2013) Influence of secondary flows on the collection efficiency of a cylindrical electrostatic precipitator. J Aerosol Sci 63:146–160
41. Le Picard R, Girshick SL (2016) The effect of single-particle charge limits on charge distributions in dusty plasmas. Journal of Physics D: Applied Physics 49(9):095201
42. Zheng C, Duan D, Chang Q et al (2019) Experiments on enhancing the particle charging performance of an electrostatic precipitator. Aerosol Air Qual Res 19(6):1411–1420
43. Vallero DA (2008) Fundamentals of air pollution, 4th Edition. Academic Press, Burlington
44. Allen M, Raabe O (1985) Slip correction measurements of spherical solid aerosol particles in an improved Millikan apparatus. Aerosol Sci Technol 4(3):269–286
45. Zheng C, Zhang X, Yang Z et al (2018) Numerical simulation of corona discharge and particle transport behavior with the particle space charge effect. J Aerosol Sci 118:22–33
46. Wang Y, Gao W, Zhang H et al (2019) Insights into the role of ionic wind in honeycomb electrostatic precipitators. J Aerosol Sci 133:83–95
47. Gao W, Wang Y, Zhang H et al (2020) A numerical investigation of the effect of dust layer on particle migration in an electrostatic precipitator. Aerosol Air Qual Res 20(1):166–179
48. Zheng C, Zhang H, Liu X et al (2020) Effect of dust layer in electrostatic precipitators on discharge characteristics and particle removal. Fuel 278:118335
49. Mertens J, Anderlohr C, Rogiers P et al (2014) A wet electrostatic precipitator (WESP) as countermeasure to mist formation in amine based carbon capture. Int J Greenhouse Gas Control 31:175–181
50. Wu H, Pan D, Bao J et al (2016) Improving the removal efficiency of sulfuric acid droplets from flue gas using heterogeneous vapor condensation in a limestone-gypsum desulfurization process 92(1):230–237
51. Yang Z, Zheng C, Zhang X et al (2018) Highly efficient removal of sulfuric acid aerosol by a combined wet electrostatic precipitator. RSC Adv 8(1):59–66
52. Zheng C, Zheng H, Shen J et al (2020) Evolution of condensable fine particle size distribution in simulated flue gas by external regulation for growth enhancement. Environ Sci Technol 54(7):3840–3848
53. Brachert L, Kochenburger T, Schaber K (2013) Facing the sulfuric acid aerosol problem in flue gas cleaning: pilot plant experiments and simulation. Aerosol Sci Technol 47(10):1083–1091

54. Yang Z, Zheng C, Li Q et al (2020) Fast evolution of sulfuric acid aerosol activated by external fields for enhanced emission control. Environ Sci Technol 54(5):3022–3031
55. Schaber K, Korber J, Ofenloch O et al (2002) Aerosol formation in gas-liquid contact devices-nucleation, growth and particle dynamics. Chem Eng Sci 57(20):4345–4356
56. Albriet B, Sartelet KN, Lacour S et al (2010) Modelling aerosol number distributions from a vehicle exhaust with an aerosol CFD model. Atmos Environ 44(8):1126–1137
57. Hinds WC (2012) Aerosol technology: properties, behavior, and measurement of airborne particles. J Aerosol Sci 31(9):1121–1122
58. Enghoff MB, Svensmark H (2008) The role of atmospheric ions in aerosol nucleation—a review. Atmos Chem Phys 8(16):4911–4923
59. Levdansky VV, Smolik J, Moravec P (2008) Joint effect of particle charge and adsorbable foreign gases on vapor condensation on fine aerosol particles. Int Commun Heat Mass Transfer 35(10):1246–1248
60. Jasper K, Joachim C, João A et al (2011) Role of sulphuric acid, ammonia and galactic cosmic rays in atmospheric aerosol nucleation. Nature 476(7361):429–433
61. Chen S, Zhao Y, Zhang R (2018) Formation mechanism of atmospheric ammonium bisulfate: hydrogen-bond-promoted nearly barrierless reactions of SO_3 with NH_3 and H_2O. ChemPhysChem 19(8):967–972
62. Sinanis S, Wix A, Ana L et al (2008) Characterization of sulphuric acid and ammonium sulphate aerosols in wet flue gas cleaning processes. Chem Eng Process 47(1):22–30
63. Mizuno A (2000) Electrostatic precipitation. IEEE Trans Dielectr Electr Insul 7(5):615–624
64. Jaworek A, Krupa A, Czech T (2007) Modern electrostatic devices and methods for exhaust gas cleaning: a brief review. J Electrostat 65(3):133–155
65. Bai M, Wang S, Chen Z et al (2010) The effects of submicrometer dust charging and coagulation on ESP efficiency by using alternating electric field. IEEE Trans Plasma Sci 38(2):127–132
66. Chang Q, Zheng C, Yang Z et al (2017) Electric agglomeration modes of coal-fired fly-ash particles with water droplet humidification. Fuel 200:134–145
67. Marshall J, Li S (2014) Adhesive particle flow. Cambridge University Press, New York
68. Hinds WC (1999) Aerosol technology: properties, behavior, and measurement of airborne particles. J Aerosol Sci 31(9):1121–1122
69. Seville JPK, Willett CD, Knight PC (2000) Interparticle forces in fluidisation: a review. Powder Technol 113(3):261–268
70. Martonen T (2011) Aerosol measurement: principles, techniques, and applications. Van Nostrand Reinhold, New York
71. Chang Q, Zheng C, Gao X et al (2015) Systematic approach to optimization of submicron particle agglomeration using ionic-wind-assisted pre-charger. Aerosol Air Qual Res 15(7):2709–2719
72. Yang Z, Zheng C, Chang Q et al (2017) Fine particle migration and collection in a wet electrostatic precipitator. J Air Waste Manag Assoc 67(4):498–506
73. Jędrusik M, Świerczok A (2009) The influence of fly ash physical and chemical properties on electrostatic precipitation process. J Electrostat 67(2–3):105–109
74. Qi L, Yuan Y (2011) Characteristics and the behavior in electrostatic precipitators of high-alumina coal fly ash from the Jungar power plant, Inner Mongolia, China. J Hazard Mater 192(1):222–225
75. Mahalik K, Sahu JN, Patwardhan AV et al (2010) Kinetic studies on hydrolysis of urea in a semi-batch reactor at atmospheric pressure for safe use of ammonia in a power plant for flue gas conditioning. J Hazard Mater 175(1):629–637
76. Shanthakumar S, Singh DN, Phadke RC (2009) The effect of dual flue gas conditioning on fly ash characteristics. J Test Eval 37(6):623–630
77. Shanthakumar S, Singh DN, Phadke RC (2008) Influence of flue gas conditioning on fly ash characteristics. Fuel 87(15–16):3216–3222
78. Shanthakumar S, Singh DN, Phadke RC (2008) Flue gas conditioning for reducing suspended particulate matter from thermal power stations. Prog Energy Combust Sci 34(6):685–695

79. Zheng C, Shen Z, Yan P et al (2017) Particle removal enhancement in a high-temperature electrostatic precipitator for glass furnace. Powder Technol 319:154–162
80. Yan J, Chen L, Yang L (2016) Combined effect of acoustic agglomeration and vapor condensation on fine particles removal. Chem Eng J 290:319–327
81. Wang Y, Gao W, Zhang H et al (2020) Enhanced particle precipitation from flue gas containing ultrafine particles through precharging. Process Saf Environ Prot 144:111–122
82. Yang Z, Zheng C, Liu S et al (2018) A combined wet electrostatic precipitator for efficiently eliminating fine particle penetration. Fuel Process Technol 180:122–129
83. Yang Z, Cai Y, Li Q et al (2021) Predicting particle collection performance of a wet electrostatic precipitator under varied conditions with artificial neural networks. Powder Technol 377:632–639
84. Zheng C, Wang Y, Zhang X et al (2018) Current density distribution and optimization of the collection electrodes of a honeycomb wet electrostatic precipitator. RSC Adv 8(54):30701–30711
85. Yang Z, Zheng C, Zhang X et al (2018) Challenge of SO_3 removal by wet electrostatic precipitator under simulated flue gas with high SO_3 concentration. Fuel 217:597–604
86. Heiredal ML, Jensen AD, Thøgersen JR et al (2013) Pilot-scale investigation and CFD modeling of particle deposition in low-dust monolithic SCR $DeNO_x$ catalysts. AIChE J 59(6):1919–1933
87. Noda N, Makino H (2010) Influence of operating temperature on performance of electrostatic precipitator for pulverized coal combustion boiler. Adv Powder Technol 21(4):495–499
88. Villot A, Gonthier Y, Gonze E et al (2012) Separation of particles from syngas at high-temperatures with an electrostatic precipitator. Sep Purif Technol 92:181–190
89. Naidis GV (2009) Simulation of streamer-induced pulsed discharges in atmospheric-pressure air. Eur Phys J Appl Phys 47(2):22803
90. Ghaleb F, Benstâali W, Belasri A (2008) Calculation of breakdown voltage in plasma display panels. Mater Sci Eng, C 28(5):791–794
91. Yuan, Z., Suzuki, K., Ehara, Y et al (2020) Corona discharge characteristics in electrostatic precipitator under high temperature. 2020 IEEE Industry Applications Society Annual Meeting, pp 1–6
92. Zheng C, Wu Z, Shen Z et al (2020) Particle capture in a high-temperature electrostatic precipitator with different electrode configurations. Powder Technol 372:84–93
93. Ma S, Zhao Y, Yang J et al (2017) Research progress of pollutants removal from coal-fired flue gas using non-thermal plasma. Renew Sustain Energy Rev 67:791–810
94. Li S, Huang Y, Wang F et al (2014) Fundamentals and environmental applications of non-thermal plasmas: multi-pollutants emission control from coal-fired flue gas. Plasma Chem Plasma Process 34(3):579–603
95. Yang Z, Zheng C, Zhang X et al (2017) Sulfuric acid aerosol formation and collection by corona discharge in a wet electrostatic precipitator. Energy Fuels 31(8):8400–8406
96. Gao X, Du X, Jiang Y et al (2010) A DFT study on the behavior of NO_2 in the selective catalytic reduction of nitric oxides with ammonia on a V_2O_5 catalyst surface. J Mol Catal A: Chem 317(1):46–53
97. Zabetta EC, Hupa M, Saviharju K (2005) Reducing NO_x emissions using fuel staging, air staging, and selective noncatalytic reduction in synergy. Ind Eng Chem Res 44(13):4552–4561
98. Chandrasekara Pillai K, Chung SJ, Raju T et al (2009) Experimental aspects of combined NO_x and SO_2 removal from flue-gas mixture in an integrated wet scrubber-electrochemical cell system. Chemosphere 76(5):657–664
99. Basfar AA, Fageeha OI, Kunnummal N et al (2008) Electron beam flue gas treatment (EBFGT) technology for simultaneous removal of SO_2 and NO_x from combustion of liquid fuels. Fuel 87(8):1446–1452
100. Shen X, Gao X, Li Z et al (2012) PLIF diagnostics of NO oxidization and OH consumption in pulsed corona discharge. Fuel 102:729–736
101. Zheng C, Shen X, Gao X et al (2013) Planar laser-induced fluorescence diagnostics for spatiotemporal OH evolution in pulsed corona discharge. IEEE Trans Plasma Sci 41(3):485–493

102. Zhu X, Zheng C, Gao X et al (2014) Experimental study of NO_2 reduction in N_2/Ar and O_2/Ar mixtures by pulsed corona discharge. J Environ Sci (China) 26(11):2249–2256
103. Tu X, Whitehead JC (2012) Plasma-catalytic dry reforming of methane in an atmospheric dielectric barrier discharge: understanding the synergistic effect at low temperature. Appl Catal B 125:439–448
104. Tu X, Gallon HJ, Whitehead JC (2011) Electrical and spectroscopic diagnostics of a single-stage plasma-catalysis system: effect of packing with TiO_2. J Phys D: Appl Phys 44 (48):482003
105. Hu X, Zhao G, Zhang J et al (2004) Nonthermal-plasma reactions of dilute nitrogen oxide mixtures: NO x-in-Argon and NO_x+CO-in-Argon. Ind Eng Chem Res 43(23):7456–7464
106. Kutasi K (2011) Modelling of NO destruction in a low-pressure reactor by an Ar plasma jet: species abundances in the reactor. J Phys D: Appl Phys 44 10:105202
107. Tsuji M, Nakano K, Kumagae J et al (2002) Decomposition of NO by microwave discharge of NO/He or NO/Ar mixtures. Bull Chem Soc Jpn 75(3):607–614
108. Arai M, Saito M, Yoshinaga S (2004) Effect of oxygen on NO_x removal in corona discharge field: NO_x behavior without a reducing agent. Combust Sci Technol 176(10):1653–1665
109. Tochikubo F, Uchida S, Yasui H et al (2009) Numerical simulation of NO oxidation in dielectric barrier discharge with microdischarge formation. Jpn J Appl Phys 48(7):076507
110. Yoshida K, Rajanikanth BS, Okubo M (2009) NO_x reduction and desorption studies under electric discharge plasma using a simulated gas mixture: a case study on the effect of corona electrodes. Plasma Sci Technol 11(3):327–333
111. López-Antón MA, Abad-Valle P, Díaz-Somoano M et al (2009) The influence of carbon particle type in fly ashes on mercury adsorption. Fuel 88(7):1194–1200
112. Deng S, Shi Y, Liu Y et al (2014) Emission characteristics of Cd, Pb and Mn from coal combustion: field study at coal-fired power plants in China. Fuel Process Technol 126:469–475
113. Zhou Z-J, Liu X-W, Zhao B et al (2015) Effects of existing energy saving and air pollution control devices on mercury removal in coal-fired power plants. Fuel Process Technol 131:99–108
114. Zhang L, Wang S, Wu Q et al (2016) Mercury transformation and speciation in flue gases from anthropogenic emission sources: a critical review. Atmos Chem Phys 16(4):2417–2433
115. Gao Y, Zhang Z, Wu J et al (2013) A critical review on the heterogeneous catalytic oxidation of elemental mercury in flue gases. Environ Sci Technol 47(19):10813–10823
116. Chen J, Wey M, Liu Z (2001) Adsorption Mechanism of Heavy Metals on Sorbents during Incineration 127(1):63–69
117. Tian H, Wang Y, Xue Z et al (2010) Trend and characteristics of atmospheric emissions of Hg, As, and Se from coal combustion in China, 1980–2007. Atmos Chem Phys 10(23):11905–11919
118. Cheng C, Hack P, Chu P et al (2009) Partitioning of mercury, arsenic, selenium, boron, and chloride in a full-scale coal combustion process equipped with selective catalytic reduction, electrostatic precipitation, and flue gas desulfurization systems. Energy Fuels 23(10):4805–4816
119. Fu Y, Ma K (2014) Application of low temperature ESP technology in thermal power plant. Adv Mater Res 1070–1072:1481–1485
120. Meij R, Vredenbregt LHJ, Ht W (2002) The fate and behavior of mercury in coal-fired power plants. J Air Waste Manag Assoc 52(8):912–917
121. Zheng C, Wang L, Zhang Y et al (2018) Co-benefit of hazardous trace elements capture in dust removal devices of ultra-low emission coal-fired power plants. J Zhejiang Univ Sci A 19(1):68–79
122. Jew AD, Rupp EC, Geatches DL et al (2015) Mercury interaction with the fine fraction of coal-combustion fly ash in a simulated coal power plant flue gas stream. Energy Fuels 29(9):6025–6038
123. Wang L, Zheng C, Zhang Y et al (2017) Speciation characteristics and mobility of trace elements across ultralow emission air pollution control devices. Energy Fuels 31(12):13963–13971

Chapter 4
Pollutant Control by Absorption Methods

4.1 Introduction

Anthropogenic activities, such as burning fossil fuels, smelting ferrous/nonferrous metals, and producing and using chemicals, produce a variety of gaseous pollutants, including SO_2, NO_x, NH_3, hydrogen halides, and organic gaseous pollutants (e.g., polycyclic aromatic hydrocarbons).

Absorption, wherein one or more components of a gas stream are selectively absorbed into an absorbent, is often used to remove gaseous pollutants from the emission of various industries. During absorption, gaseous pollutants are chemically absorbed via mass transfers and reactions between gas and liquid phases. Moreover, in some systems, such as a limestone–gypsum wet flue gas desulfurization (WFGD) system, the dissolution and crystallization of absorbents and by-products are involved [1, 2].

Most research on the removal of pollutants via absorption has focused on the removal of SO_2. Key technologies for SO_2 removal includes pH adjustment, desulfurization additives and scrubber optimization. Furthermore, the simultaneous removal of other pollutants, including Hg and NO_x, during SO_2 absorption is a globally promising technology [3–6].

To meet the increasingly strict emission regulations, a higher removal efficiency of SO_2 absorption technology, already used in mature industrial applications, is required either by enhancing mass transfer and two-phase reaction mechanisms or by developing more efficient scrubbers for different absorption characteristics. Depending on the application, different absorbents can be used, including calcium-based [7], magnesium-based [8], seawater [9], and other absorbents [10]. Systems capable of efficiently removing pollutants simultaneously can allow for simplified and more cost-effective pollutant control methods.

In this chapter, the principles of pollution control via absorption are introduced in detail. To accomplish this, the fundamentals and key factors of SO_2 absorption are discussed in the first section, including process chemistry, mass transfer, and reaction kinetics. Based on these key factors, SO_2 absorption enhancement methods

X. Gao et al., *Multi-Pollutant Control for Flue Gases*, Advanced Topics in Science and Technology in China 63, https://doi.org/10.1007/978-981-16-1518-4_4

are discussed in the second section, including pH adjustment, desulfurization additives, and scrubber optimization. In the third section, owing to the urgent need for multi-pollutant control, technologies for simultaneous reduction of other pollutants, such as NO_x and Hg, in the SO_2 absorption process are introduced.

4.2 Fundamentals of Pollutant Control by Absorption

4.2.1 Process Chemistry

Several types of materials can be used as an absorbent for SO_2 removal from flue gas, such as limestone ($CaCO_3$), calcium hydroxide ($Ca(OH)_2$), magnesium hydroxide ($Mg(OH)_2$), and caustic soda (NaOH). Among them, limestone is commonly used owing to its low cost and high SO_2 removal efficiency [11]. SO_2 absorption is a complex process that involves three main steps, including: (1) SO_2 absorption, (2) dissolution, and (3) oxidation and crystallization. During the first step, taking limestone-based WFGD as an example, the SO_2 is absorbed in water and exists in ion forms, e.g., HSO_3^- and SO_3^{2-}. Next, the $CaCO_3$ is dissolved into Ca^{2+}, CO_3^{2-}, HCO_3^- and $CO_2(aq)$. During oxidation and crystallization, HSO_3^- and SO_3^{2-} are oxidized by oxygen in the air to form SO_4^{2-}; Ca^{2+}, SO_4^2, and H_2O react to form $CaSO_4$, which can be recycled for further use. The reactions involved in these three main steps during limestone WFGD are detailed below.

(1) SO_2 absorption

$$SO_2(g) \leftrightarrow SO_2(aq) \tag{4.1}$$

$$SO_2(aq) + H_2O \leftrightarrow H^+ + HSO_3^- \tag{4.2}$$

$$HSO_3^- \leftrightarrow H^+ + SO_3^{2-} \tag{4.3}$$

(2) Limestone dissolution

$$CaCO_3(s) \leftrightarrow Ca^{2+} + CO_3^{2-} \tag{4.4}$$

$$CO_3^{2-} + H^+ \leftrightarrow HCO_3^- \tag{4.5}$$

$$HCO_3^- + H^+ \leftrightarrow CO_2(aq) + H_2O \tag{4.6}$$

$$CO_2(aq) \leftrightarrow CO_2(g) \tag{4.7}$$

(3) Oxidation and crystallization

$$2HSO_3^- + O_2 \leftrightarrow 2H^+ + 2SO_4^{2-} \tag{4.8}$$

$$2SO_3^{2-} + O_2 \leftrightarrow 2SO_4^{2-} \tag{4.9}$$

$$2Ca^{2+} + 2SO_3^{2-} + H_2O \leftrightarrow 2CaSO_3 \cdot 1/2H_2O(s) \tag{4.10}$$

$$Ca^{2+} + SO_4^{2-} + 2H_2O \leftrightarrow CaSO_4 \cdot 2H_2O(s) \tag{4.11}$$

4.2.1.1 SO_2 Absorption

Absorption involves the mass transfer of a soluble gas-phase component to a solvent liquid in a device that promotes intimate contact between the gas and liquid phases. The removal of one or more selected components from a mixture of gases via absorption into a liquid is based on interphase mass transfer and is majorly controlled by diffusion rates. Absorption processes can be divided into two groups: those involving solely physical processes and those involving chemical reactions. For example, Methanol (HCHO) steam can be absorbed in water in a physical process; no chemical reaction occurs in this case. Conversely, the absorption of SO_2 in a slurry of limestone involves chemical reactions, the nature of which influence the rate of absorption.

When flue gas containing SO_2 is brought into contact with water, SO_2 molecules that come into contact with the water's surface dissolve rapidly. However, SO_2 molecules are dispersed throughout the gas phase; to reach the water's surface, they must diffuse through the gas phase. Furthermore, SO_2 that has dissolved at the water's surface is distributed throughout the water phase through diffusion. Consequently, the rate of absorption is determined by the rates of diffusion in both the gas and liquid phases.

a. *SO_2 Absorption Mechanism*

Gas–liquid phase equilibrium: In addition to absorption, desorption, which is the mass transfer of a solute from the liquid phase to the gas phase, occurs. At a certain temperature and pressure, the absorption and desorption rate eventually become equal, i.e., the mass transfer between the gas and liquid phase achieves dynamic equilibrium, as shown in Fig. 4.1, wherein the net mass transfer flux is zero.

Two concepts can be derived: gas solubility and equilibrium partial pressure; they represent the concentration of solute in the liquid phase at equilibrium and the partial

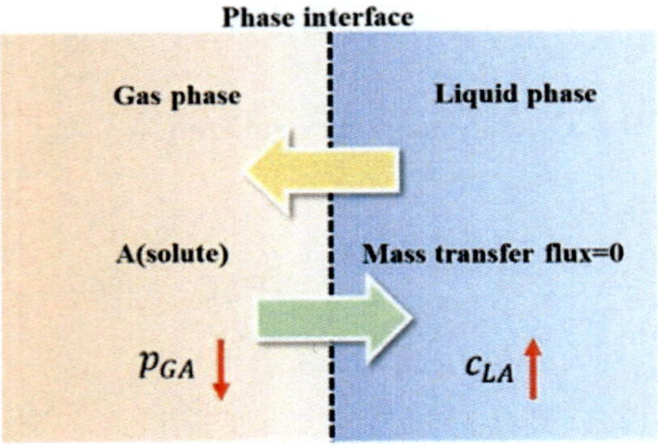

Fig. 4.1 Diagram of dynamic equilibrium, where the mass transfer of solute A between the gas and liquid phases is zero

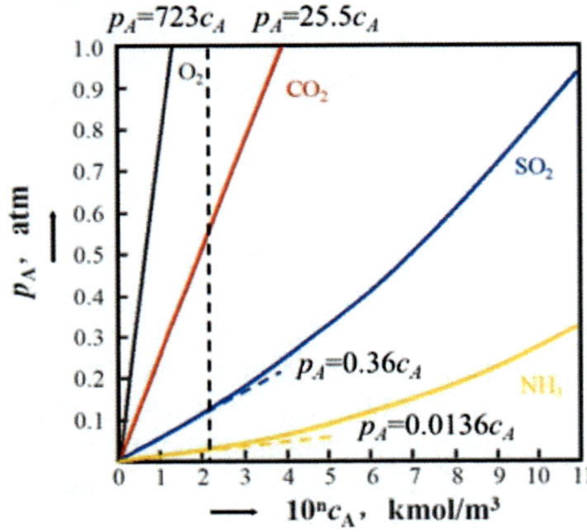

Fig. 4.2 Solubility of some common gases in 20 °C water (p_A, partial pressure of solute in the gas phase; c_A, concentration of solute in the liquid phase)

pressure of solute in the gas phase at equilibrium, respectively. The solubility of a given gas is affected by the temperature, pressure, and solute concentration in the gas phase; the solubility of several gases in water is shown in Fig. 4.2.

When the liquid phase is an ideal solution, the gas phase is an ideal gas and the temperature of the solution is higher than the critical temperature of the gas. The gas–liquid equilibrium relationship can be described by Henry's Law:

$$p_A^* = Ex_A, \tag{4.12}$$

where p_A^* represents the equilibrium partial pressure of the solute in the gas phase, kPa, x_A represents the mole fraction of the solute in the liquid phase, and E represents Henry's law coefficient, kPa.

At pressures below 5 atm, the solubility curve of a dilute solution at a given temperature can be approximated as a straight line; i.e., the solubility of a solute in the liquid phase is proportional to its partial pressure in the gas phase. In general, the higher the temperature, the higher the Henry's law coefficient. Moreover, the higher the Henry's law coefficient, the lower the gas solubility. Henry's law coefficients of common gases in water are shown in Table 4.1.

Table 4.1 Henry's law coefficients of common gases in water [12]

Gas	Temperature (°C)															
	0	5	10	15	20	25	30	35	40	45	50	60	70	80	90	100
	$E \times 10^{-6}$ (kPa)															
H_2	5.87	6.16	6.44	6.70	6.92	7.16	7.39	7.52	7.61	7.70	7.75	7.75	7.71	7.65	7.61	7.55
N_2	5.35	6.05	6.77	7.48	8.15	8.76	9.36	9.98	10.5	11.0	11.4	12.2	12.7	12.8	12.8	12.8
Air	4.38	4.94	5.56	6.15	6.73	7.30	7.81	8.34	8.82	9.23	9.59	10.2	10.6	10.8	10.9	10.8
CO	3.57	4.01	4.48	4.95	5.43	5.88	6.28	6.68	7.05	7.39	7.71	8.32	8.57	8.57	8.57	8.57
O_2	2.58	2.95	3.31	3.69	4.06	4.44	4.81	5.14	5.42	5.70	5.96	6.37	6.72	6.96	7.08	7.10
CH_4	2.27	2.62	3.01	3.41	3.81	4.18	4.55	4.92	5.27	5.58	5.85	6.34	6.75	6.91	7.01	7.10
NO	1.71	1.96	2.21	2.45	2.67	2.91	3.14	3.35	3.57	3.77	3.95	4.24	4.44	4.45	4.58	4.60
C_2H_6	1.28	1.57	1.92	2.90	2.66	3.06	3.47	3.88	4.29	4.69	5.07	5.72	6.31	6.70	6.96	7.01
$E \times 10^{-5}$ (kPa)																
C_2H_4	5.59	6.62	7.78	9.07	10.3	11.6	12.9	–	–	–	–	–	–	–	–	–
N_2O	–	1.19	1.43	1.68	2.01	2.28	2.62	3.06	–	–	–	–	–	–	–	–
CO_2	0.738	0.888	1.05	1.24	1.44	1.66	1.88	2.12	2.36	2.6	2.87	3.46	–	–	–	–
C_2H_2	0.73	0.85	0.97	1.09	1.23	1.35	1.48	–	–	–	–	–	–	–	–	–
Cl_2	0.272	0.334	0.399	0.461	0.537	0.604	0.669	0.74	0.8	0.86	0.9	0.97	0.99	0.97	0.96	–
H_2S	0.272	0.319	0.372	0.418	0.489	0.552	0.617	0.686	0.755	0.825	0.689	1.04	1.21	1.37	1.46	1.50
$E \times 10^{-4}$ (kPa)																
SO_2	0.167	0.203	0.245	0.294	0.355	0.413	0.485	0.567	0.661	0.763	0.871	1.11	1.39	1.7	2.01	–
$E \times 10^{-3}$ (kPa)																

(continued)

Table 4.1 (continued)

Gas	Temperature (°C)															
	0	5	10	15	20	25	30	35	40	45	50	60	70	80	90	100
	$E \times 10^{-6}$ (kPa)															
HCl	0.247	0.255	0.263	0.271	0.279	0.287	0.293	–	0.303	–	–	0.299	–	–	–	–
NH_3	0.208	0.224	0.240	0.257	0.277	0.297	0.321	–	–	–	–	–	–	–	–	–

Henry's law can also be expressed by the partial pressure p_A of a species and its aqueous-phase concentration c_A or the molar mixing ratio in the gas phase y_A^* and molar mixing ratio in the aqueous phase x_A:

$$p_A^* = \frac{c_A}{H}, \quad (4.13)$$

where c_A is the molality of the solute in the liquid phase, kmol/m^3, and H is the solubility coefficient, kmol/(m^3 kPa).

$$y_A^* = mx_A, \quad (4.14)$$

where y_A^* is the mole fraction of the solute in the gas phase and m is the phase equilibrium constant.

Mass transfer in the gas or liquid phase: Gas absorption involves the mass transfer of mixed gas components from the gas to liquid phase owing to a species concentration gradient. Mass transfer involves both diffusion and convection, wherein diffusion is the macroscopic result of random molecular motion on a microscopic scale and convection is the transport of material between a boundary interface and a moving fluid or between two relatively immiscible moving fluids.

Diffusion: Diffusion is dependent on molecular motion and occurs in a stationary, laminar fluid. During desulfurization, SO_2 diffuses from the gas phase to the phase interface, and then from the phase interface to the liquid phase. Thus, gas diffusion occurs in both the gas and liquid phases. Diffusion occurs owing to the gradient of the concentration; the rate of diffusion can be measured through the diffusion flux (the amount of material transferred per unit area per unit time). Fick's law describes the diffusion flux for a homogeneous mixture comprising A and B as follows:

$$J_{A,Z} = -D_{AB}\frac{\partial c_A}{\partial z}, \quad (4.15)$$

where $J_{A,Z}$ represents the diffusion flux of A in B (kmol/(m^2 s)), $\partial c_A/\partial z$ represents the concentration gradient of A on the z-axis ((kmol/m^3)/m), and D_{AB} represents the diffusion coefficient of A in B, m^2/s. Similarly, the diffusion flux of B in A can be described using Eq. (4.16). Commonly, the $J_{A,Z}$ and $J_{B,Z}$ are opposite to each other with the same absolute values:

$$J_{B,Z} = -D_{BA}\frac{\partial c_B}{\partial z}. \quad (4.16)$$

Convective mass transfer: Mass transfer by convection involves the transport of material between a boundary surface and a moving fluid or between two relatively immiscible, moving fluids. Convective mass transfer occurs in turbulent fluids, as shown in Fig. 4.3.

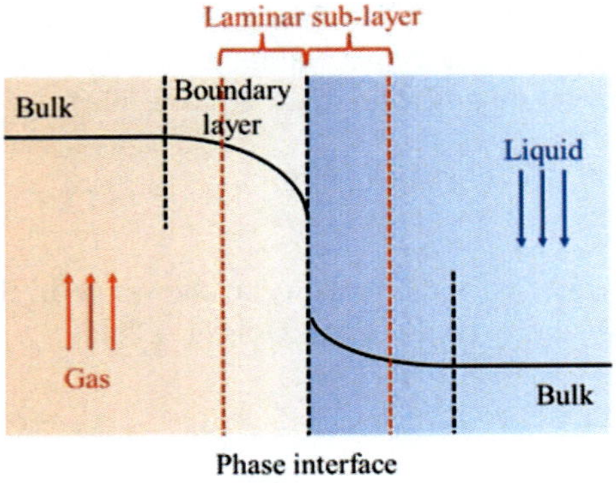

Fig. 4.3 Diagram of convective mass transfer

The main body of turbulent flow is uniform in concentration, so the mass transfer resistance is very small. The mass transfer in the laminar sub-layer is molecular diffusion with great resistance. According to boundary layer theory, in convective mass transfer, the mass transfer flux of A (N_A, kmol/(m^2 s)) can be calculated as follows:

$$N_A = k(c_{A1} - c_{A2}), \tag{4.17}$$

where k is the mass transfer coefficient, m/s, which can be calculated using the Sherwood number (Sh)

$$\mathrm{Sh} = \frac{kd}{D_A} = f(\mathrm{Re}, \mathrm{Sc}). \tag{4.18}$$

The Reynolds number (Re) can be obtained using

$$\mathrm{Re} = \frac{du\rho}{\mu}. \tag{4.19}$$

The Schmidt number (Sc) can be obtained using

$$\mathrm{Sc} = \frac{\mu}{\rho D_A}. \tag{4.20}$$

Mass transfer between the gas and liquid: Over the past few decades, considerable researches have been conducted for the development and analysis of mass transfer models. The models demonstrated to best represent mass transfer within a WFGD system include the two-film, penetration, and surface renewal theories.

Two-film theory: The two-film theory, as proposed by Whiteman [13], is based on the following three points.

(1) Mass transfer resistance occurs exclusively in two thin, stagnant films near the gas–liquid interface. Gas passes from the gas film to the phase interface, then to the liquid film, and finally into the liquid bulk.
(2) At the phase interface, the mass transfer resistance is zero and phase equilibrium is reached.
(3) The solute concentration in the gas and liquid bulk is considered to be uniform; the concentration gradients exist only in the two stagnant films.

Thus, mass transfer resistance comprises gas- and liquid-phase resistance and is used as the basis for scrubber design. Generally, the absorption rate can be obtained using the conditions at the gas–liquid interface and the edge of the film as boundary conditions as

$$N_A = -D_A \frac{dc_A}{dz}|_{z=0} = \frac{D_A}{z_b}(c_{Ai} - c_A) \quad (4.21)$$

where c_{Ai} and c_A represent the SO_2 concentration at the gas–liquid interface and in liquid bulk (kmol/m^3).

Thus, the mass transfer coefficient k_l in the liquid phase is proportional to the first power of the diffusion coefficient as follows:

$$k_l = \frac{D_A}{z_b}. \quad (4.22)$$

The diffusion rate is thus directly proportional to both the surface area and concentration difference and is inversely proportional to the thickness of the film.

According to the convective mass transfer model, the mass transfer flux can be obtained as follows:

$$N_A = k_g(p_A - p_{Ai}) = k_l(c_{Ai} - c_A). \quad (4.23)$$

As Henry's law is satisfied at the gas–liquid interface,

$$N_A = \frac{p_A - \frac{c_A}{H}}{\frac{1}{k_g} + \frac{1}{k_l \cdot H}}. \quad (4.24)$$

The total resistance of the two phases can be expressed in terms of the overall gas-phase mass transfer coefficient, K_g, kmol/(m^2 s kPa), or overall liquid-phase mass transfer coefficient, K_l, m/s as

$$K_g = \frac{1}{\frac{1}{k_g} + \frac{1}{k_l \cdot H}}, \quad (4.25)$$

$$K_l = \frac{1}{\frac{H}{k_g} + \frac{1}{k_l}}, \quad (4.26)$$

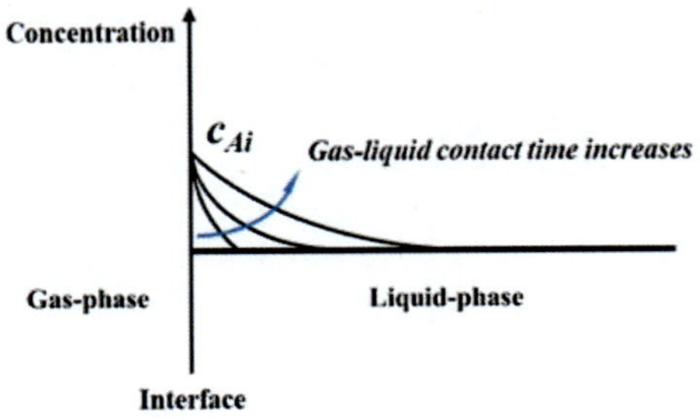

Fig. 4.4 Diagram of the penetration theory

where k_g represents the gas-phase mass transfer coefficient, kmol/(m^2 s kPa), and k_l represents the liquid-phase mass transfer coefficient, m/s.

The two-film theory only focuses on the mass transfer in each phase and ignores the accumulation of diffused components in the film. A stable interface between the gas and liquid phases is assumed; mass transfer resistance is primarily attributed to the molecular diffusion resistance in the gas and liquid phases. The flux through each small area is constant. This model is suitable for representing mass transfer between two fluids with a fixed phase interface and low velocity.

Penetration theory: Penetration theory, proposed by Higbie [14] and based on the two-film theory, states that the gas–liquid contact time is insufficient, as shown in Fig. 4.4. Here, the mass transfer is described as a process in which a solute gradually penetrates from the phase boundary toward the center of the liquid to form a stable concentration gradient.

However, this theory assumes that no chemical reactions exist in the system and represents the one-dimensional unsteady diffusion of solute A in the liquid phase. The concentration of microfluid units is uniform, and the concentration at the phase interface remains constant. The molecular mass transfer differential equation can then be represented as follows:

$$\frac{\partial c_A}{\partial t} = D_A \frac{\partial^2 C_A}{\partial z^2}. \tag{4.27}$$

The average mass transfer flux of component A through the interface is expressed as follows:

$$N_A = 2(c_{Ai} - c_A)\sqrt{\frac{D_A}{\pi\tau}}. \tag{4.28}$$

The liquid phase mass transfer coefficient is expressed as follows:

$$k_l = 2\sqrt{\frac{D_A}{\pi\tau}}, \tag{4.29}$$

where τ represents the gas–liquid contact time (i.e., penetration time), s.

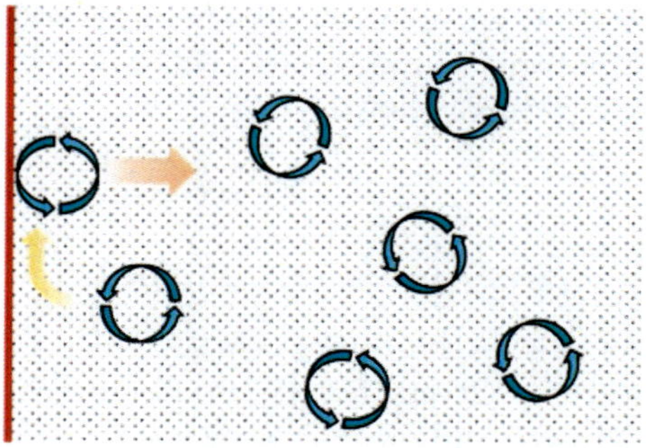

Fig. 4.5 Diagram of the surface renewal theory

Molecules penetrate from the interface to the bulk of the liquid phase. The instantaneous concentration distribution and corresponding interfacial diffusion rate differ at every time instant. Until time τ, the fluid in the liquid film completely mixes with the main flow; therefore, the penetration and two-film theories are the same at an infinite penetration time.

Surface renewal theory: Surface renewal theory comprises a modified penetration theory that considers that the liquid surface comprises liquid units with varying exposure times, as proposed by Danckwerts [15] and shown in Fig. 4.5. Here, the mean time mass transfer coefficient of the liquid phase is expressed as follows:

$$k_l = \sqrt{D_A S}, \tag{4.30}$$

where S represents the surface renewal rate.

The assumptions made in surface renewal theory more closely represent reality. When the surface renewal rate S of all liquid units is the same, i.e., the contact time of each element on the interface is the same, the surface renewal theory can be simplified as the permeation theory.

Chemical absorption: Many industrial applications of absorption employ chemical absorption, wherein chemical reactions are used to increase the absorption rate. Chemical reactions in the liquid phase can improve the mass transfer rate via three aspects: mass transfer force, mass transfer coefficient, and effective mass transfer area. In general, chemical reactions increase the mass transfer coefficient of the liquid phase. To express the degree of enhancement of a chemical reaction as a function of the mass transfer rate [7], an enhancement factor E can be defined as the ratio of the mass transfer flux with a chemical reaction to the mass transfer flux without a chemical reaction under the same driving force as follows:

$$E = \frac{k_l}{k_l^0}, \tag{4.31}$$

where k_l^0 represents the mass transfer coefficient under physical absorption, m/s. The corresponding mass transfer equation considering the chemical reaction can thus be expressed as follows:

$$N_A = k_l(c_{Ai} - c_A) = Ek_l^0(c_{Ai} - c_A). \tag{4.32}$$

b. *Factors Affecting SO_2 Absorption*

As SO_2 absorption is a complex process that involves mass transfer and chemical reactions. Several factors can affect the absorption rate. Taking SO_2 absorption into the slurry of limestone as an example, the SO_2 absorption rate is affected by the concentration of SO_2 in the gas phase, the solution pH, and the relative velocity between the droplet and flue gas. The SO_2 absorption rate can be obtained using the model described in Sect. 4.2.2.1.

SO_2concentration in the gas phase and pH: The SO_2 absorption rate and overall gas-phase mass transfer coefficient under different SO_2 concentrations and pH values are shown in Fig. 4.6a. The rate of absorption increases as the SO_2 concentration increases, whereas the overall gas-phase mass transfer coefficient shows an inverse trend. Thus, at high SO_2 concentrations, it is more difficult to absorb SO_2, making the desulfurization of flue gas having high SO_2 concentration more difficult. As the pH increases, the overall gas-phase mass transfer coefficient and SO_2 absorption rate increase. Furthermore, the content of S(IV) in the droplets decreases as the pH increases, as shown in Fig. 4.6b.

Relative velocity between droplets and flue gas: The relative velocity between the gas and the liquid is also an important factor. Figure 4.7 shows that as the relative velocity between the gas and liquid increases, both the SO_2 absorption rate and overall gas-phase mass transfer coefficient increase. However, for flue gas having

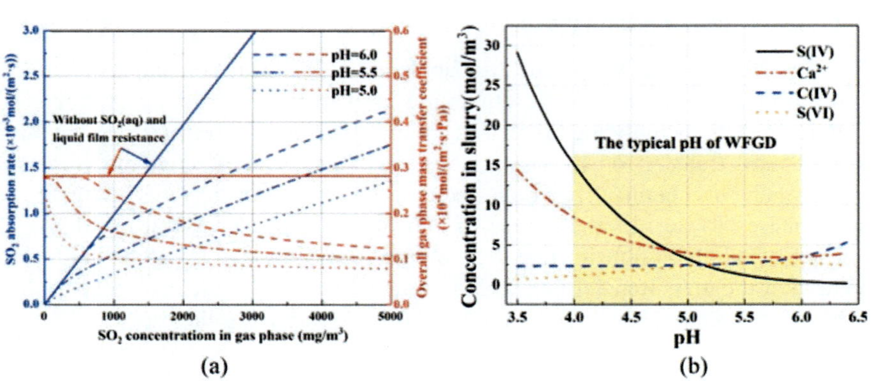

Fig. 4.6 Effect of pH and SO_2 concentration on the **a** SO_2 absorption rate, overall mass transfer coefficient, and **b** concentration of S(IV), Ca^{2+}, C(IV), and S(VI) in the slurry. Temperature = 50 °C; total pressure of flue gas in scrubber = 10^4 kPa; partial pressure of CO_2 = 12 kPa; initial velocity of droplets = 4 m/s [16]

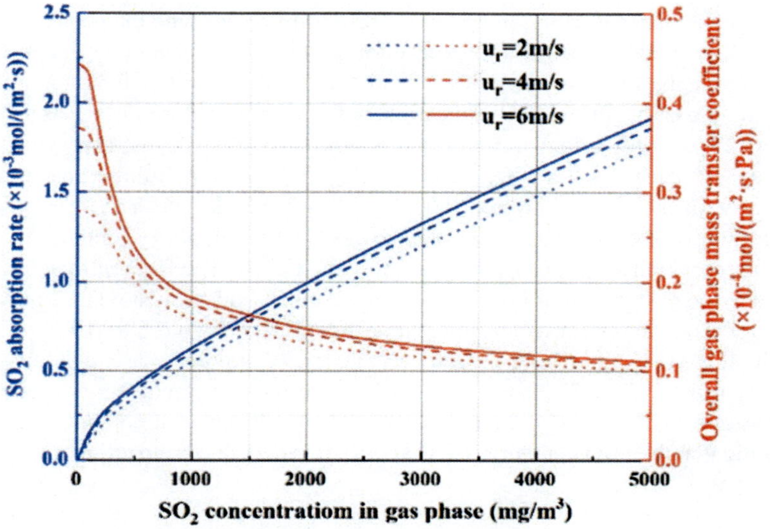

Fig. 4.7 Effect of the relative velocity between the gas and liquid phase on the SO_2 absorption rate and overall mass transfer coefficient. Temperature = 50 °C; total pressure of flue gas in scrubber = 10^4 kPa; partial pressure of CO_2 = 12 kPa; pH = 5.5

high SO_2 concentration, increasing the gas–liquid relative velocity cannot effectively increase the mass transfer coefficient because the liquid-phase mass transfer coefficient decreases.

Parameters such as the temperature and composition of slurry or flue gas can also influence the SO_2 absorption rate; for example, upon an increase in temperature, the velocity of diffusion and reaction increase, which promotes SO_2 absorption, whereas solubility coefficient will decrease, which hinders SO_2 absorption. Therefore, an optimal temperature for the absorption of SO_2 likely exists. Furthermore, the presence of oxidizing ions such as NO_3^{2-}, MnO_4^-, or ClO^- can increase the SO_2 absorption rate owing to the oxidation of SO_3^{2-} to SO_4^{2-}.

4.2.1.2 Dissolution Rate of Limestone

In limestone-based WFGD—an effective and commonly used SO_2 absorption—the dissolution of the limestone absorbent is a rate-limiting step; further development and optimization of this method thus requires understanding the dissolution rate of limestone [17].

Natural limestone is a sedimentary rock. Biological debris comprising calcium carbonate gradually sinks into the seabed, forming a rock layer composed of shells, carapace, and bones, which, over many years, transforms into dense, hard rocks. The chemical composition, mineral composition, and physical properties of the resulting limestone vary considerably; In particular, its density varies with the formation time. The quality of limestone used as a desulfurization agent depends on its structure, composition, impurity content, and uniformity of the distribution of these impurities. Pure calcium carbonate ($CaCO_3$), the primary component of limestone, has a density ranging 2700–2950 kg/m^3, a molecular weight of 100.09, and an average specific heat capacity of 0.59 kJ/(kg °C). Furthermore, it is a white or colorless crystal with extremely low solubility in water (0.0038 g/100 g water at 50 °C).

The quality of limestone is determined by its CaO and impurity contents, the most common of which are $MgCO_3$, SiO_2, Al_2O_3, and Fe_2O_3. The form of magnesium is related to the dissolution of limestone; intercrystalline-substituted magnesium (e.g., $MgCO_3$) promotes the dissolution of limestone, whereas dolomite ($CaMg(CO_3)_2$) inhibits it [18]. The purity of limestone is generally expressed as $CaCO_3$ in limestone/gypsum WFGD. The content of CaO is determined by analyzing the chemical composition of limestone, which is then used to calculate the content of $CaCO_3$, assuming that the CaO content is derived only from the decomposition of $CaCO_3$ in limestone, as follows:

$$M_{CaCO_3} = M_{CaO} \times \frac{100.09}{56.08}, \tag{4.33}$$

where M_{CaCO_3} and M_{CaO} represent the predicted and determined mass percentage of $CaCO_3$ and CaO in limestone, respectively.

The composition, purity, activity, and grindability coefficient of limestone are commonly used as performance indicators of WFGD in coal-fired power plants. Limestone purity requirements are generally above 90%, fineness is above 325 mesh, and the sieve residual rate is less than 5%. However, a high-purity absorbent with a fine particle size increases the cost of absorbent preparation and the operating cost of the system. The components of 42 limestone samples from different regions in China are presented in Table 4.2.

a. *Dissolution Model*

Many chemical reactions occur during dissolution, including the dissolution of calcium carbonate,

$$CaCO_3(s) \overset{K_{CP}}{\longleftrightarrow} Ca^{2+} + {CO_3}^{2-}, \tag{4.34}$$

the first ionization of carbonic acid,

$$CO_2(aq) + H_2O \overset{K_{C1}}{\longleftrightarrow} H^+ + {HCO_3}^-, \tag{4.35}$$

the second ionization of carbonic acid,

$${HCO_3}^- \overset{K_{C2}}{\longleftrightarrow} H^+ + {CO_3}^{2-}, \tag{4.36}$$

Table 4.2 Component contents of limestone samples [19]

No.	Region	SiO_2 (wt%)	Al_2O_3 (wt%)	Fe_2O_3 (wt%)	CaO (wt%)	MgO (wt%)
1	An Shun	3.11	0.52	0.31	49.68	1.79
2	Ban Qiao	2.14	0.62	0.19	54.19	0.70
3	Bei Lun	0.89	0.69	0.40	54.28	0.19
4	Cang Zhou	1.15	0.84	0.12	53.82	0.94
5	Chang Xing	0.49	0.38	0.12	55.01	0.19
6	Chi Zhou	2.47	0.74	0.37	52.29	0.69
7	Chong Qing	3.73	1.41	0.23	51.78	1.35
8	Dai Hai	1.98	0.69	0.46	53.44	0.83
9	Dian Dong	3.33	0.50	0.32	52.16	0.15
10	Fang Chenggang	0.43	0.78	0.38	54.81	0.38
11	Feng Cheng	4.72	1.28	0.37	50.74	0.93
12	Fu Yang	3.64	0.77	0.27	50.70	2.00
13	Han Cheng	5.27	0.71	0.67	52.69	2.36
14	He Chuan	5.60	1.22	0.54	50.31	0.56
15	Huai Yin	0.86	0.08	0.12	55.68	0.12
16	Jia Yuguan	1.21	0.75	0.31	53.79	0.90
17	Ke Mengang	3.53	0.81	0.16	52.47	0.49
18	Mao Ming	0.87	0.62	0.12	54.69	0.31
19	Nan Tong	4.32	0.85	0.19	53.30	0.78
20	Ning Hai	0.56	0.37	0.28	55.23	0.57
21	Qi Shuyan	3.44	0.50	0.20	54.04	0.12
22	Shang Du	0.93	0.17	0.28	55.46	0.40
23	She Yang	2.47	0.92	0.85	51.35	2.28
24	Shi Jingshan	6.01	1.94	0.91	50.15	1.14
25	Ta Shan	1.29	0.38	0.25	52.46	2.05
26	Tai Cang	3.28	1.15	0.62	53.14	1.50
27	Tai Shan	1.85	0.56	0.18	53.37	1.80
28	Wai Gaoqiao	0.99	0.43	0.19	54.19	0.20
29	Wen Zhou	0.74	0.36	0.20	54.41	0.25
30	Wu Shashan	2.96	0.63	0.76	52.69	0.68
31	Xia Guan	1.97	1.14	0.77	54.02	1.21
32	Xiao Shan	3.45	0.44	0.25	52.14	1.18
33	Xin Mi	1.75	0.32	1.38	51.77	1.64

(continued)

Table 4.2 (continued)

No.	Region	SiO_2 (wt%)	Al_2O_3 (wt%)	Fe_2O_3 (wt%)	CaO (wt%)	MgO (wt%)
34	Xin Yang	1.29	0.31	0.25	51.42	3.74
35	Yan Cheng	1.96	0.45	0.15	55.06	0.52
36	Yang Cheng	1.72	0.31	0.25	50.50	3.29
37	Yang Zhou	2.70	0.84	0.16	54.20	0.39
38	Yi Yang	0.99	0.61	0.17	54.44	0.27
39	Yu Huan	3.82	1.08	0.12	53.05	0.92
40	Zhang Jiagang	2.10	0.73	1.08	53.81	1.26
41	Zhu Zhou	2.13	0.18	0.20	54.39	0.74
42	Zun Yi	3.36	0.48	0.16	54.51	0.12

hydrolysis,

$$H_2O \overset{K_W}{\longleftrightarrow} H^+ + OH^-, \tag{4.37}$$

the dissolution of gypsum when gypsum is present in the slurry,

$$CaSO_4(s) \overset{K_{SP}}{\longleftrightarrow} Ca^{2+} + SO_4^{2-}, \tag{4.38}$$

and the formation of HSO_4^-,

$$H^+ + SO_4{}^{2-} \overset{K_{S1}}{\longleftrightarrow} HSO_4{}^-, \tag{4.39}$$

where K_{CP} represents the solubility product of $CaCO_3$, $(mol/m^3)^2$; K_{C1} and K_{C2} represent the equilibrium constants of the first and second ionization of carbonic acid, mol/m^3; K_W represents the ionic product of water, $(mol/m^3)^2$; K_{SP} represents the solubility product of gypsum, $(mol/m^3)^2$; and K_{S1} represents the equilibrium constant of $SO_4{}^{2-}$ hydrolysis, mol/m^3.

The limestone dissolution model, summarized in Fig. 4.8, is based on the application of the two-film theory to the mass transfer system between the fluid and solid.

Some contents in the WFGD system, such as $SO_3{}^{2-}$, may affect the dissolution rate of limestone. Limestone dissolution in the presence of sulfite is controlled by a combined surface kinetics/mass transfer mechanism. The dissolution flux of limestone per unit area can be expressed as follows:

$$Flux = k_c \frac{\left(\left[CaCO^0_{3eq}\right]_s - \left[CaCO^0_3\right]_s\right)^{0.5}}{\left[CaSO^0_3\right]_s \times \left[CaCO^0_3\right]_s}, \tag{4.40}$$

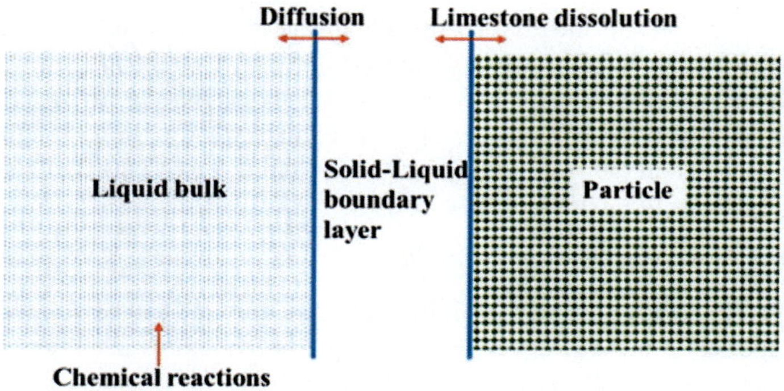

Fig. 4.8 Schematic of limestone particle dissolution [20]

where *Flux* represents the dissolution flux of limestone per unit area, mol/(m^2 s); k_c represents the surface reaction rate constant, $mol^{\frac{5}{2}}/mol^{\frac{13}{2}}$ s; $\left[CaCO^0_{3eq}\right]_s$ represents the equilibrium concentration of $CaSO^0_3$ on the surface of limestone particles in the absence of $SO_3{}^{2-}$, mol/m^3; $[CaSO^0_3]_s$ represents the equilibrium concentration of $CaSO^0_3$ on the surface of limestone particles in the presence of $SO_3{}^{2-}$, mol/m^3; and $[CaSO^0_3]_s$ represents the equilibrium concentration of $CaSO^0_3$ on the surface of limestone particles in the presence of $SO_3{}^{2-}$, mol/m^3.

The dissolution of limestone is assumed to only occur during the main liquid phase in the absorption zone. At this point, the liquid film on the surface of the limestone particles is stagnant. Therefore, for limestone particles with a particle size of d_j,

$$Sh = \frac{k_{Lj} \cdot d_j}{D_{Ca^{2+}}} = 2, \tag{4.41}$$

where k_{Lj} represents the liquid mass transfer coefficient of dissolved limestone particles for a particle size range of *j*.

The total dissolution rate of limestone particles of all size distributions can be expressed as follows:

$$r_{CaCO_3} = \sum_{j=1}^{n} r_{CaCO_3,j} A_j. \tag{4.42}$$

The dissolution flux of limestone particles in the particle size range of *j* of the absorption zone can be expressed as follows:

$$r_{CaCO_3,j} = Flux_j \cdot A_j, \tag{4.43}$$

$$A_j = \frac{6m}{\rho_L} \sum \frac{\Phi_j}{d_j}, \tag{4.44}$$

where $r_{CaCO_3,j}$ represents the dissolution rate of limestone, mol/(m^3 s); A_j represents the total surface area of limestone particles in the particle size range of j per unit slurry volume, m^2/m^3; m represents the mass of limestone, kg/m^3; ρ_L represents the true density of the limestone, kg/m^3; Φ_j represents the volume fraction of the limestone particles in the particle size range of j; and d_j represents the mean size of limestone particles in the particle size range of j, m.

b. *Factors Affecting Limestone Dissolution*

The factors affecting limestone dissolution are numerous and can be divided into two major categories: limestone characteristics (e.g., surface pores, chemical composition, microstructure, and particle size distribution) and the physical and chemical environment of limestone particles (e.g., slurry pH, temperature, stirring rate, organic acid additives, SO_3^{2-}, metal cations, Cl^-, F^-, and SO_4^{2-}). Several typical factors are elaborated below.

Microstructure: Calcium carbonate has an ABO_3 polytype crystal structure and can comprise three homogeneous allotypes, including calcite, aragonite, and vaterite, which have slightly different stacking modes [21]. The typical crystalline structures are cubic, needle type, and spherical, among which calcite is the most stable crystalline phase. Under the same conditions, the dissolution rate of calcite, aragonite, and globular aragonite increases successively.

Particle size distribution: The particle size distribution of the limestone determines the surface area of the reaction between limestone and slurry, which has an important effect on the limestone's dissolution rate. The dissolution rate of limestone can be studied using a pH-stat apparatus, as shown in Fig. 4.9. Under the same conditions, the dissolution rate decreases with increasing particle size, as shown in Fig. 4.10, wherein F represents the dissolution fraction of limestone during the reaction.

pH: The effect of pH on the limestone dissolution rate in the presence/absence of sulfite is shown in Fig. 4.11, wherein Flux (mol/m^2 s) represents the dissolution flux

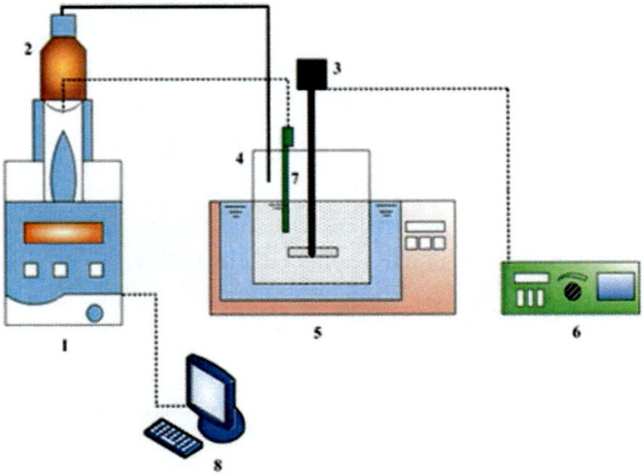

Fig. 4.9 Schematic of experimental apparatus: (1) autotitrator, (2) HCl bottle, (3) stirrer, (4) reactor, (5) water bath, (6) tachometer, (7) pH electrode, and (8) data collector [22]

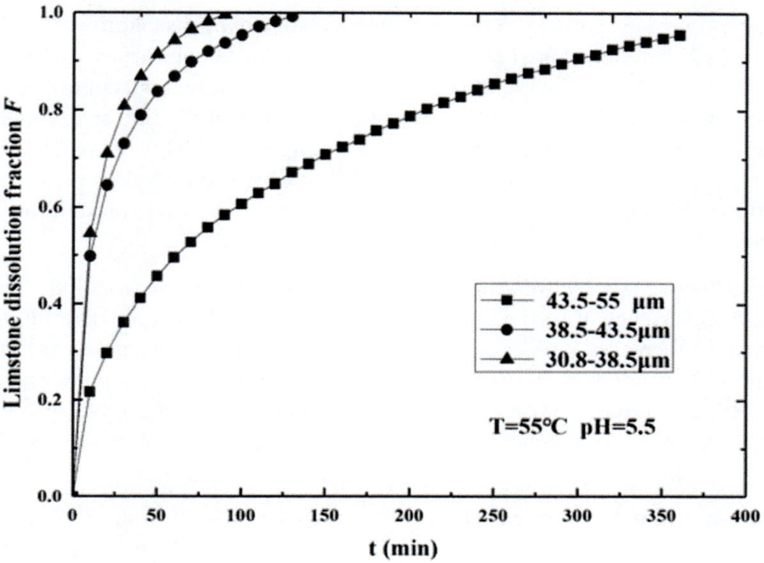

Fig. 4.10 Influence of particle size on dissolution rate [19]

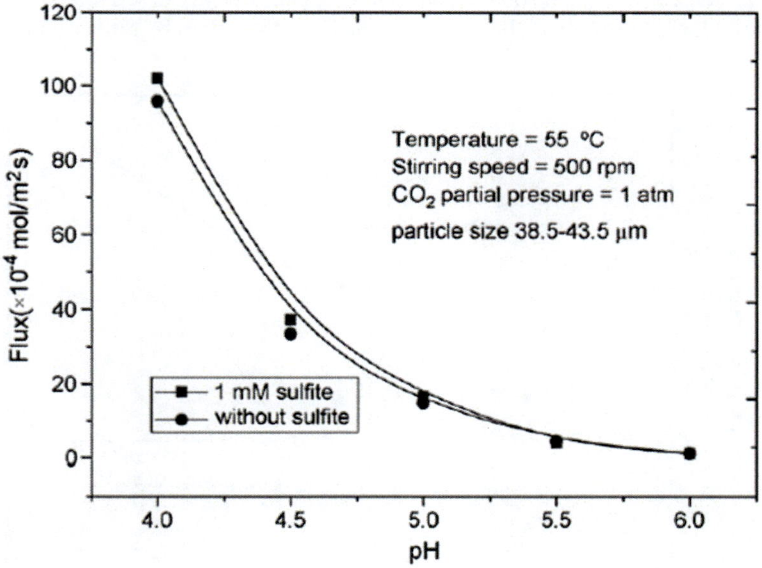

Fig. 4.11 Effect of pH on the limestone dissolution rate in the presence/absence of sulfite [17]

of a single limestone particle. A lower pH is favorable to the dissolution of limestone because the H^+ gradient between the liquid bulk and limestone surface is the driving force, although limestone dissolution in the presence of sulfite is not controlled solely by mass transfer. At pH < 5.5, sulfite promotes limestone dissolution by providing H^+ consumed during dissolution via the buffer effect of HSO_3^-/SO_3^{2-} at the limestone's surface. At pH > 5.5, the promotion and inhibition effects of sulfite on limestone dissolution offset each other. Sulfite has little effect on limestone dissolution in the pH range of 5.5–6.0 under these conditions.

Temperature: The effect of reaction temperature on limestone dissolution rate in the presence/absence of sulfite is shown in Fig. 4.12. The limestone dissolution rate significantly increases with increasing temperature. Surface reactions are thus in dissolution, because in general, diffusion-controlled processes are slightly dependent on temperature and chemical reaction-controlled processes are strongly dependent on temperature. That is to say, surface reactions play an important role in the dissolution processes in this experiment.

Fluorine-aluminum complex: During WFGD, Al^{3+} from fly ash or limestone is enriched in the scrubber and reacts with F^- in the flue gas, which negatively affects the dissolution rate [23], as shown in Fig. 4.13. In the presence of both Al^{3+} and F^-, the dissolution rate of limestone decreases significantly because F^- has a strong coordination ability and forms a coordinated complex with Al^{3+}. This coordinated complex adheres strongly to the surface of the limestone, thereby hindering the mass

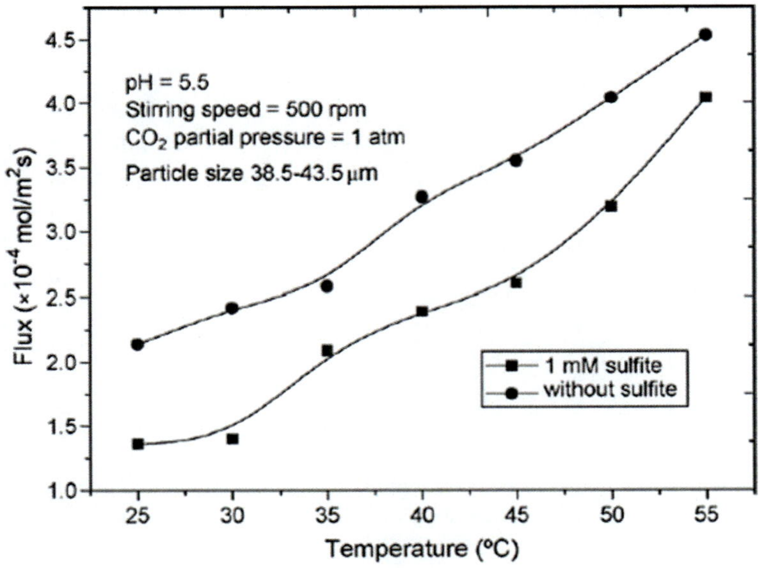

Fig. 4.12 Effect of temperature on the limestone dissolution rate in the presence/absence of sulfite [17]

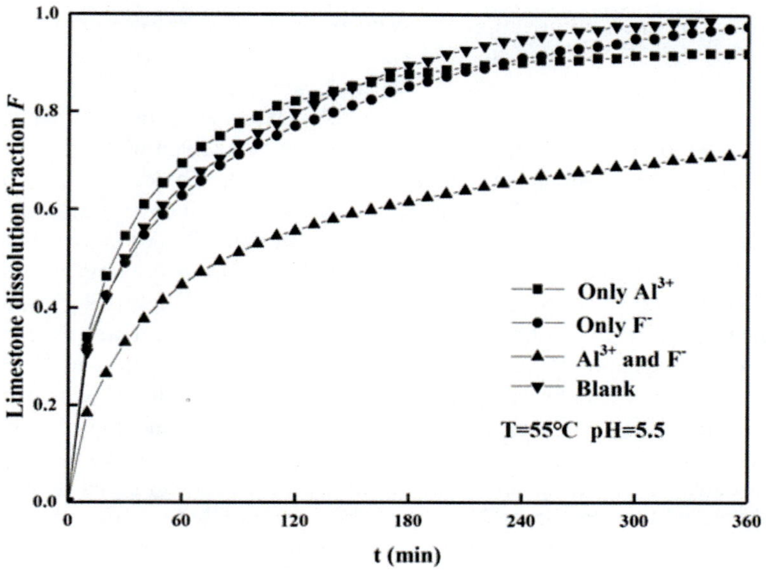

Fig. 4.13 Influence of Al^{3+} and F^{-} on the limestone dissolution rate [19]

Fig. 4.14 SEM images of the **a** active limestone surface and **b** limestone surface covered with aluminum–fluoride complex [19]

transfer of H^+ from the liquid bulk to it. Scanning electron microscopy images of the active limestone surface and the limestone surface covered by the fluorine–aluminum complex are shown in Fig. 4.14.

4.2.1.3 Oxidation and Crystallization

a. *Oxidation*

The oxidation of desulfurization by-products during desulfurization affects the stability of WFGD and the quality of these by-products. Early installations of limestone-based FGD processes experienced operating problems caused by the homogeneous nucleation and precipitation of solid reaction products on equipment surfaces [24]. In modern systems, this problem can be avoided with proper design and operation of the system, particularly the extent of sulfite oxidation. If the oxidation fraction is less than 0.15, the reaction products precipitate as a solid solution of calcium sulfite and calcium sulfate hemihydrate, which under most conditions, do not form scale inside the equipment. If oxidation is nearly complete (i.e., an oxidation fraction of ≥ 0.9), the reaction product primarily comprises gypsum that precipitates preferentially on seed crystals maintained in the circulating slurry rather than as scale. Gypsum scale most often forms at sulfite oxidation fractions of approximately 0.15–0.3, where the calcium sulfite crystal lattice is saturated with respect to sulfate, leading sulfate to be precipitated as gypsum. However, incomplete oxidation of sulfite to sulfate is in this range leads to insufficient gypsum seed surface area to prevent homogeneous nucleation. As a result, gypsum scale is formed. In practice, most lime- and limestone-based FGD processes are operated in a forced-oxidation mode; other processes (e.g., ammonia or magnesium methods) may also be operated to inhibit oxidation, as discussed in the following subsections.

Natural oxidation: Under natural oxidation conditions, the fraction of sulfite oxidized to sulfate is not deliberately controlled but is instead determined by the inherent rates of the oxidation reactions in the system and the ratio of SO_2 to O_2 that is absorbed from the flue gas. Some FGD processes can be operated without additives while maintaining low oxidation if the coal's sulfur content is high and the excess air in the flue gas is low. Under these conditions, the oxidation fraction is low because the ratio of absorbed SO_2 to O_2 is high. Conversely, near-complete oxidation in an FGD process can be obtained without forced air if the rate at which the sulfite is oxidized is nearly as high as the rate at which SO_2 is absorbed. Normally, this is possible only if the coal's sulfur content is low and the slurry pH is controlled in the range of 5.0–5.5.

Forced oxidation: Forced oxidation is obtained by pumping air into the slurry in the reaction tank, thereby converting the dissolved sulfite and bisulfite to sulfate and causing the gypsum to precipitate. The extent of oxidation that can be obtained depends on the reaction tank depth and quantity of air. A typical design for complete ($> 99\%$) oxidation uses approximately two moles of O_2 per mole of absorbed SO_2. Nearly complete oxidation is required if commercial-quality gypsum is produced as a by-product; complete oxidation is not required if the by-product is landfilled.

Oxidation rate model: The oxidation rate is affected by variety of factors, including the sulfite concentration, oxygen concentration, airflow rate, slurry pH, slurry temperature, presence of trace amounts of catalysts or inhibitors, stirring speed, oxidation mode (i.e., spray gun or pipe network) and operating parameters.

The overall oxidation rate includes the intrinsic oxidation rate, oxygen mass transfer rate, and calcium sulfite dissolution (which occurs in the presence of calcium

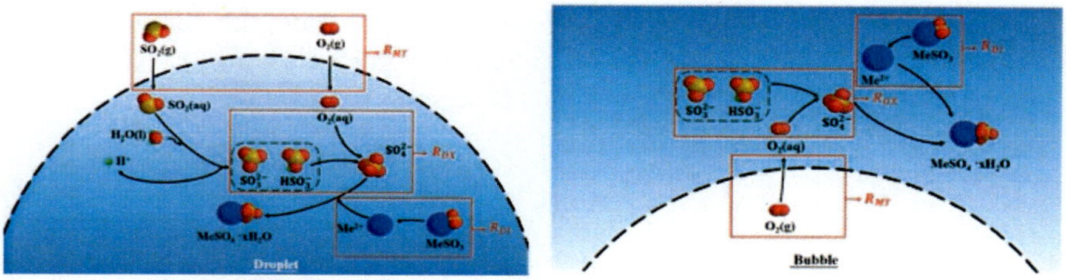

(a) (b)

Fig. 4.15 Schematic of **a** natural oxidation and **b** forced oxidation processes [25]

sulfite precipitation) rate. A process diagram of oxidation is shown in Fig. 4.15. Here, O_2 transfers from flue gas to a droplet (Fig. 4.15a) in the absorption zone, whereas oxygen transfers from a bubble to the liquid phase (Fig. 4.15b) in the slurry tank. The actual reaction rate is determined by the lowest of the three rates

$$R = \min(R_{OX}, R_{MT}, R_{DI}), \tag{4.45}$$

where R, R_{OX}, R_{MT}, and R_{DI} represent the actual reaction rate, chemical reaction oxidation rate, oxygen transfer rate, and calcium sulfite dissolution rate, respectively (mol m^{-3} s^{-1}).

Chemical reaction kinetics: The oxidation reaction process can be divided into noncatalytic and catalytic reaction processes according to the presence or absence of a catalyst. However, the reaction orders reported in the literature are inconsistent. The intrinsic reaction rate R_{OX} of the noncatalytic reaction process can be expressed as Eq. (4.46), wherein m and n represent the reaction orders of S(IV) and dissolved oxygen, respectively.

$$R_{OX} = k_{ox}^{nc} C_{S(IV)}^{m} C_{O_2}^{n}, \tag{4.46}$$

where k_{ox}^{nc} represents the reaction rate constant ($\text{mol}^{-\frac{1}{2}}\ \text{mol}^{-\frac{3}{2}}\ \text{s}^{-1}$) and $C_{S(IV)}$ and C_{O_2} represent the sulfite and O_2 concentration, respectively (mol/m^3).

The intrinsic chemical reaction rate of the catalytic reaction process can be expressed as [26]

$$R_{OX} = k_{ox}^{c} C_{S(IV)}^{m} C_{O_2}^{n} C_{Me^{2+}}^{p}, \tag{4.47}$$

where $C_{Me^{2+}}$ represents the catalytic metal ion concentration, mol/m^3, and k_{ox}^{c} represents the rate constant of the intrinsic noncatalytic or catalytic chemical reaction, which is closely related to the temperature and properties (e.g., pH, composition) of the S(IV)-loaded solution.

Calcium sulfite dissolution rate: In the case of nonagitation, the diffusion rate of sulfite ions in the liquid phase can be described through Fick's law; the diffusion rate of a single calcium sulfite particle can thus be described as follows:

$$R'_{SO_3{}^{2-}} = -D_{SO_3{}^{2-}} - 4\pi r^2 \left(\frac{\partial C_{SO_3{}^{2-}}}{\partial r} \right), \tag{4.48}$$

where $D_{SO_3{}^{2-}}$ represents the diffusion coefficient of sulfite, m^2/s.

At unsteady conditions, molecular and eddy diffusion occur simultaneously during mass transfer. The mass transfer rate of a single calcium sulfite particle can thus be written as

$$R'_{SO_3{}^{2-}} = -(D_{SO_3{}^{2-}} + D_e) 4\pi r^2 \left(\frac{\partial C_{SO_3{}^{2-}}}{\partial r} \right), \tag{4.49}$$

where D_e represents the eddy diffusion coefficient (related to turbulence intensity), m^2/s.

According to the conservation of mass, the sum of masses of dissolved identical spherical calcium sulfite pellets is equal to the increase in the solute concentration in the slurry. The dissolution rate of calcium sulfite can be expressed as

$$V \frac{\partial C_{SO_3{}^{2-}}}{\partial t} = -4\pi r_j^2 (D_{SO_3{}^{2-}} + D_e) N \frac{\partial C_{SO_3{}^{2-}}}{\partial t}, \tag{4.50}$$

where V, m^3, represents the volume of the slurry tank, N represents the number of calcium sulfite pellets, $C^S_{SO_3{}^{2-}}$, mol/m^3 and $C^b_{SO_3{}^{2-}}$, mol/m^3, represent the saturated concentration of calcium sulfite and concentration of calcium sulfite in solution, respectively, and the boundary conditions are $r = r_j$, $C_{SO_3{}^{2-}} = C^S_{SO_3{}^{2-}}$ and $r = +\infty$, $C_{SO_3{}^{2-}} = C^b_{SO_3{}^{2-}}$.

Oxygen mass-transfer rate: Assuming that the transfer of oxygen in the slurry is equal to the diffusion of oxygen molecules in the gas and liquid films, it can be represented by the two-film theory. The stationary phase interface between the gas and liquid contacts with each other, and the gas and liquid are in equilibrium at the phase interface. The concentration gradient in the phase bulk is zero; all composition changes are concentrated in the two effective films. According to diffusion theory, the diffusion flux of O_2, N_{O_2}, $mol/(m^2\ s)$, on the surface of the droplet is expressed as

$$N_{O_2} = K_g \left(p_{O_2} - \frac{c_{O_2}}{H_{O_2}} \right), \tag{4.51}$$

where K_g represents the overall mass transfer coefficient, $mol/(m^2\ s\ Pa)$, and H_{O_2} represents the solubility coefficient of O_2, $mol/(m^3\ Pa)$.

The total mass transfer coefficient K_g and the partial pressure of O_2, p_{O_2}, can be obtained, respectively, as

$$\frac{1}{K_g} = \frac{1}{k_g} + \frac{1}{H_{O_2} \cdot k_l \cdot E_{O_2}}, \tag{4.52}$$

$$p_{O_2} \cdot Q = n_{O_2} \cdot R \cdot T. \tag{4.53}$$

According to the Frössling equation [27], the gas-phase mass transfer coefficient of O_2, k_g, can be obtained using Eqs. (4.54) to (4.57).

$$k_g = \frac{\mathrm{Sh} \cdot D_{O_2}}{d}, \tag{4.54}$$

$$\mathrm{Sh} = 2 + 0.55\mathrm{Re}_g^{0.5}\mathrm{Sc}^{\frac{1}{3}}, \tag{4.55}$$

$$\mathrm{Sc} = \frac{\mu_g}{\rho_g D_{O_2}}, \tag{4.56}$$

$$\mathrm{Re}_g = \frac{d|\mu_g + \mu_l|\rho_g}{\mu_g}. \tag{4.57}$$

Here, the gas diffusion coefficient, D_{O_2} (m^2/s), represents the diffusion capacity of the gas [28] and can be obtained as [29]

$$D_{O_2} = \frac{9.86 \times 10^{-9} \cdot T^{1.75} \cdot \left|\frac{1}{M_{\mathrm{air}}} + \frac{1}{M_{O_2}}\right|^{\frac{1}{2}}}{0.000001 \cdot p \cdot \left(v_{\mathrm{air}}^{\frac{1}{3}} + v_{O_2}^{\frac{1}{3}}\right)^2}, \tag{4.58}$$

where the total pressure p (Pa) in the scrubber is inversely proportional to the gas diffusion coefficient of O_2 and v_{air} and v_{O_2} are the diffusion volume of air and O_2, respectively.

According to the penetration theory, the mass transfer coefficient of the liquid phase can be determined as

$$k_l = \sqrt{\frac{4D_l}{\pi t_p}}, \tag{4.59}$$

where D_l represents the diffusion coefficient of O_2 in the slurry, m^2/s, and t_p represents the penetration time, s, which is determined through the fluid dynamics characteristics inside the droplet.

This oxidation rate model can be clarified using an example of a 300 MW unit with an inlet SO_2 concentration of 2000 mg/m^3 and a flue gas flow rate of 1,000,000 m^3/h.

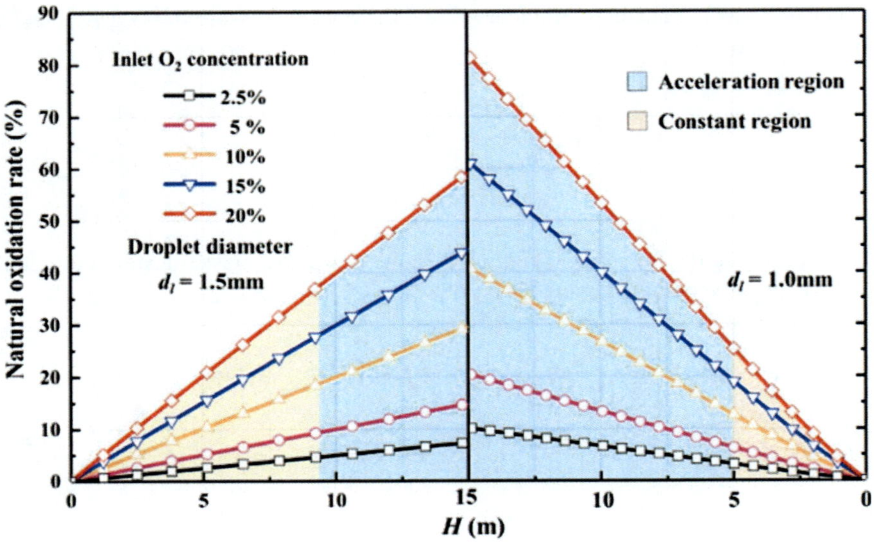

Fig. 4.16 Natural oxidation rate for different inlet O_2 concentrations [25]

As the concentration of O_2 at the inlet increases, a single droplet absorbs more O_2, thereby increasing the natural oxidation rate, as shown in Fig. 4.16. The oxidation process thus is controlled by O_2 diffusion. Once O_2 passes through the liquid film, it quickly combines with HSO_3^-. An increase in the inlet O_2 concentration from 2.5 to 20% leads increases the natural oxidation rate from 7.40 to 59.18%, with a droplet diameter of 1.5 mm. Smaller droplets have a higher oxidation rate. Moreover, increasing the absorption height in certain conditions improves the natural oxidation because a linear relationship exists between the natural oxidation rate and absorption height. Under forced oxidation conditions in the slurry tank, as shown in Fig. 4.17, increasing the airflow rate from 1200 to 2100 m^3/h increases the oxidation rate from 24.81 to 43.42%, with a droplet diameter of 1.5 mm. As previously observed, smaller bubbles have a higher oxidation rate.

b. *Crystallization*

When the concentration of ions (Ca^{2+}, CO_3^{2-}, and SO_4^{2-}) in the slurry is higher than the equilibrium concentration, the crystal nuclei grow and can become visible. In this process, ion pairs are first formed owing to the charge interaction between anions and cations in solution. At a certain degree of saturation, ion pairs form larger molecules. A solid lattice is formed as the ion-pair concentration continues to increase and then aggregates to form larger particles, as shown in Fig. 4.18a. This aggregation process is in dynamic equilibrium with the solution. When the aggregates reach a critical size (determined by the supersaturation of the solution), solid particles nucleate and these larger nucleated particles are less likely to be redissolved. This nucleation process mainly occurs via homocovalent nucleation in the liquid solution, as shown

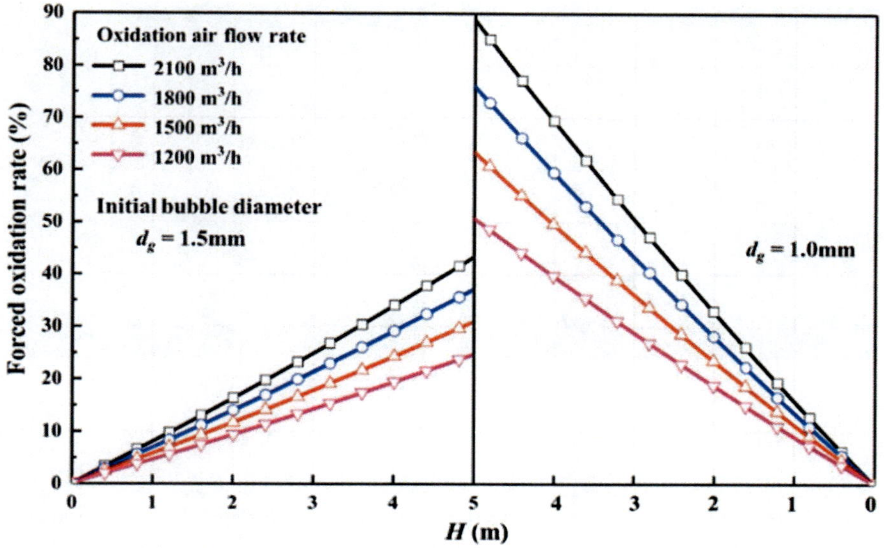

Fig. 4.17 Forced oxidation rate for different oxidation airflow rates [25]

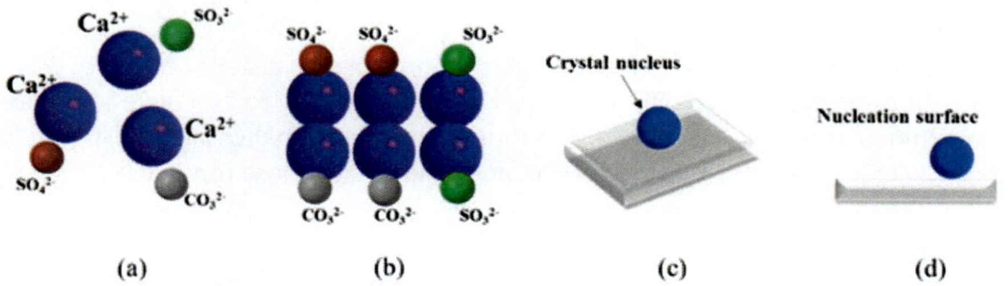

Fig. 4.18 Schematic of gypsum crystallization process: **a** solid lattice **b** nucleation **c** heterogeneous nucleation **d** gypsum crystal

in Fig. 4.18b, but can also occur via heterogeneous nucleation on some surfaces, as shown in Fig. 4.18c. When the relative saturation reaches a certain value, a crystal nucleus is formed. At a certain level of supersaturation, gypsum crystals grow on both the surface of existing gypsum crystals in the suspension liquid and the surface of various other objects, as shown in Fig. 4.18d.

Ca^{2+} and $SO_4{}^{2-}$ crystallize to calcium sulfate dihydrate at the slowest rate in the desulfurization process. In the oxidation tank, their crystallization rate is affected by $CaSO_4 \cdot 2H_2O$ supersaturation, slurry temperature, acidity, and mixing mode. The shape and size of the gypsum crystal can directly affect the operation of other systems. Thick, dense columnar (short rod-shaped) gypsum, a reasonable particle size distribution, and high compressive strength are preferred. The larger the particle size of the gypsum crystal is, the better it will separate $CaSO_4 \cdot 2H_2O$ from the

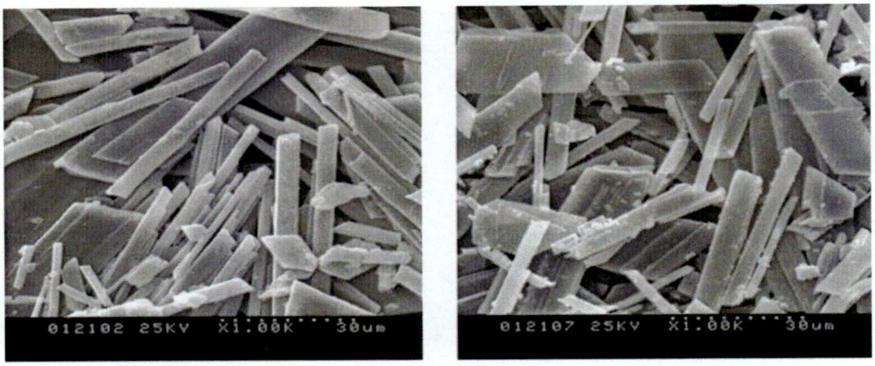

(a) (b)

Fig. 4.19 SEM imaging of $CaSO_4 \cdot 2H_2O$ crystals formed at **a** 40 °C and at **b** 55 °C

unreacted $CaCO_3$, thereby improving the quality of gypsum and the utilization rate of the desulfurizer. Moreover, crystal growth is closely related to the supersaturation of the solution. Ideal crystallization can be obtained in an isothermal crystallization process with constant oversaturation and low supersaturation. Further, the presence of Mg^{2+} and Fe^{3+} also reduces the growth rate of the gypsum crystal and leads to the formation of needle-shaped crystals [30].

SEM images of $CaSO_4 \cdot 2H_2O$ at different temperatures are shown in Fig. 4.19. Here, most of the $CaSO_4 \cdot 2H_2O$ crystals are columnar; at 55 °C, there are more short columnar crystals and their distribution is more uniform than at 40 °C. Crystals formed at 55 °C thus have better crystal morphology than those formed at 40 °C.

4.2.2 *Mass Transfer and Kinetic Reaction Modeling*

SO_2 can be removed from flue gas via absorption using an alkaline slurry or solution including limestone ($CaCO_3$), calcium hydroxide ($Ca(OH)_2$), magnesium hydroxide ($Mg(OH)_2$), or caustic soda (NaOH). This method of removing SO_2 from flue gas is commonly known as WFGD; WFGD systems commonly employ limestone as an absorbent owing to its low cost and high efficiency. In this section, the mass transfer and reaction model of a scrubber in a typical WFGD system using limestone is introduced.

4.2.2.1 SO_2 Absorption in the Scrubber

a. *Process Description*

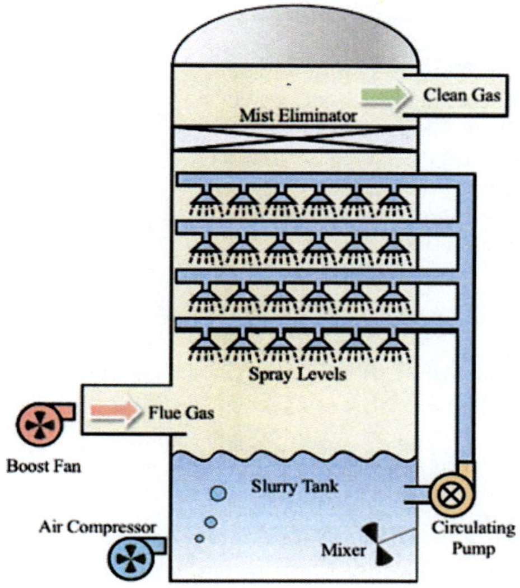

Fig. 4.20 Diagram of a typical WFGD scrubber

The main device of WFGD, such as in the typical limestone WFGD system shown in Fig. 4.20, is the scrubber. The flue gas enters the scrubber from the inlet, contacts the limestone slurry, goes through the mist eliminator, and then finally exits the scrubber. During this process, SO_2 in the flue gas is absorbed by the slurry droplets that then fall into the slurry tank, where the sulfite is oxidized to gypsum by air injected by an air compressor at the bottom of the tank. Several circulating pumps carry the slurry from the slurry tank up to the nozzles again.

b. *Assumptions*

WFGD is a complex process with many simultaneous complicated processes, such as collisions between droplets, gas–liquid distribution in the scrubber, and collisions between droplets and scrubber sidewalls. To simplify the process and allow overall processes to be calculated, the following assumptions are made [16].

- Assumption 1: The flue gas is an ideal gas whose volume is uninfluenced by mass transfer. Furthermore, the temperature of the gas and liquid phases are equal.
- Assumption 2: All slurry droplets are spherical, identical in size, and evenly distributed in the horizontal plane. There is no collision between droplets.
- Assumption 3: The ion and charge equilibrium are reached instantaneously in the droplets.

c. *Model Development*

In Sect. 4.2.1.1, we introduced the absorption theory. In this section, we will introduce how to establish the SO_2 removal model for the whole scrubber, so as to obtain the influence of key parameters on the absorption efficiency of the scrubber.

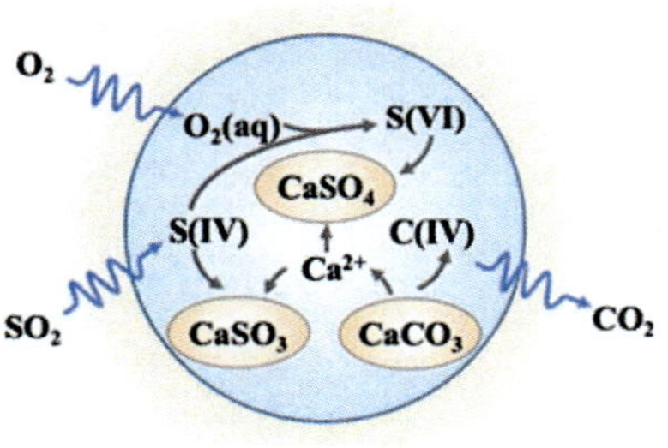

Fig. 4.21 Mass transfer and reaction around and within a slurry droplet during WFGD (the semicircle arrows mean the reaction inside the droplet, the squiggle arrows mean the absorption or desorption process of the gas phase solute)

Motion of droplets: The motion of droplets in the scrubber can be described by the balance of forces as

$$\frac{\pi}{6}d_1^3\rho_1\frac{du_1}{dt} = \frac{\pi}{6}d_1^3(\rho_1 - \rho_g)g - C_D\rho_g\frac{\pi}{4}d_1^2\frac{(u_g - u_1)^2}{2}, \tag{4.60}$$

where C_D represents the drag coefficient and can be calculated as

$$C_D = \frac{24}{\text{Re}_1}\left(1 + 0.125\text{Re}_1^{0.72}\right), \tag{4.61}$$

$$\text{Re}_1 = \frac{d_1|u_g - u_1|\rho_g}{u_g}. \tag{4.62}$$

Absorption or desorption of SO_2, CO_2, and O_2: The slurry absorbs SO_2 and O_2 and releases CO_2, as shown in Fig. 4.21. The absorption or desorption rate of SO_2, O_2, and CO_2 can be defined by calculating the flux of SO_2, O_2, and CO_2 as

$$N_A = k_{g,A}(p_A - p_{Ai}), \tag{4.63}$$

where A can represent SO_2, O_2, or CO_2.

The gas-phase coefficient $k_{g,A}$ is calculated using the Frössling correlation [28],

$$\text{Sh} = \frac{k_{g,A}d_1RT}{D_{SO_2}} = 2 + 0.55\text{Re}_1^{0.5}\text{Sc}^{\frac{1}{3}}, \tag{4.64}$$

where Sc is the Schmidt number,

$$\text{Sc} = \frac{\mu_g}{\rho_g D_A}. \tag{4.65}$$

The gas-phase diffusion coefficient ($D_{g,A}$) is predicted by Fuller et al. [29] as

$$D_{g,A} = \frac{9.86 \times 10^{-9} \cdot T^{1.75} \cdot \left[\frac{1}{M_{\text{air}}} + \frac{1}{M_A}\right]^{\frac{1}{2}}}{0.000001 \cdot p \cdot \left(v_{\text{air}}^{\frac{1}{3}} + v_A^{\frac{1}{3}}\right)^2}. \tag{4.66}$$

The partial pressure of the gas and the concentration of SO_2, CO_2, and O_2 at the gas–liquid interface follow Henry's law:

$$p_{i,A} = \frac{c_{i,A}}{H}. \tag{4.67}$$

Mass transfer and reaction in the liquid phase: After determining the concentration of SO_2, CO_2, and O_2 at the gas–liquid interface, the concentration distribution of the total S(IV), S(VI), C(IV), Ca, and O_2 in the droplet can be calculated. According to Fick's law, this unsteady mass transfer process including chemical reactions can be described as [31]

$$\begin{aligned}\frac{\partial c_{\text{S(IV)}}}{\partial t} = {} & D_{\text{SO}_2(\text{aq})}\left(\frac{\partial^2 c_{\text{SO}_2(\text{aq})}}{\partial b^2} + \frac{2}{b}\frac{\partial c_{\text{SO}_2(\text{aq})}}{\partial b}\right) + D_{\text{HSO}_3^-}\left(\frac{\partial^2 c_{\text{HSO}_3^-}}{\partial b^2} + \frac{2}{b}\frac{\partial c_{\text{HSO}_3^-}}{\partial b}\right) \\ & + D_{\text{SO}_3^{2-}}\left(\frac{\partial^2 c_{\text{SO}_3^{2-}}}{\partial b^2} + \frac{2}{b}\frac{\partial c_{\text{SO}_3^{2-}}}{\partial b}\right) - r_{c,\text{CaSO}_4} - r_{ox},\end{aligned} \tag{4.68}$$

$$\begin{aligned}\frac{\partial c_{\text{S(IV)}}}{\partial t} = {} & D_{\text{HSO}_4^-}\left(\frac{\partial^2 c_{\text{HSO}_4^-}}{\partial b^2} + \frac{2}{b}\frac{\partial c_{\text{HSO}_4^-}}{\partial b}\right) + D_{\text{SO}_4^{2-}}\left(\frac{\partial^2 c_{\text{SO}_4^{2-}}}{\partial b^2} + \frac{2}{b}\frac{\partial c_{\text{SO}_4^{2-}}}{\partial b}\right) \\ & - r_{c,\text{CaSO}_4} + r_{ox},\end{aligned} \tag{4.69}$$

$$\begin{aligned}\frac{\partial c_{\text{S(IV)}}}{\partial t} = {} & D_{\text{CO}_2(\text{aq})}\left(\frac{\partial^2 c_{\text{CO}_2(\text{aq})}}{\partial b^2} + \frac{2}{b}\frac{\partial c_{\text{CO}_2(\text{aq})}}{\partial b}\right) + D_{\text{HSO}_3^-}\left(\frac{\partial^2 c_{\text{HCO}_3^-}}{\partial b^2} + \frac{2}{b}\frac{\partial c_{\text{HCO}_3^-}}{\partial b}\right) \\ & + D_{\text{CO}_3^{2-}}\left(\frac{\partial^2 c_{\text{CO}_3^{2-}}}{\partial b^2} + \frac{2}{b}\frac{\partial c_{\text{CO}_3^{2-}}}{\partial b}\right) - r_{c,\text{CaCO}_3},\end{aligned} \tag{4.70}$$

$$\frac{\partial c_{\text{Ca}}}{\partial t} = D_{\text{Ca}^{2+}}\left(\frac{\partial^2 c_{\text{Ca}^{2+}}}{\partial b^2} + \frac{2}{b}\frac{\partial c_{\text{Ca}^{2+}}}{\partial b}\right) - r_{d,\text{CaCO}_3} - r_{c,\text{CaSO}_4} - r_{c,\text{CaSO}_3}, \tag{4.71}$$

$$\frac{\partial c_{\text{O}_2}}{\partial t} = D_{\text{O}_2}\left(\frac{\partial^2 c_{\text{O}_2}}{\partial b^2} + \frac{2}{b}\frac{\partial c_{\text{O}_2}}{\partial b}\right) - \frac{1}{2}r_{ox}. \tag{4.72}$$

Dissolution or crystallization of $CaCO_3$, $CaSO_3$, and $CaSO_4$: The dissolution or crystallization of $CaCO_3$, $CaSO_3$, and $CaSO_4$ can be described as [31–33]

$$r_{d,\text{CaCO}_3} = k_{d,\text{CaCO}_3}\left(RS_{\text{CaCO}_3} - 1\right), \tag{4.73}$$

$$RS_{\text{CaCO}_3} = \frac{c_{\text{Ca}^{2+}} \cdot c_{\text{CO}_3^{2-}}}{k_{SP,\text{CaCO}_3}}, \tag{4.74}$$

Table 4.3 Equilibrium constants of S(IV), C(IV), S(VI), and H_2O

Symbol	Explanation	Formula
k_{s1}	Equilibrium constant of $SO_2(aq)$	$c_{H^+}c_{HSO_3^-}/c_{SO_2(aq)}$
k_{s2}	Equilibrium constant of HSO_3^-	$c_{H^+}c_{SO_3^{2-}}/c_{HSO_3^-}$
k_{c1}	Equilibrium constant of CO_2 (aq)	$c_{H^+}c_{HCO_3^-}/c_{CO_2(aq)}$
k_{c2}	Equilibrium constant of HCO_3^-	$c_{H^+}c_{CO_3^{2-}}/c_{HCO_3^-}$
k_{s42}	Equilibrium constant of HSO_4^-	$c_{H^+}c_{SO_4^{2-}}/c_{HSO_4^-}$
k_w	Equilibrium constant of H_2O	$c_{H^+} \cdot c_{OH^-}$

$$r_{c,\mathrm{CaSO_3}} = 1.62 \times 10^{-4} \exp\left(-\frac{5153}{T}\right)\left(RS_{\mathrm{CaSO_3}} - 1\right)^n \frac{A_1}{RS_{\mathrm{CaSO_4}}}, \tag{4.75}$$

$$RS_{\mathrm{CaSO_3}} = \frac{c_{\mathrm{Ca^{2+}}} \cdot c_{\mathrm{SO_3^{2-}}}}{k_{SP,\mathrm{CaSO_3}}}, \tag{4.76}$$

$$r_{c,\mathrm{CaSO_4}} = 1.1 \times 10^{-4} A_2\left(RS_{\mathrm{CaSO_4}} - 1\right), \tag{4.77}$$

$$RS_{\mathrm{CaSO_4}} = \frac{c_{\mathrm{Ca^{2+}}} \cdot c_{\mathrm{SO_4^{2-}}}}{k_{SP,\mathrm{CaSO_4}}}. \tag{4.78}$$

Ion and charge equilibrium: According to the ionic equilibrium theory, the equilibrium constants of S(IV), C(IV), S(VI), and H_2O as calculated by Pasiuk-Bronikowska and Rudziński [34] are shown in Table 4.3. Clearly, the composition of the slurry is closely related to the concentration of H^+. The distribution of S(IV) at different pH values is shown in Fig. 4.22.

According to assumption 3, the charge equilibrium can be described as

$$c_{\mathrm{H^+}} - c_{\mathrm{OH^-}} + 2c_{\mathrm{Ca^{2+}}} - 2c_{\mathrm{SO_3^{2-}}} - c_{\mathrm{HSO_3^-}} - 2c_{\mathrm{CO_3^{2-}}} - c_{\mathrm{HCO_3^-}} - 2c_{\mathrm{SO_4^{2-}}} - c_{\mathrm{HSO_4^-}} = 0. \tag{4.79}$$

d. *Calculation Process*

To calculate the desulfurization efficiency of the scrubber, the spray zone is divided into n horizontal cells and each droplet is divided into m layers, as shown in Fig. 4.23. Since the absorption in the scrubber is countercurrent, it is impossible to know the parameters of the flue gas and the slurry on the same boundary. It is thus necessary to assume a range of outlet SO_2 partial pressure and calculate from top to bottom, and then continuously amend outlet SO_2 partial pressure by the dichotomy method. First, the SO_2 absorption rate and the rate of each reaction are calculated according to

Fig. 4.22 Distribution of S(IV) at different pH values

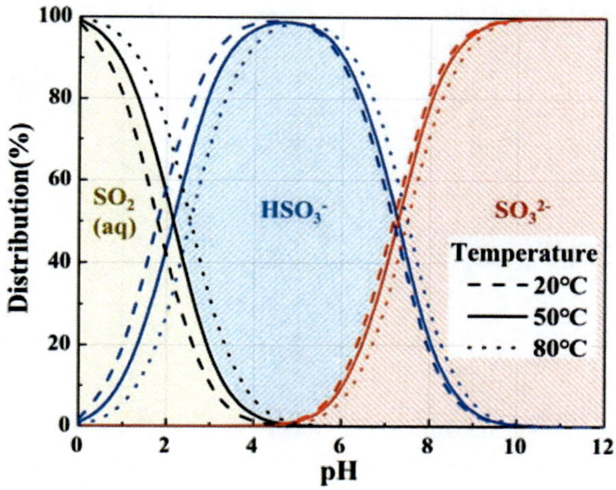

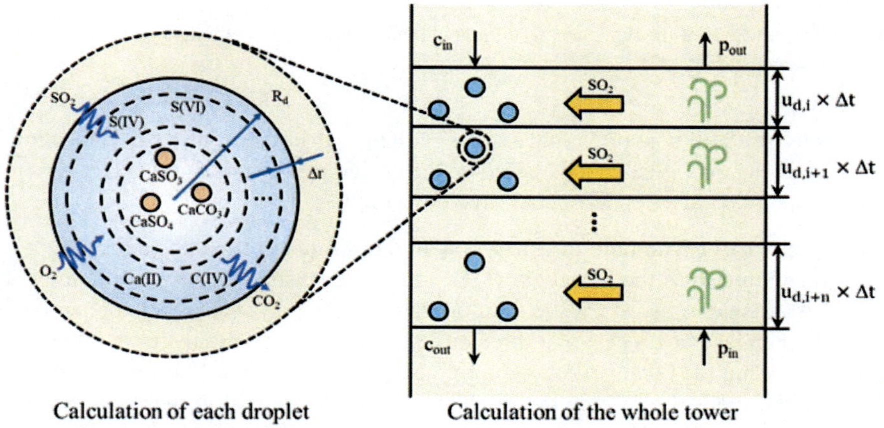

Fig. 4.23 Diagram of calculation process when solving the WFGD model

Eqs. (4.60) to (4.67) and Eqs. (4.73) to (4.78), respectively. The droplet composition and concentration distribution can then be obtained according to Fick's law and the ion and charge equilibrium with Eqs. (4.68) to (4.72), Eq. (4.79) and the equilibrium constants mentioned in Table 4.3.

4.2.2.2 Key Factors Affecting Desulfurization Efficiency

According to the model mentioned above, the key factors affecting desulfurization efficiency are explored. Knowing these factors and how they are interrelated with the process chemistry of each system will provide an understanding of how each process

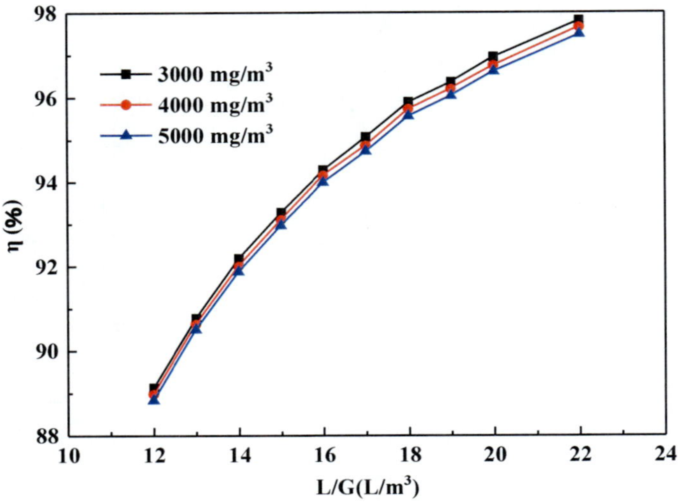

Fig. 4.24 Influence of the liquid/gas (L/G) ratio on the desulfurization efficiency at different inlet concentrations of SO_2 [31]

functions, in addition to providing a set of guidelines to be used during an inspection. The following data is calculated based on the design parameters of a WFGD system which is applied in a 300 MW coal-fired power unit.

Liquid/gas ratio: The ratio of slurry flow in the scrubber to the quenched flue gas flow is referred to as the liquid/gas (L/G) ratio and is usually expressed in units of L/m^3. As shown in Fig. 4.24, the desulfurization efficiency increases exponentially with the increase of L/G ratio at three different SO_2 concentrations in the flue gas. Although a high L/G ratio allows for high SO_2 removal, it also reduces scalability because the spent slurry from the scrubber is more dilute with respect to absorbed SO_2. Increasing the L/G ratio can also increase the capital and operating costs owing to the higher capacity requirements of the reaction tank and associated hold tanks, dewatering equipment, higher pumping requirements, slurry preparation and storage requirements, and reagent and utility necessities.

Flow rate and concentration of SO_2in the flue gas: As the SO_2 concentration in the flue gas increases, the desulfurization efficiency gradually decreases, as shown in Fig. 4.25. Higher inlet SO_2 depletes the available liquid-phase alkalinity more rapidly, increasing the liquid-film resistance to absorption.

The flue gas flow rate is an important parameter in the WFGD system because it is closely related to the exposure time of the flue gas and the total interfacial area of mass transfer in the scrubber. Increasing the gas velocity decreases the exposure time of the flue gas while increasing the total interfacial area of mass transfer in the scrubber; increasing the gas velocity decreases the relative velocity of droplets and increases the mass transfer coefficient. As shown in Fig. 4.25, increasing the

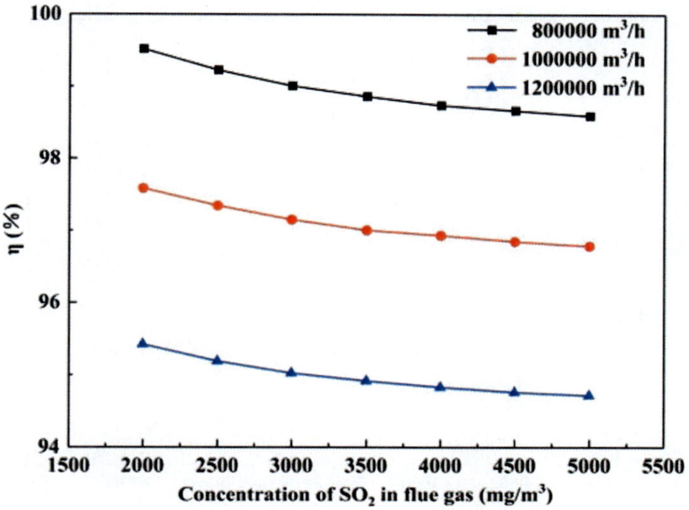

Fig. 4.25 Influence of inlet SO_2 concentration on the desulfurization efficiency at different flue gas flow rate [31]

flue gas flow rate under the given conditions decreases the desulfurization efficiency, indicating that the effect of exposure time on the desulfurization efficiency is greater than the other factors.

Combinations of different spray layers: Scrubbers generally comprise several spray layers; the combination of spray layers used affects the desulfurization efficiency of the WFGD system. For a typical scrubber layers (1, 2, 3, and 4, from bottom to top of the scrubber), the possible spray layer combinations using two or three spray layers are shown in Table 4.4.

The resulting desulfurization efficiency when three spray layers are used (i.e., schemes 1–4 from Table 4.4) is shown in Fig. 4.26. The desulfurization efficiencies of schemes 2 and 4 are higher than those of schemes 1 and 3, indicating that higher spray layers can increase the desulfurization efficiency because the slurry dropping from higher spray layer travels further, thereby increasing the contact time between the flue gas and slurry droplets.

Table 4.4 Combination of different spray layers [31]

Scheme	1	2	3	4	5	6	7	8	9	10
1st spray layer	+	+	+	–	+	+	+	–	–	–
2nd spray layer	+	–	+	+	+	–	–	+	+	–
3rd spray layer	+	+	–	+	–	+	–	+	–	+
4th spray layer	–	+	+	+	–	–	+	–	+	+

Note + represents the spray layers are in operation, – represents the spray layers are out of operation

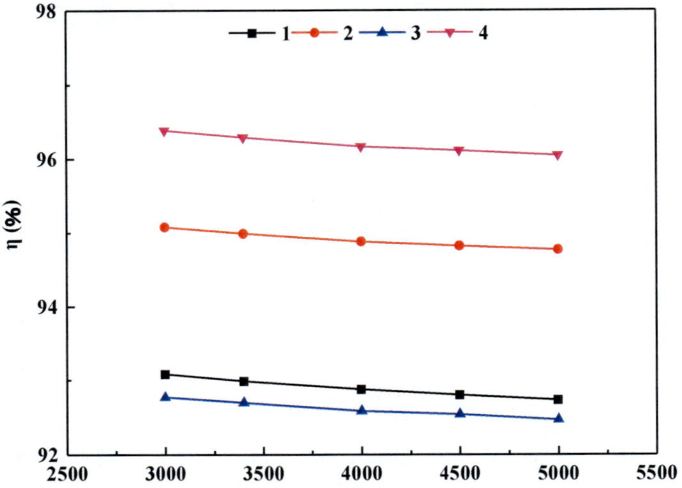

Fig. 4.26 Influence of spray layer combinations using three layers on the desulfurization efficiency (100% load) [31]

The power generation unit often operates at the partial load mode during the night, and the flue gas flux typically decreases to approximately 800,000 m^3/h. The influence of different spray layer schemes and SO_2 concentration on desulfurization efficiency in the part load of boiler is shown in Fig. 4.27. Based on the calculation

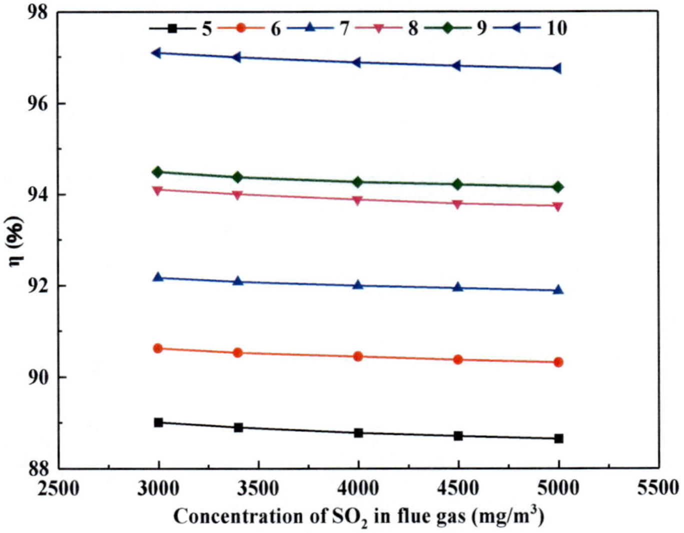

Fig. 4.27 Influence of spray layers combination on the desulfurization efficiency (80% load) [31]

results of the combination of spray layers, it can be concluded that combinations of higher spray layers are more effective in cleaning the flue gas. When the variation of concentration of SO_2 is high, the scheme of combinations of higher spray layers is effective. But the energy consumption of higher spray layers is higher than low spray layers. As at full load, combinations including higher spray layers more effectively cleaned the flue gas at partial load than those with lower spray layers; however, higher spray layers consume more energy than lower spray layers due to the added consumption of the circulating pump.

4.3 SO_2 Control Enhancement and Applications

Increasingly strict emission requirements have increased the necessity for low-cost, highly efficient WFGD systems. In this section, some of the recently developed methods to improve the desulfurization performance of WFGD based on the above-mentioned parameters affecting the SO_2 absorption process (e.g., slurry pH, the specific area between the gas and liquid phases, and S(IV) concentration) are described.

4.3.1 PH Adjustment

4.3.1.1 Realization of a Dual-Loop System with Different pH Values

The slurry pH is an important parameter for WFGD. A high pH can promote the absorption of SO_2, whereas a low pH can promote the oxidation of S(IV) and the dissolution of solid absorbents, such as limestone. To enhance the performance of WFGD, the system can be divided into two zones with different pH values, thereby allowing SO_2 absorption to occur at a high pH and S(IV) oxidation and absorbent dissolution to occur at a low pH. This so-called dual-loop technology comprises two slurry loops: one with a lower slurry pH for limestone dissolution and S(IV) oxidation and one with a higher slurry pH for SO_2 absorption. Several dual-loops methods have been proposed, as summarized in Table 4.5. The main mechanism behind these four methods are similar. In this section, the fourth method is introduced for example to discuss the mechanism behind dual-loop technology and how it can be used to enhance SO_2 removal.

4.3.1.2 Performance of pH Dual-Loop Technology

Effect of slurry residence time on pH: A simple titration experiment using the slurry from three power plants can be used to demonstrate the relationship between

Table 4.5 Different methods of dual-loop technology

Strategy	Brief introduction	Characteristics
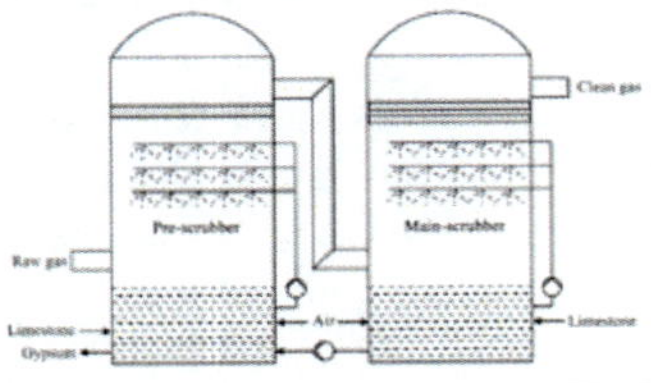	◆ Two scrubbers ◆ Lower pH in prescrubber ◆ Higher pH in main scrubber	◆ High efficiency ◆ High cost ◆ Large covering areas
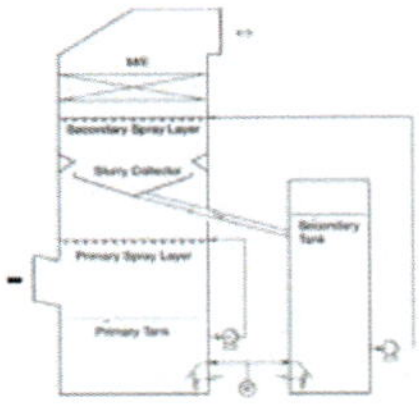	◆ Slurry collector ◆ Higher pH in upper loop ◆ Lower pH in lower loop ◆ Primary tank and secondary tank	◆ Relatively high cost ◆ High efficiency ◆ Special design for the slurry collector ◆ Relatively large pressure drop
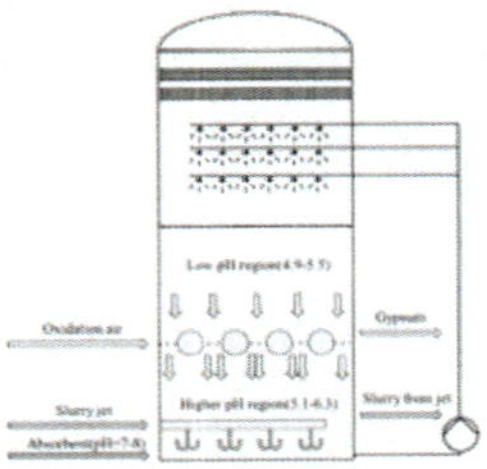	◆ Slurry tank with double zone ◆ Higher pH in upper loop ◆ Lower pH in lower loop	◆ Low cost Relatively high efficiency ◆ Less area occupancy ◆ Special design in the slurry tank
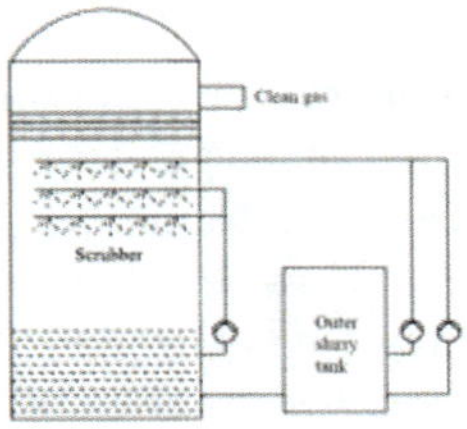	◆ Main/auxiliary slurry tank ◆ Higher pH in upper loop ◆ Lower pH in lower loop	◆ Low cost ◆ High efficiency ◆ Convenient for scrubber upgrade

the slurry residence time and pH; as shown in Fig. 4.28. First, hydrochloric acid is added into the slurry by titration until pH = 3.8, and then slurry pH increases quickly within a short time after titration. As H^+ is consumed, the increase rate of pH gradually decreases. The slurry pH increases as the residence time increases in a certain range, demonstrating that the residence time can significantly affect the slurry pH. In the fourth method, the slurry residence time is different between the main and auxiliary slurry tank, and thus the pH of the slurry in the main and auxiliary slurry tank is different.

pH difference between the main and auxiliary slurry tanks: A pilot-scale experimental system with a gas flow rate of 50,000 m^3/h and four circulating pumps (A,

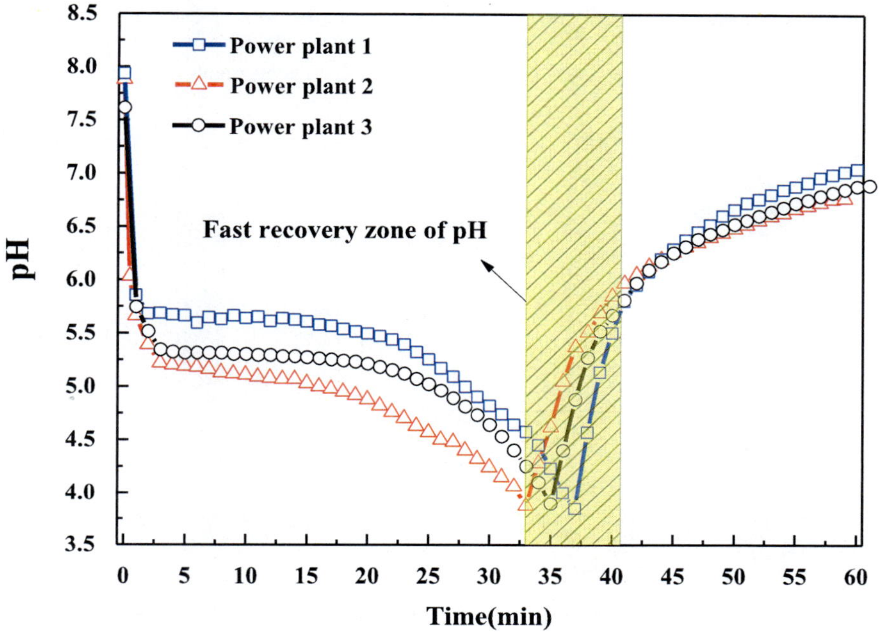

Fig. 4.28 pH variation in the titration and recovery process [35]

B, C, and D, from top to bottom) can be used to clarify the impact of slurry pH, flue gas flow rate, SO_2 concentration, and spray density. Here, the pH of the main and auxiliary slurry tank is 3.5–5.3 and 5.5–6.0, respectively; the term pH difference is used to describe the difference in pH between these two tanks.

As shown in Fig. 4.29, the pH difference increases with increasing inlet SO_2 concentration. Increasing the number of circulating pumps in operation increased the pH difference; furthermore, as clarified in Fig. 4.30, the pH difference decreased as the pH of the main slurry tank increased. Such results is helpful for the designing of the dual-loop technology in order to achieve a stable and enough pH difference between the main and auxiliary slurry tanks.

Desulfurization performance of dual-loop technology: The following results are achieved by the same pilot-scale experimental system mentioned above. The influence of unit load on desulfurization efficiency and outlet SO_2 concentration is shown in Fig. 4.31. Here, as the unit load increased from 50 to 100%, the desulfurization efficiency decreased from 99.97 to 99.26%, increasing the outlet SO_2 concentration from 0.2 to 5.7 mg/m^3. Furthermore, increasing the number of circulating pumps from two to four increased the desulfurization efficiency from 93.1 to 99.97% and decreased the outlet SO_2 concentration from 61.76 to < 1 mg/m^3, as shown in Fig. 4.32. Finally,

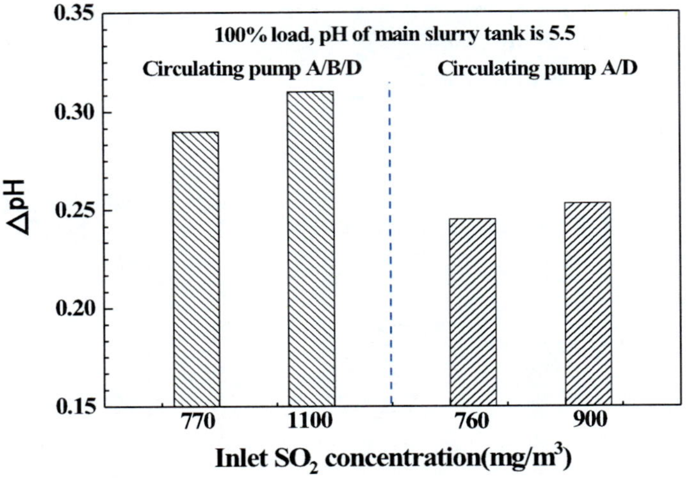

Fig. 4.29 Effect of the inlet SO_2 concentration and circulating pumps on pH difference. 100% load, slurry pH of main tank = 5.0

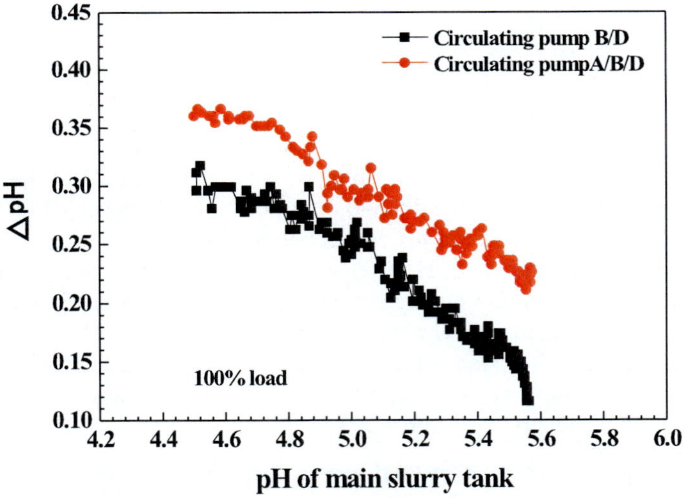

Fig. 4.30 Effect of the pH of the main slurry tank and circulating pumps on the pH difference. 100% load

increasing the pH of the main tank slurry from 5.0 to 6.0 increased the desulfurization efficiency from 98.9 to 99.86% and decreased the outlet SO_2 concentration from 9.07 to < 1 mg/m^3, as shown in Fig. 4.33.

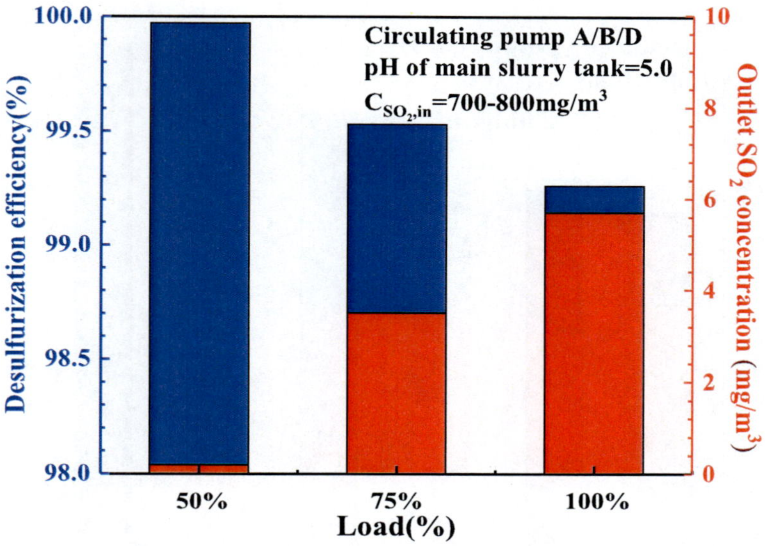

Fig. 4.31 Influence of unit load on the desulfurization efficiency and outlet SO_2 concentration using three circulating pumps [A/B/D], a main tank slurry pH of 5.0, and inlet SO_2 concentration of 700–800 mg/m^3

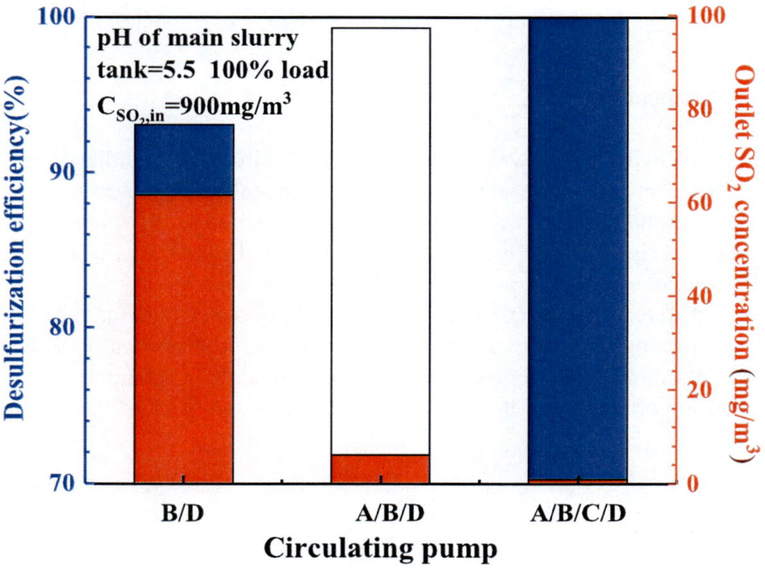

Fig. 4.32 Influence of circulating pump usage on the desulfurization efficiency and outlet SO_2 concentration at a 100% load, main tank slurry pH of 5.0, and inlet SO_2 concentration of 900 mg/m^3

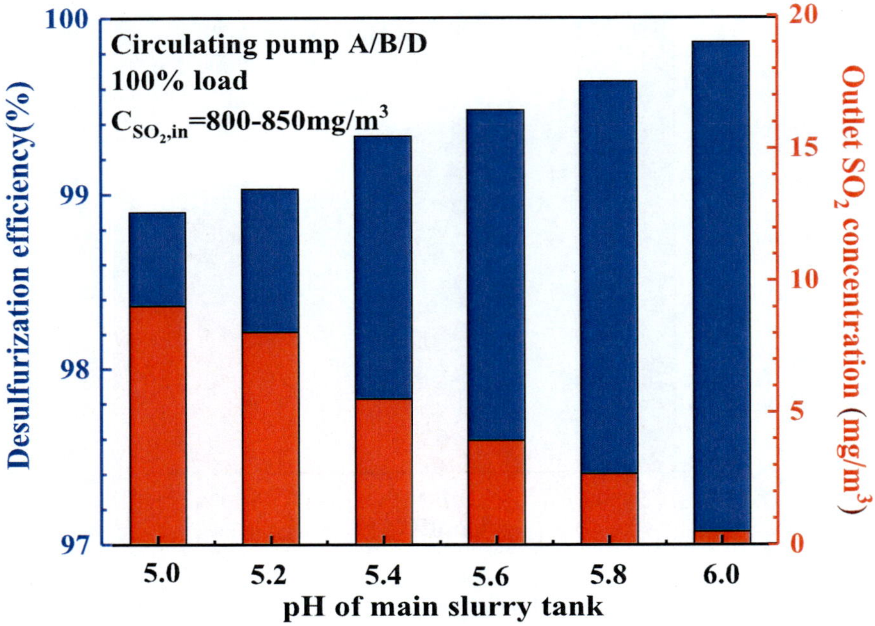

Fig. 4.33 Influence of slurry pH in the main tank on the desulfurization efficiency and outlet SO_2 concentration. 3 circulating pumps [A/B/D], 100% load, inlet SO_2 concentration = 800–850 mg/m^3

4.3.1.3 Application

A dual-loop WFGD system was applied to a 220-t/h circulating fluidized bed (CFB) unit that uses coal and wastewater sludge as fuel. The inlet SO_2 concentration varied from 1000 to > 4000 mg/m^3 according to the mass fraction of the sludge in the fuel. A schematic of the scrubber is shown in Fig. 4.34, and parameters of the inlet flue gas and scrubber are summarized in Table 4.6.

The pH difference between the main and auxiliary slurry tank ranged from 0.15 to 0.42 (increasing with decrease in the pH in the main slurry tank), as shown in Table 4.7. At an inlet SO_2 concentration of approximately 3000 mg/m^3, the overall desulfurization efficiency of this dual-loop WFGD system was > 99.2%, as shown in Fig. 4.35.

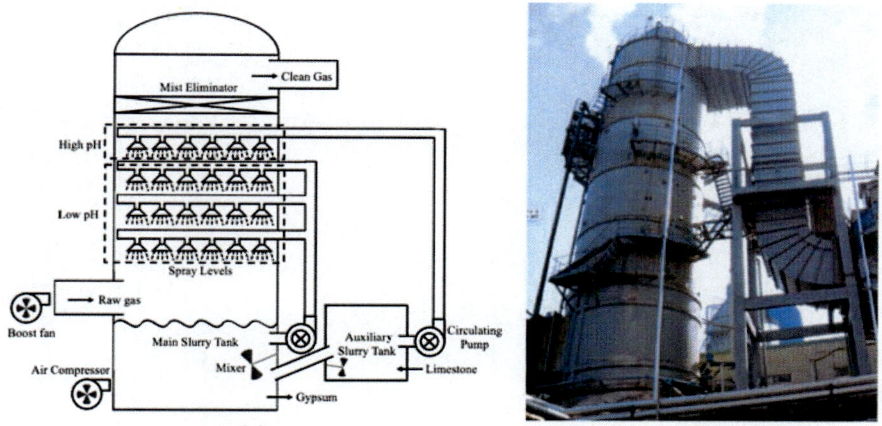

(a) (b)

Fig. 4.34 **a** Schematic and **b** photograph of a dual-loop WFGD system applied to a 220-t/h circulating fluidized bed (CFB) unit

Table 4.6 Designed parameters of WFGD system

Parameters	Unit	Value
Load	t/h	220
Flue gas flow rate	m^3/h	300,000
Inlet SO_2 concentration	mg/m^3	3500
Slurry flow rate	m^3/h	1600×4
Liquid–gas ratio	/	20.1
Height of first spray layer	/	24.2
Height of second spray layer	/	22.4
Height of third spray layer	/	20.6
Height of fourth spray layer	/	18.8

Table 4.7 pH difference between main and auxiliary slurry tank

pH of main slurry tank	5.68	5.36	5.24	4.91	4.85	4.48
pH of auxiliary slurry tank	5.83	5.52	5.44	5.2	5.12	4.9
ΔpH	0.15	0.16	0.2	0.29	0.27	0.42

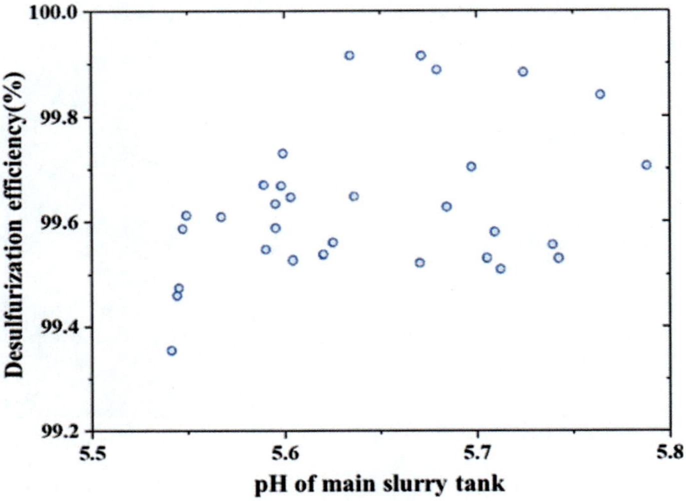

Fig. 4.35 Desulfurization performance of the pH dual-loop WFGD system at an inlet SO_2 concentration of approximately 3000 mg/m^3

4.3.2 *Desulfurization Additives*

4.3.2.1 Mechanism of Desulfurization Additives

Organic desulfurization additives, such as dicarboxylic acids in the form of an adipic or dibasic acid, are commonly used to enhance SO_2 absorption in WFGD; in particular, adipic acid is an effective scrubber additive, the mechanism of which is shown in Fig. 4.36. Acting as buffers, these additives tend to neutralize acid-generated hydrogen ions to prevent the decrease in the system pH and SO_2 removal.

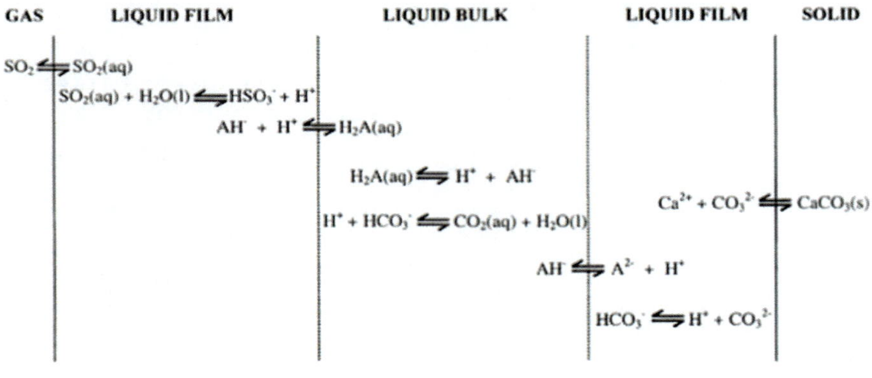

Fig. 4.36 Mechanism for the action of adipic acid in a WFGD system [36]

When SO_2 dissociates in the liquid film between the gas and liquid phases, H^+ ions are produced and taken up by the adipate (A^{2-}) and hadipate (AH^-) ions and transported to the liquid bulk phase. Here, they may be released and may react with, e.g., HCO_3^-, producing CO_2 (aq) and H_2O (l). The transport of H^+ from the liquid film to the bulk phase promotes the dissociation of SO_2, increasing the overall absorption rate. The dissolution of limestone is dependent on the diffusion of H^+ to the surface of the particles. Adipic acid can provide additional transport of H^+, as AH^- or H_2A, from the liquid phase to the liquid film surrounding the limestone particles. The AH^- ion releases H^+ near the surface of the limestone where it can react with CO_3^{2-}, thereby increasing the limestone dissolution rate. In limestone-based WFGD systems, additives can promote the dissolution of limestone and the absorption of SO_2, as discussed in the following subsection.

4.3.2.2 Effects of Desulfurization Additives

On limestone dissolution: An experiment similar to the titration experiment mention in the Sect. 4.2.1.2 can be used to elucidate the effects of additives on limestone dissolution; as shown in Fig. 4.37. Additives can cause the dissolution rate of limestone

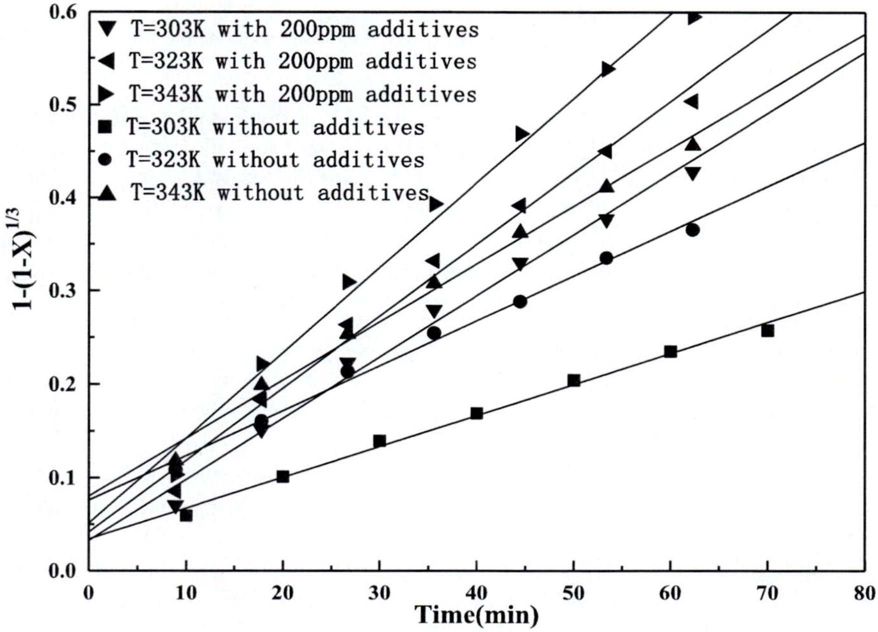

Fig. 4.37 Effect of the temperature and additives on the dissolution of limestone

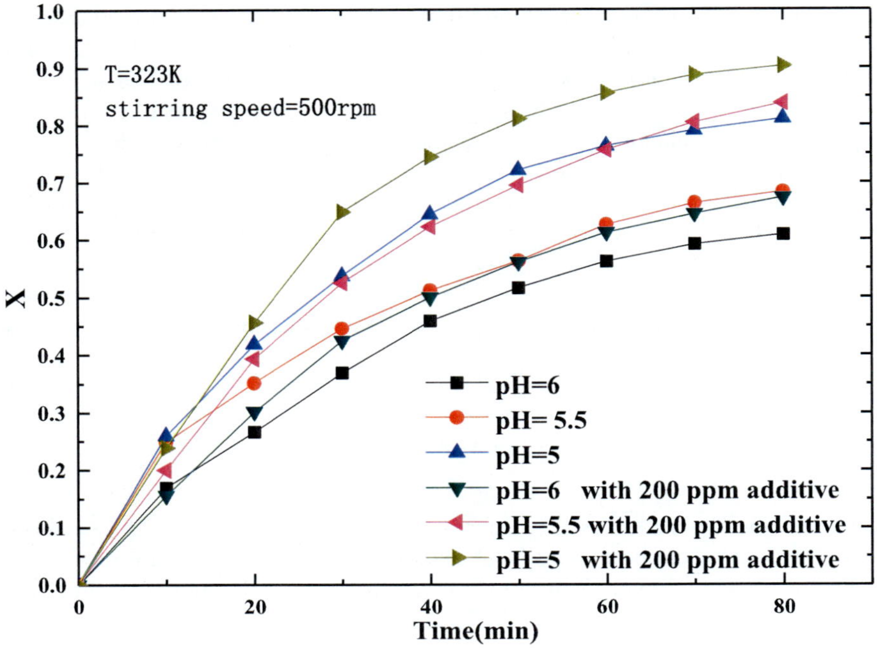

Fig. 4.38 Effect of additives on the dissolution of limestone under different pH values

to increase significantly, because the dissolution of limestone is related to the diffusion of H^+ on the surface of limestone particles. Organic acid additives promote the diffusion of H^+ from the liquid bulk to the surface of the limestone particles where H^+ can react with CO_3^{2-}, thereby increasing the dissolution rate of limestone. The effect of additives on the dissolution of limestone under different pH values is shown in Fig. 4.38. Additives could significantly increase the dissolution rate of limestone under different pH values, and the effect is most significant when pH = 5.5.

On SO_2absorption: Using a double stirred-cell absorber with a $CaCO_3$ concentration of 1% and temperature of 50 °C, the presence of additives significantly increased the absorption rate of SO_2, particularly under high SO_2 concentration, as shown in Fig. 4.39.

In a lab-scale WFGD used to study the effect of desulfurization additives in the absorber, additives were shown to increase the SO_2 absorption rate, as shown in Fig. 4.40. Here, the organic acid additives promoted the diffusion of H^+ from the gas–liquid interface to the liquid bulk, thereby promoting the conversion of SO_2(aq) into HSO_3^- and SO_3^{2-} and increasing the absorption rate of SO_2. Among the several additives studied, succinic acid showed the best result.

As a result of the improved limestone dissolution and SO_2 absorption, the overall desulfurization efficiency of the limestone slurry was increased to 99.6% using the

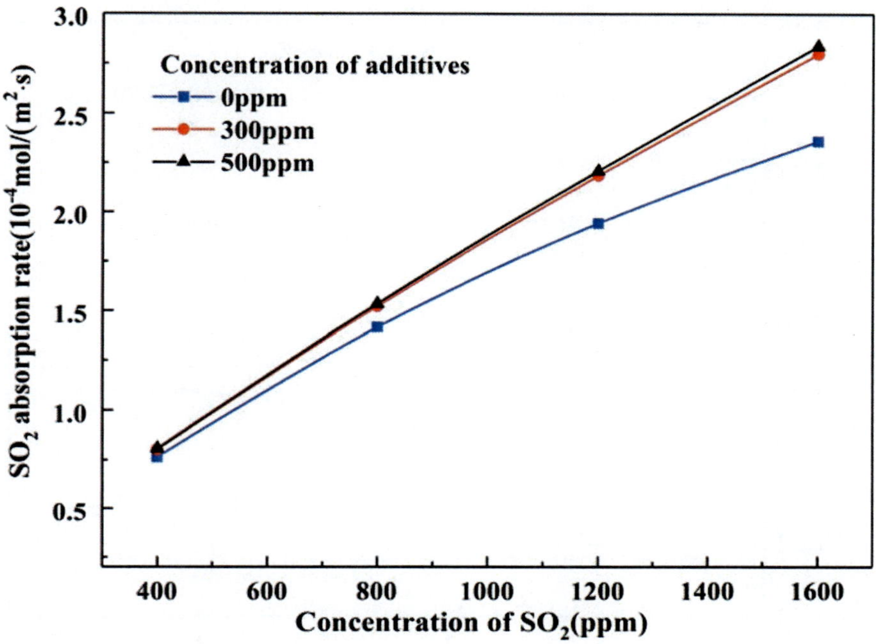

Fig. 4.39 Effect of the SO_2 concentration and additives on the SO_2 absorption rate

compound additive X, which is a mixture of succinic acid, adipic acid and glutaric acid, as shown in Fig. 4.41.

4.3.2.3 Application

At a low inlet SO_2concentration: To demonstrate the effects of additives on WFGD, a system with desulfurization additives was applied to a 600-MW coal fired power unit, shown in Fig. 4.42, with an inlet SO_2 concentration of 1000–2000 mg/m^3. The detailed parameters of the inlet flue gas and scrubber are shown in Table 4.8.

As shown in Fig. 4.43, the use of additives decreased the outlet SO_2 concentration from 117 to 20 mg/m^3; the average desulfurization efficiency increased by 4.2% from 93.6 to 97.8%, decreasing the average outlet SO_2 concentration from 85.8 to 31.5 mg/m^3. The additives using here is a mixture of succinic acid, adipic acid and glutaric acid.

At a high inlet SO_2concentration: Similarly, a WFGD system with desulfurization additives, as shown in Fig. 4.44, was applied to a 300-MW coal fired power unit with an inlet SO_2 concentration of 4500–4800 mg/m^3 and parameters shown in Table 4.9. The additives using here is a mixture of succinic acid, adipic acid and glutaric acid.

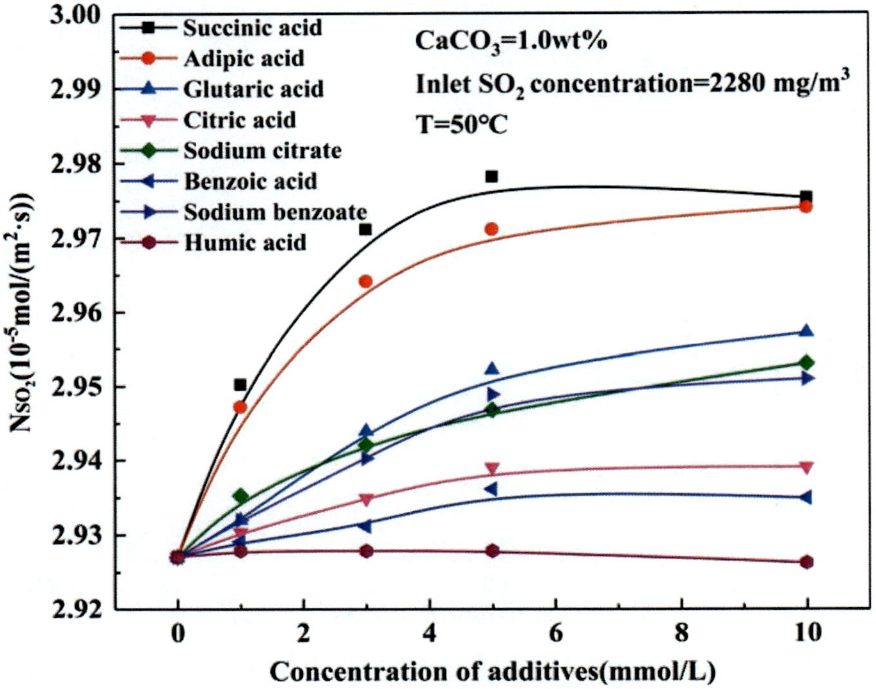

Fig. 4.40 Effect of additives on the SO_2 absorption rate [37]

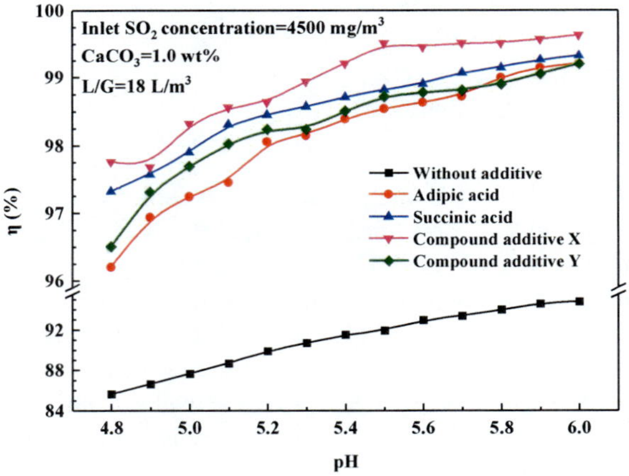

Fig. 4.41 Effect of additives on the desulfurization efficiency [37] (compound additive X and Y are mixtures of succinic acid, adipic acid and glutaric acid with different proportion)

Fig. 4.42 WFGD system with desulfurization additives

Table 4.8 Parameters of inlet flue gas of WFGD system at BMCR (boiler maximum continuous rating)

Parameters	Unit	Value
SO_2 concentration	mg/m^3	1148
Flow rate	m^3/s	571
Temperature	°C	125

Fig. 4.43 Effect of additives on the desulfurization efficiency

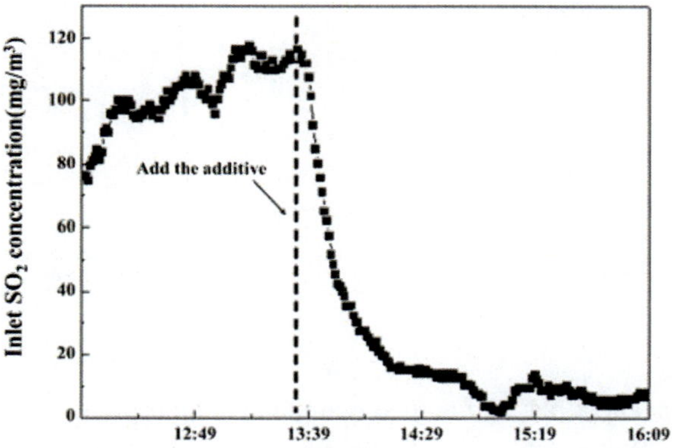

The addition of additives helped stabilize the pH of the system; as shown in Fig. 4.45, the pH fluctuated from 4.5 to 6.0 with an average of 5.28 before the addition of additives, whereas it fluctuated from 5.0 to 5.5 with an average of 5.32 once the additives were added to the system. Thus, under a steady inlet SO_2 concentration and limestone slurry supply, additives reduced the pH fluctuation, increased the

Fig. 4.44 a WFGD system and b scrubber with desulfurization additives

Table 4.9 Parameters of inlet flue gas of WFGD system at BMCR

Parameters	Unit	Value
SO_2	mg/m^3	4554
Flow rate	m^3/s	335
Temperature	°C	132

Fig. 4.45 Effect of additives on pH fluctuation

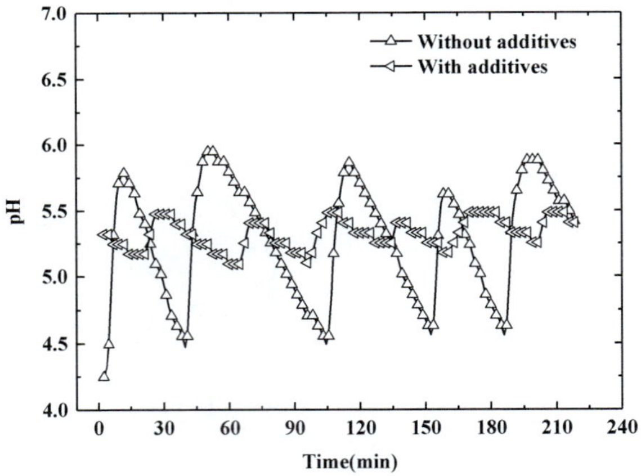

average pH, and helped maintain an appropriate scrubber slurry pH, thereby facilitating system stability and increasing SO_2 removal. As a result, the desulfurization efficiency was increased by 3–4%, as shown in Fig. 4.46; using four circulation pumps allowed for an efficiency of 99.4% for an average SO_2 outlet concentration of 26 mg/m^3 to be achieved, as shown in Fig. 4.47.

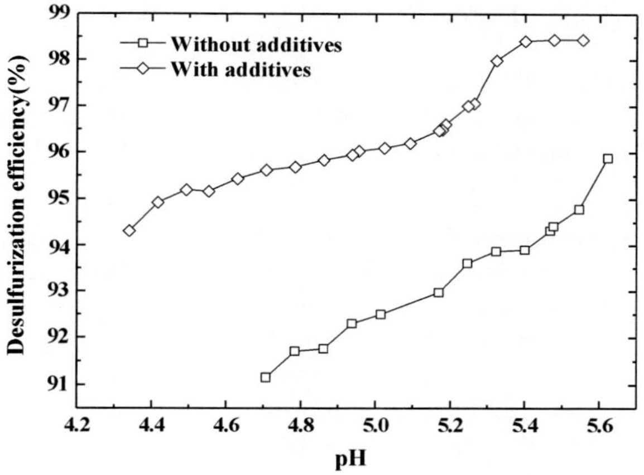

Fig. 4.46 Effect of additives on the desulfurization efficiency

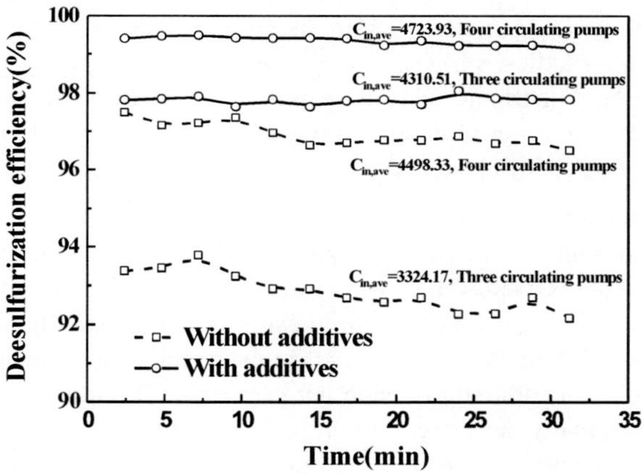

Fig. 4.47 Effect of additives on the variation of desulfurization efficiency

4.3.3 Scrubber Optimization

The SO_2 absorption process can be divided into the mass transfer across the phase boundary and the chemical reactions in the aqueous phase. In addition to pH regulation, the SO_2 absorption process can be enhanced by promoting mass transfer via improving the contact and mixing state of the gas and liquid phases. Highly efficient scrubbers encourage immediate contact between the gas phase and the absorbents; this contact can be enhanced in a scrubber by optimizing its geometric structure and introducing internal components.

4.3.3.1 Typical Scrubbers for SO_2 Absorption

Each scrubber must be customized to its application (i.e., flue gas condition, targeted gaseous pollutants); scrubber designs include a spray tower, packed tower, bubble column, liquid-column tower, and rotating packed bed, as shown in Fig. 4.48a–e, respectively, and detailed below.

Spray tower (Fig. 4.48a): Here, liquid absorbents are sprayed as droplets from the top to contact an upward-flowing gas stream. The dispersion of the liquid phase within the gas phase allows for a large gas–liquid interfacial contact area. Spray towers have been widely applied for SO_2 absorption owing to their simple structure, low investment cost, and small pressure drop. The nozzles require regular replacement because they are susceptible to wear and blockage owing to their small nozzle aperture fused to decrease the droplet diameter.

Packed tower (Fig. 4.48b): Here, the gas–liquid interfacial area is enhanced by the contact of gas rising through the void space between packing with a liquid film flowing downward, where the packing includes solid objects such as ceramic rings and saddles, which are used to ensure mixing between the gas and liquid phases. A liquid distributor is usually used to evenly distribute the absorbents and increase the effective wetting area of the packing. Despite reaching SO_2 absorption efficiencies of up to 99%, packed towers often face large pressure drops, scaling, and blockage.

Bubble column (Fig. 4.48c): Here, gas bubbles are generated in a sparger and rise through a downward-flowing liquid stream. Contrary to the spray tower, the gas phase is dispersed in a continuous liquid phase, thus eliminating the need for a spray system. A bubbling column can provide the simultaneous, high-efficiency removal of SO_2 and PM owing to the unique contact that is induced between the gas and liquid phases, although passing the gas stream through the thick liquid layer can cause a relatively large pressure drop. The bubble size considerably influences the removal efficiency of SO_2: smaller bubbles provide higher desulfurization efficiency.

Liquid-column tower (Fig. 4.48d): Here, liquid absorbents are dispersed from the bottom in the form of a liquid column. Liquid-column towers can provide sufficient gas–liquid mass transfer and high desulfurization efficiency as the liquid column

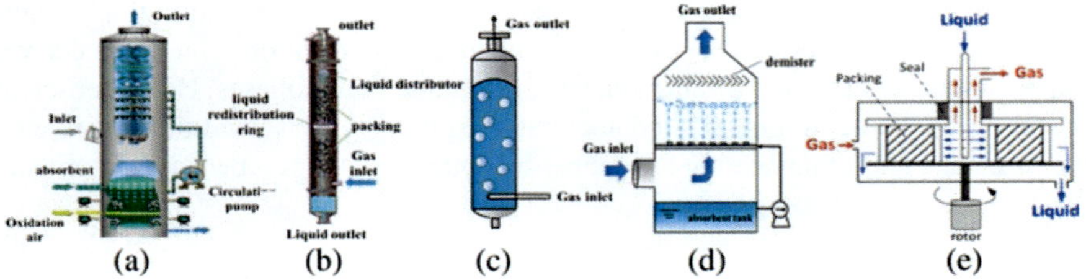

Fig. 4.48 Five types of scrubbers: **a** spray tower **b** packed tower **c** bubble column **d** liquid-column tower **e** rotating packed bed

rises and falls repeatedly. Large-diameter nozzles are commonly used and have a better blocking resistance and higher operational stability than their smaller-diameter counterparts. However, the investment and operation costs of a liquid-column tower are relatively high owing to high pressure drop.

Rotating packed bed (Fig. 4.48e): As the most recently developed model, rotating packed bed is a novel and efficient device designed to enhance mass transfer between the gas and liquid phases and thus gaseous pollutant removal. Here, a liquid is injected from the liquid distributors into the inner periphery of the rotor, flows uniformly through the rotor packing, and then flows from the rotor into the machine case. The flue gas flows through the packing zone to contact the liquid within the rotor via countercurrent, cocurrent, and cross flows. This system can provide a mean acceleration 100 or even 1000 times greater than the force of gravity, causing the liquid to spread and split into micro/nanodroplets and thin films [38]. This allows for volumetric mass transfer coefficients 1–3 orders of magnitude higher than those in a conventional packed bed. Moreover, rotating packed beds provide a high hydraulic and flooding capacity, low liquid hold-up, small reactor size, and better rotor self-cleaning.

Overall, the crucial factor influencing SO_2 absorption, and thus the design basis for scrubber optimization, is the mixing between the gas and liquid phases. The internal and external geometric structures of a scrubber considerably influence the flow and distribution characteristics of the gas and liquid phases. Therefore, in the subsequent sections, the methods and parameters used to optimize the external and internal scrubber structure are discussed in detail.

4.3.3.2 Optimization of External Structure

Of the many structural parameters that can be used to describe a scrubber and can affect the flow field, the diameter, entrance angle, entrance deflector, spacing of spray layers, and spray angle are described here [39].

a. *Scrubber Diameter*

At a constant flue gas flow rate, decreasing the scrubber diameter increases the velocity of the gas phase, as shown in Fig. 4.49. As the gas velocity increases, the relative velocity between the gas and liquid phases increases, thereby promoting mass transfer across the phase interfaces. A smaller diameter can also increase the liquid mass per unit volume and enlarge the mass transfer area. However, as the diameter decreases, the pressure drop increases owing to the enhanced gas–liquid interaction, leading to higher operating costs.

b. *Entrance Angle*

The entrance angle of the flue gas influences the uniformity of the flow field. As shown in Fig. 4.50, increasing the entrance angle can improve velocity distribution uniformity by weakening the vortex in the bottom of the scrubber and the gas backflow

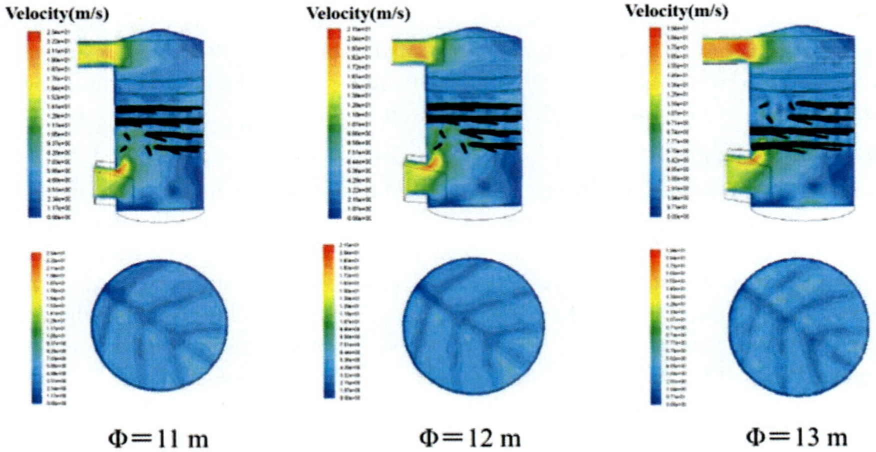

Fig. 4.49 Effect of the scrubber diameter on the flow field [38]

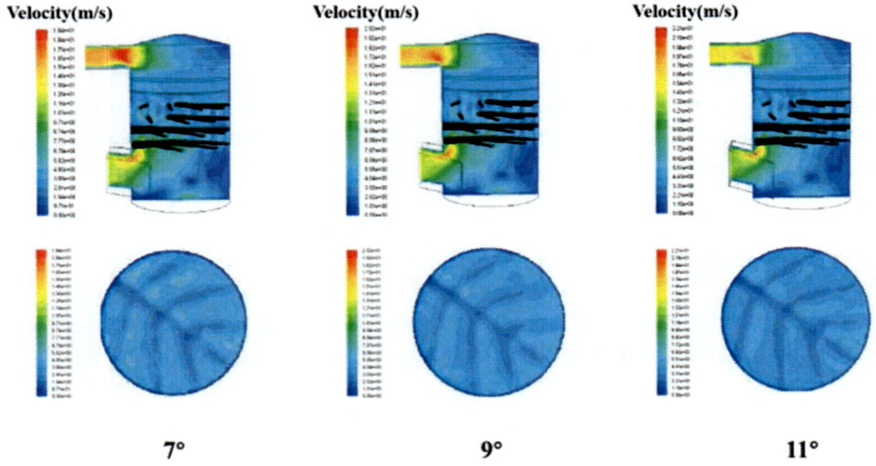

Fig. 4.50 Effect of the entrance angle on the flow field [39]

above the entrance. Furthermore, the downward inclined angle of the entrance can help prevent liquid from entering and corroding the entrance.

c. *Entrance Deflector*

The deflector is usually installed near the entrance of the scrubber to improve the uneven flow field caused by the inertia of the gas stream entering the scrubber by guiding it toward the center of the scrubber, as shown in Fig. 4.51. A deflector can also help prevent corrosion caused by liquid entering the entrance duct. However, increasing the length of the deflector increases the pressure drop.

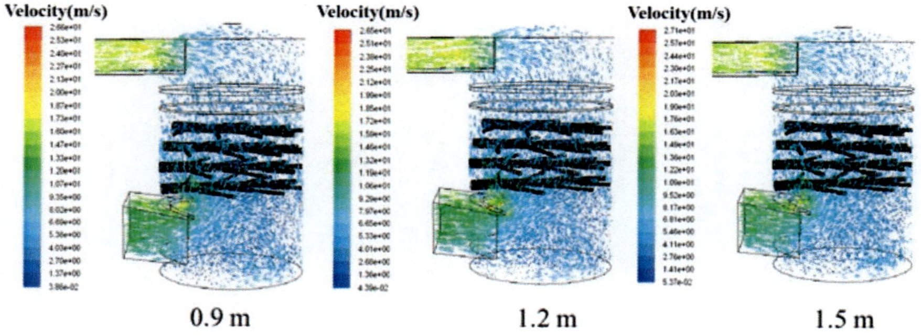

Fig. 4.51 Effect of the deflector length on the flow field [39]

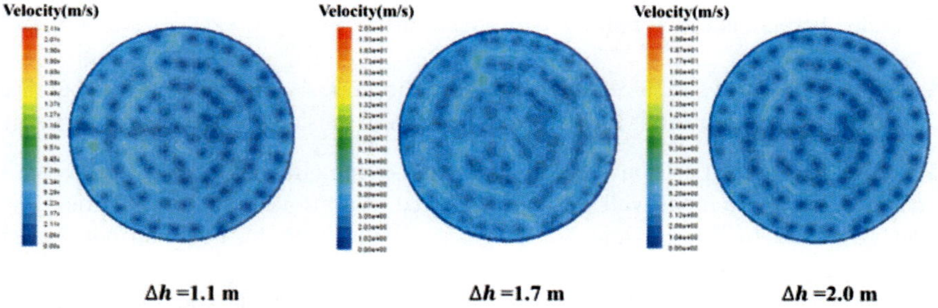

Fig. 4.52 Effect of the spray layer spacing on the liquid distribution [39]

d. *Spacing of the Spray Layers*

Although the spray system influences the distribution of the liquid phase in the scrubber, the spacing of the spray layer has no obvious effect on the coverage area of the liquid phase, as shown in Fig. 4.52. However, as the spacing increases, the contact time between the gas and liquid phases increases, thereby enhancing mass transfer. Still, increasing the spacing increases the pressure drop and operating costs. Therefore, these factors need to be balanced when designing a suitable arrangement of spray layers.

e. *Spray Angle of Nozzles*

The nozzle spray angle can influence the coverage area of the liquid. Spray angles $\geq$ 90° generally allow for adequate coverage, as shown in Fig. 4.53, where no obvious variation in the coverage area is seen between angles of 120° and 90°. Although a larger spray angle can enhance mass transfer between phases, it is accompanied by a trade-offs of larger pressure drops, operating costs and desulfurization efficiency.

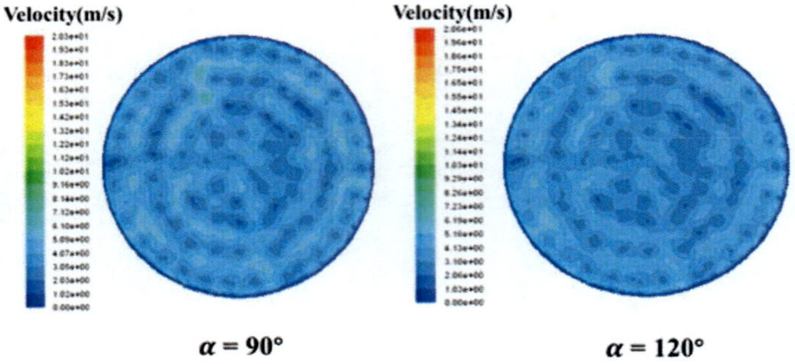

Fig. 4.53 Effect of the spray angle of nozzles on the liquid distribution [39]

4.3.3.3 Internals for Scrubber Enhancement

Introducing internal structures, such as distribution rings, sieve trays, and turbulence flow grids, can improve mixing between phases and thus the overall SO_2 absorption.

Distribution rings: Distribution rings are center-symmetrical ring-shaped structures arranged along the inner wall of the tower used to improve scrubber flow distribution. As shown in Fig. 4.54a and b, the gas stream flowing through distributor rings converges toward the center of the scrubber rather than remaining in a high-velocity gas stream near the wall. Distribution rings can also help make the droplets hitting wall return to the center of the scrubber (Fig. 4.54c) and thus increase the concentration of active absorbent liquid, as shown in Fig. 4.54d. As a result, installing distribution rings leads to an enhanced gas–liquid mixing zone in the center of the scrubber, thereby enhancing SO_2 absorption; as shown in Fig. 4.55, SO_2 concentration decreases more quickly in the distribution ring zone than in the scrubber without distribution rings.

Sieve tray: A sieve tray is introduced, as shown in Fig. 4.56, to modify the flow characteristics and contact state of the gas and liquid phases. Above each tray, a liquid holding layer forms; as the gas stream flows through the sieve tray, it is dispersed in the liquid phase, thereby causing a bubbling region, as is seen in a bubbling column. As shown in Fig. 4.57, gas–liquid contact above the sieve tray goes through five stages as the velocity of the gas stream increases: a thin liquid layer, a froth region, bubble vibration, vibration intensification, and no-liquid shed. Among them, bubble vibration is optimal, as it allows for strong gas–liquid mixing and turbulence intensification by bubble formation and breaking, resulting in a large gas–liquid interfacial area and a long contact time.

The effect of sieve trays on desulfurization of a scrubber under different working conditions was demonstrated using a pilot-scale platform compromising SO_2 generation, flue gas production, absorption, oxidized air production, limestone slurry preparation, and by-product processing systems with a flue gas flow rate of 5000 m^3/h, a scrubber 0.9 m in diameter and 14.5 m in height, and four spray layers. At a pH of 5.8,

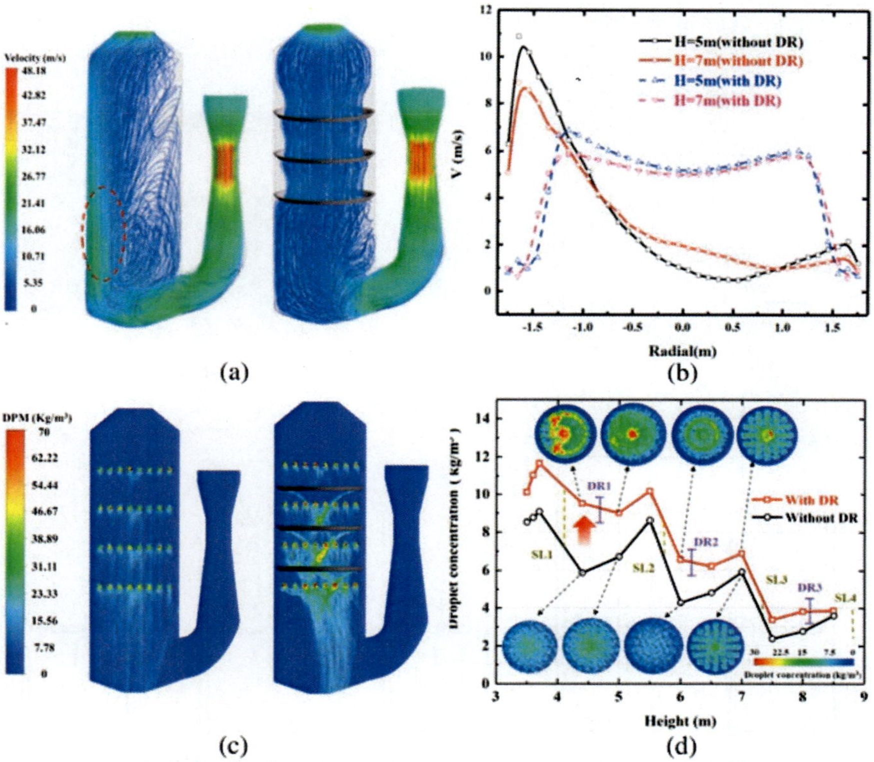

Fig. 4.54 Improvement of the gas–liquid distribution and SO_2 removal using distribution rings: **a** comparison of flow field **b** velocity variation along radial direction **c** comparison of droplets concentration distribution **d** droplets concentration variation as height variation

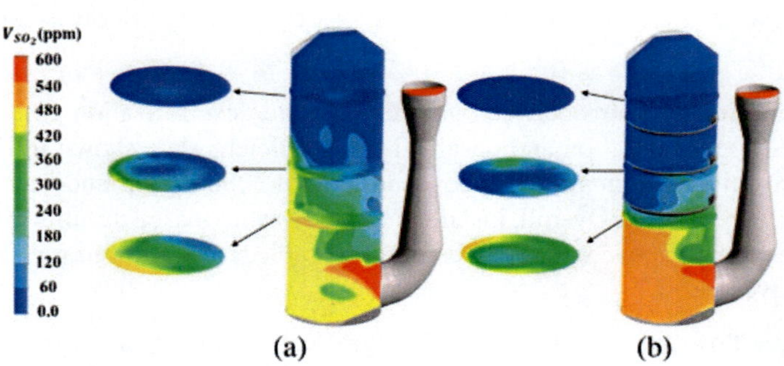

Fig. 4.55 Comparison of the SO_2 distribution in a scrubber **a** with and **b** without a distribution ring

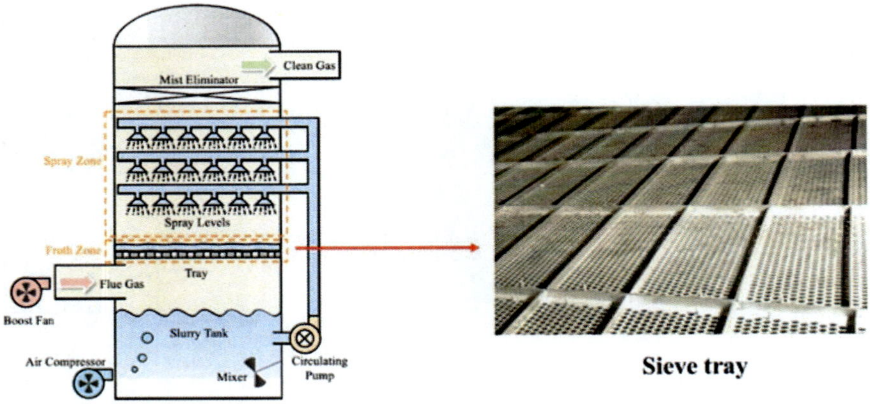

Fig. 4.56 Sieve tray in a scrubber

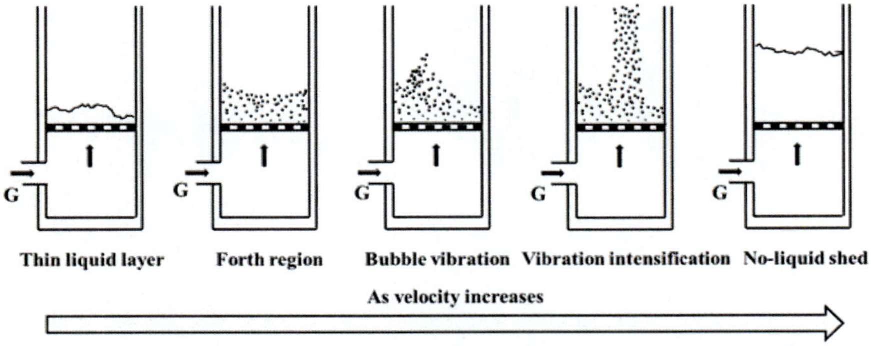

Fig. 4.57 Contact state of the gas–liquid above the sieve tray as the gas velocity increases [40]

the resulting relationships between gas velocity and desulfurization efficiency and between SO_2 concentration and desulfurization efficiency are shown in Fig. 4.58a and b, respectively, at an inlet SO_2 concentration of 3000 mg/m^3 and a gas velocity of 3.7 m/s, respectively. Overall, installing a sieve tray increased desulfurization efficiency (remaining above 99% at all conditions studied) and allowed for a reduction in spray layers used.

Turbulence flow grid: A turbulence flow grid comprises a row of tubes arranged side by side, as shown in Fig. 4.59, and can be classified into several types according to the tube shape and arrangement. As shown in Fig. 4.60a, installing a turbulence glow grid improved the uniformity of the velocity field in the scrubber owing to improved contact and mixing between the gas and liquid phases. Further, the SO_2 concentration is more evenly distributed in the scrubber, leading to a lower outlet concentration as the uniformity of the velocity field increases, as shown in Fig. 4.60b.

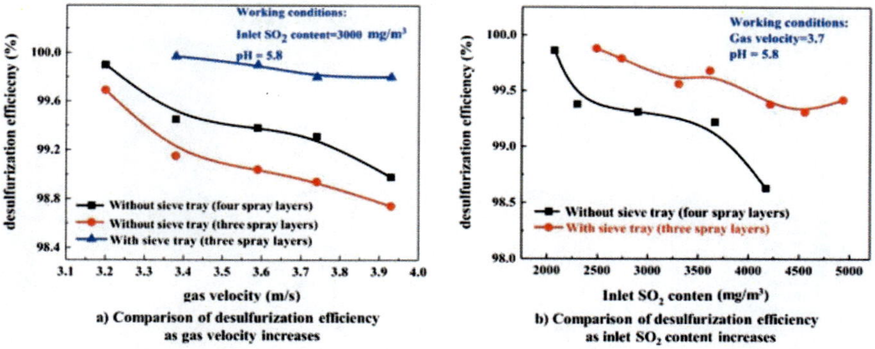

Fig. 4.58 SO_2 removal enhancement using a sieve tray under various operating conditions

Fig. 4.59 Turbulence flow grid in a scrubber

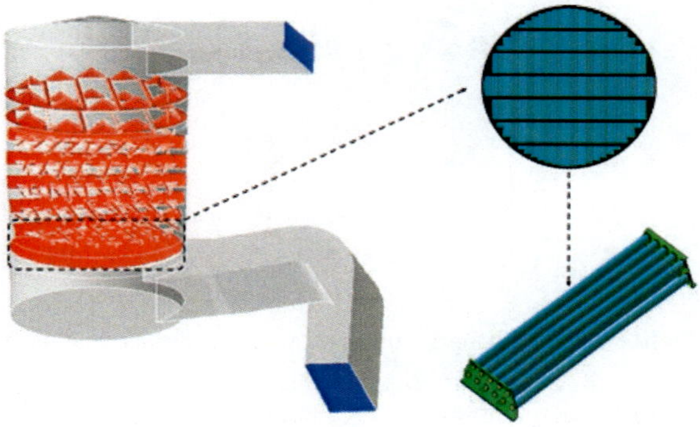

A liquid holding area can form above the turbulence flow grid, where SO_2 in the gas stream is absorbed in both the liquid holding area and spray zone, further improving the desulfurization efficiency. This effect has also been demonstrated experimentally on a pilot-scale platform, as shown in Fig. 4.61, where the slurry pH is 5.8, the gas velocity is 3.7 m/s, and three spray layers are used. Although the desulfurization efficiency decreased as inlet SO_2 concentration increased, the turbulence flow grid mitigated this effect and allowed for better SO_2 removal even under increased inlet SO_2 concentrations.

4.3.3.4 Applications

Case 1. Enhanced scrubber with sieve tray: The scrubber of the 660 MW coal-fired power unit shown in Fig. 4.62 was improved by adding a sieve tray; the design parameters are shown in Table 4.10. On-site operating data confirmed an SO_2 removal efficiency of > 98.5% under different working conditions, as shown in Table 4.11.

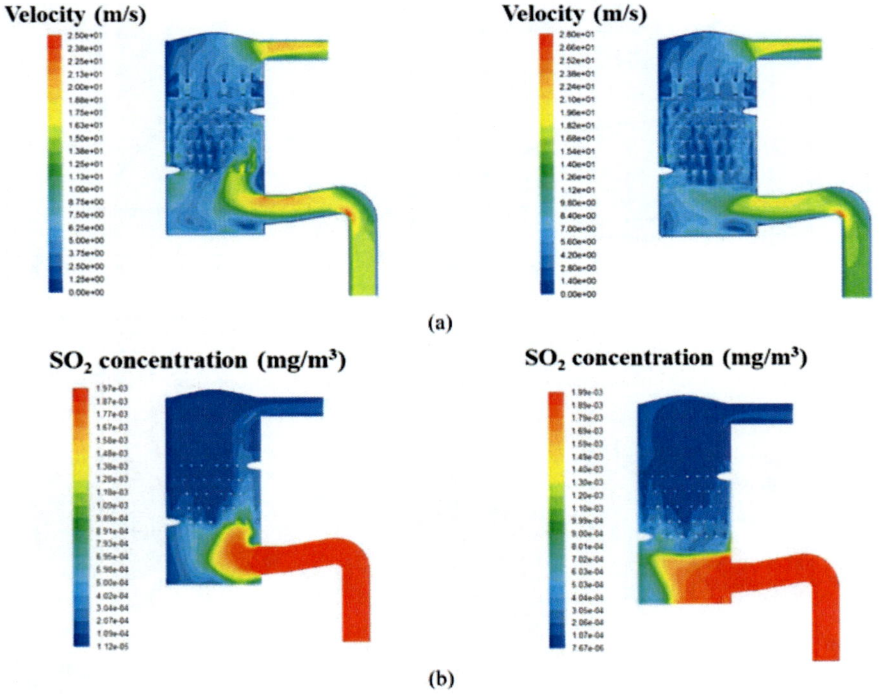

Fig. 4.60 Effect of the turbulence flow grid on the distribution of **a** velocity and **b** SO_2 concentration

Fig. 4.61 SO_2 removal enhancement with turbulence flow grid

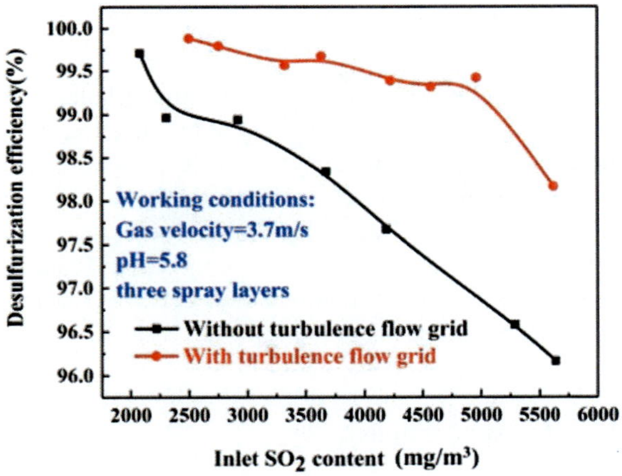

Fig. 4.62 Enhanced scrubber with sieve tray in a 600 MW unit

Table 4.10 Design parameters of scrubber used in case 1

Item	Parameter	Unit
Unit capacity	660	MW
Flue gas treatment	1,920,945	m^3/h, dry, 6% O_2
Inlet SO_2 concentration	2317	m^3/h, dry, 6% O_2
Slurry flow rate	10,400 × 4	m^3/h
L/G	19.1	L/m^3
Number of spray layers	4	
Height of sieve tray	17.6	m

Table 4.11 On-site operating data from case 1

Working conditions	Load	Spray layers	pH	Inlet SO_2 concentration	Outlet SO_2 concentration	Desulfurization efficiency
Unit	MW			mg/m^3	mg/m^3	%
1	534	3	5.33	1494.17	13.50	99.10
2	597	3	5.29	1650.07	19.79	98.8
3	512	3	5.28	1552.55	17.24	98.89
4	576	3	5.33	1405.45	14.07	99.00
5	584	3	5.35	1426.23	13.56	99.05
6	547	3	5.36	1749.27	14.31	99.18
7	618	3	5.34	1870.65	19.85	98.94
8	612	3	5.26	1831.43	24.18	98.68
9	667	4	4.92	1896.22	20.26	98.93

Case 2. Enhanced scrubber with sieve tray: A sieve tray was added to the scrubber of a 300 MW coal-fired power unit as shown in Fig. 4.63. From the on-site operating data shown in Table 4.12, the SO_2 removal efficiency remained above 98.5% at an inlet SO_2 concentration of approximately 2000 mg/m^3.

Case 3. Enhanced scrubber with turbulence flow grid at low inlet SO_2 concentration: A turbulence flow grid was installed under the first spray layer of the scrubber of the 600 MW coal-fired power unit shown in Fig. 4.64 with a relatively low SO_2

Fig. 4.63 Enhanced scrubber with sieve tray in a 300 MW unit

Table 4.12 On-site operating data of scrubber used in case 2

Load (MW)	Flow rate (m^3/h)	Inlet SO_2 concentration (mg/m^3)	Outlet SO_2 concentration (mg/m^3)	Desulfurization efficiency (%)
301.0	1,069,331.6	2038.1	16.8	99.2
300.3	1,064,544.9	2004.3	16.8	99.1
300.5	1,060,478.8	2043.9	18.6	99.0
300.5	1,062,590.6	2045.2	18.7	99.0

Fig. 4.64 Enhanced scrubber with the turbulence flow grid in a 600 MW unit

inlet concentration. A comparison between the desulfurization efficiency before and after installing the turbulence flow grid under an inlet SO_2 concentration of 900–1000 mg/m^3 and L/G of 7.6 is shown in Table 4.13. Here, installing the turbulence flow grid increased the desulfurization efficiency by 6.55% and 3.66% under 75% and 100% loads, respectively.

Case 4. Enhanced scrubber with turbulence flow grid at high inlet SO_2 concentration: A turbulence flow grid was installed in the scrubber of the 330 MW coal-fired power unit shown in Fig. 4.65 with a relatively high SO_2 inlet concentration. The resulting desulfurization efficiency under different loads is listed in Table 4.14. Under 80% and 100% loads, the desulfurization efficiency reached

Table 4.13 Desulfurization efficiency improvement of the enhanced scrubber

Items	Unit	600 MW (100% load)		450 MW (75% load)	
		Unmodified	Modified	Unmodified	Modified
Inlet gas flow rate	m^3/h	1,955,085	2,012,396	1,590,178	1,636,715
L/G	L/m^3	7.6	7.6	7.6	7.6
Desulfurization efficiency	%	90.83	94.49	90.65	97.2

Fig. 4.65 Enhanced scrubber with the turbulence flow grid in a 330 MW unit

Table 4.14 Desulfurization efficiency of the enhanced scrubber under different loads

Item	Unit	330 MW (100% load)	270 MW (80% load)
Inlet gas flow rate	m^3/h	1,104,097	895,623
Inlet SO_2 concentration	mg/m^3	4489	4595
Outlet SO_2 concentration	mg/m^3	29.6	27
Desulfurization efficiency	%	99.35	99.42

99.42% and 99.35%, respectively, for a respective outlet SO_2 concentration of 27 and 29.6 mg/m^3, thereby satisfying the ultra-low emission limits in both cases.

4.4 Simultaneous Control of Multiple Pollutants

To efficiently and stably meet the increasingly strict emission standards, researchers have aimed to develop integrated technologies capable of simultaneously eliminating multiple pollutants [41]. In this section, technologies aiming to simultaneously remove SO_2 and NO_x or trace heavy metals are summarized.

4.4.1 *Simultaneous Control of NO_x*

The solubility of NO, which accounts for 90–95% of the total NO_x in flue gas [42], is extremely low (0.032 g/dm^3 in water) [43]. However, higher-order nitrogen species, such as NO_2, NO_3, and N_2O_5, can react with water to form NO_2^- and NO_3^-, which can be removed during WFGD [44]. Preoxidation processes, such as selective catalytic oxidation, corona discharge oxidation, and ozone oxidation, can be used to transform insoluble NO into highly soluble NO_2 and N_2O_5, which can then be removed with SO_2 during the wet absorption process. However, achieving a high NO oxidation efficiency in these preoxidation processes requires high operating costs. Researchers have thus aimed to develop techniques to either allow NO absorption in the presence of NO_2 during limestone-based WFGD or enhance NO reduction, a few of which are discussed in this section.

4.4.1.1 Absorption and Reaction Mechanism

The results discussed in this section from a series of NO_x absorption experiments performed in a cylindrical double-stirred tank reactor with a volume of 1250 mL (100 mm in diameter and 160 mm in height) and two nested rotating axial impellers installed on stirring rods for gas- and liquid-phase mixing. Each stirring rod was locked to a pulley, and each pulley was connected to a drive pulley with a belt.

Effect of $CaCO_3$ concentration on NO_x absorption: As NO is insoluble in water, the effect of $CaCO_3$ concentration on the NO absorption rate was trivial, as shown in Fig. 4.66; the NO concentration between the inlet and outlet was reduced by only 4–7 ppm, corresponding to a NO absorption rate of approximately 2.885×10^{-7} mol/(m^2 s) at equilibrium.

Reaction products of NO_x absorption: To clarify the absorption mechanism of NO_x in the $CaCO_3$ slurry, the presence of nitrite and nitrate ions (i.e., NO_2^- and NO_3^-, respectively) can be analyzed. The absorbed NO_x in a fixed period represents the

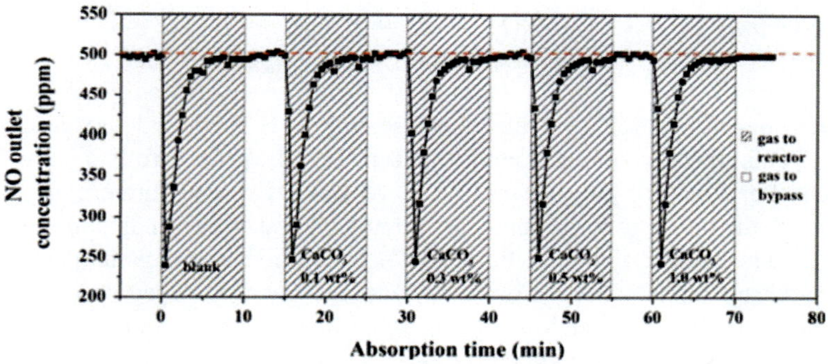

Fig. 4.66 Effect of $CaCO_3$ concentration on NO absorption. NO inlet concentration = 500 ppm, $T = 50$ °C [45]

area under the NO_x absorption rate curve and thus can be calculated by integrating the NO_x absorption rate as follows:

$$n_{NO_x} = \int_0^{t^*} N_{NO_x}(t) \cdot S dt, \tag{4.80}$$

where n_{NO_x} represents the total NO_x absorbed in the slurry, mol; $N_{NO_x}(t)$ represents the NO_x absorption rate at time t, mol/(m^2 s); and t represents the total absorption time, s. The concentration of nitrate and nitrite ions in the $CaCO_3$ slurry increase as the absorption proceeds and the calculated total nitrate and nitrite concentrations are consistent with the experimental values, as shown in Fig. 4.67a. The mole percentage of nitrite ions, i.e., the molar ratio of NO_2^- to ($NO_2^- + NO_3^-$) decreases gradually during the absorption time. The mole percentage of nitrite was approximately one-third when the absorption reaches equilibrium owing to the decomposition of HNO_2, as shown in Fig. 4.67a.

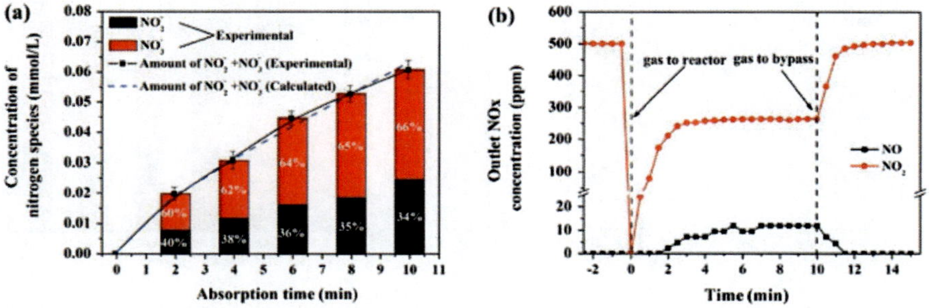

Fig. 4.67 NO_2 absorption in $CaCO_3$ slurry. **a** Experimental and calculated concentration of absorption products; **b** NO and NO_2 concentrations in the outlet [45]

$$3HNO_2(aq) \leftrightarrow 2NO(g) + NO_3^- + H^+ + H_2O(l),\ \Delta_f G^0 = -35.2\ \text{kJ/mol}. \tag{4.81}$$

HNO_2 decomposition can thus increase nitrate content and decrease nitrite content. This phenomenon is demonstrated in Fig. 4.67b, where NO starts being released after 2 min of NO_2 absorption in the $CaCO_3$ slurry. Furthermore, as NO is released, the outlet concentrations of NO and NO_2 stabilize, indicating that NO_2 uptake reaches equilibrium and further NO_2 absorption is inhibited. This implies that the release of NO does not occur as soon as NO_2 is absorbed into the $CaCO_3$ slurry but is necessary for HNO_2 accumulation. In addition, the liquid-phase oxidation of NO_2^- by dissolved oxygen (DO), as described by Eq. (4.82), contribute to the reduced mole percentage of nitrite lower more than 50%.

$$2NO_2^- + O_2(aq) \leftrightarrow 2NO_3^-,\ \Delta_f G^0 = -158.2\ \text{kJ/mol}. \tag{4.82}$$

Effect of NO_x composition on NO_x absorption in limestone slurry: The following reactions occur when NO and NO_2 coexist in the gas stream.

$$NO(g) + NO_2(aq) \leftrightarrow N_2O_3(g),\ \Delta_f G^0 = +3.5\ \text{kJ/mol}, \tag{4.83}$$

$$N_2O_3(g) + H_2O(aq) \leftrightarrow 2HNO_2(aq,\ \text{undissociated}),\ \Delta_f G^0 = +2.7\ \text{kJ/mol} \tag{4.84}$$

The existence of NO_2 may promote NO absorption. As the molar ratio of NO_2 to $(NO + NO_2)$ (OR) increased from 0 to 0.5, the NO absorption rate increased from 2.885×10^{-7} to its maximum at 1.057×10^{-6} mol/(m^2s), as shown in Fig. 4.68a. This is because N_2O_3, which is produced with the existence of NO and NO_2, has a higher solubility than NO does. Hence, N_2O_3 formation significantly increases NO absorption rate. Moreover, as OR increased from 0.2 to 1.0, the absorption rate of

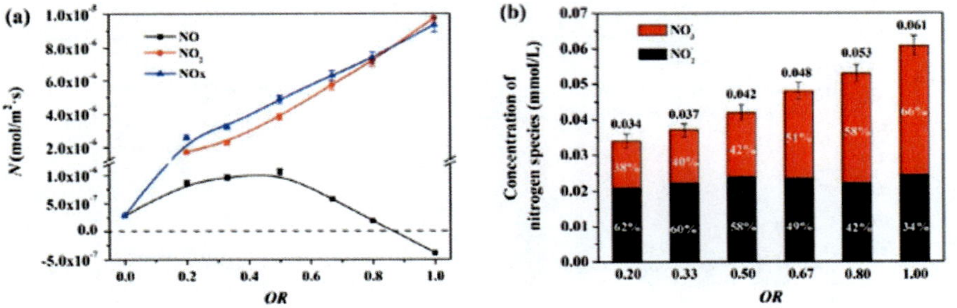

Fig. 4.68 **a** Effect of NO_x compositions in the gas stream on NO_x absorption rates; **b** nitrite/nitrate concentrations in the slurry for different molar ratios of NO_2 to $(NO + NO_2)$ (ORs; total NO_x inlet concentration = 500 ppm, T = 50 °C) [45]

NO_2 was much higher than that of NO under the same condition and increased from 1.720×10^{-6} to 9.741×10^{-6} mol/(m^2 s), indicating that the total NO_x absorption rate increases with increases in the OR.

The concentration of nitrite/nitrate ions at equilibrium in the $CaCO_3$ slurry increases as the OR increases, as shown in Fig. 4.68b. Here, NO_x is more soluble and reactive when more oxidized NO_x species are present in the gas stream, leading to a higher total dissolved NO_x dissolved.

Effect of SO_2 on NO_x absorption: The influence of SO_2 concentration on the NO_x absorption rate at an inlet NO_x concentration of 500 ppm with different ratios of NO and NO_2 at 50 °C is shown in Fig. 4.69a. When OR = 0, the NO_x absorption rate remained nearly constant at 2.887×10^{-7} mol/(m^2 s) as the SO_2 concentration was increased from 0 to 2000 ppm; at an OR of 0.5 or 1, however, the NO_x absorption rate increased with increasing SO_2 concentrations. This suggests that SO_2 does not promote NO absorption but promotes the overall NO_x absorption rate when inlet NO_x is a mixture of NO and NO_2. SO_2 that promoted NO_x absorption may be related to S(IV) ions, the hydrolysis products of SO_2.

A breakdown of the nitrate and nitrite ions in the $CaCO_3$ slurry with different inlet SO_2 concentrations is shown in Fig. 4.69b, where cases 1 and 2 represent ORs of 1 and 0.5, respectively. Overall, the concentration of nitrogen species increased in the presence of SO_2. At an inlet OR of 1, the mole percentage of nitrite increased from 34 to 41% as the SO_2 concentration increased from 0 to 2000 ppm. However, at an OR of 0.5, the mole percentage of nitrite decreased from 58 to 51% with increased SO_2 concentration because of the reactions between nitrite and S(IV).

Effect of NO_x composition on NO_x absorption in ammonia-based WFGD system: Ammonia-based wet flue gas desulfurization is a widely used desulfurization method using ammonia to absorb SO_2. NO_x absorption in $(NH_4)_2SO_3$ is strongly related to the NO_2/NO_x concentration ratio, as shown in Fig. 4.70, with an $(NH_4)_2SO_3$ concentration of 0.1 mol/L, inlet NO_x concentration of 500 ppm, and a temperature of 30 °C. As the NO_2/NO_x ratio increased, the NO_x absorption rate and equilibrium content

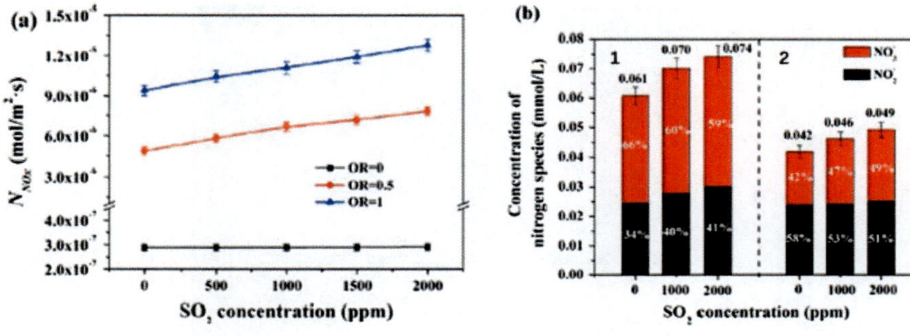

Fig. 4.69 Effect of SO_2 concentration on **a** NO_x absorption rates for different ORs and **b** nitrite/nitrate concentration in the liquid phase as when (1) OR = 1, NO_2 inlet concentration = 500 ppm and (2) total NO_x inlet concentration = 500 ppm and OR = 0.5 (T = 50 °C) [45]

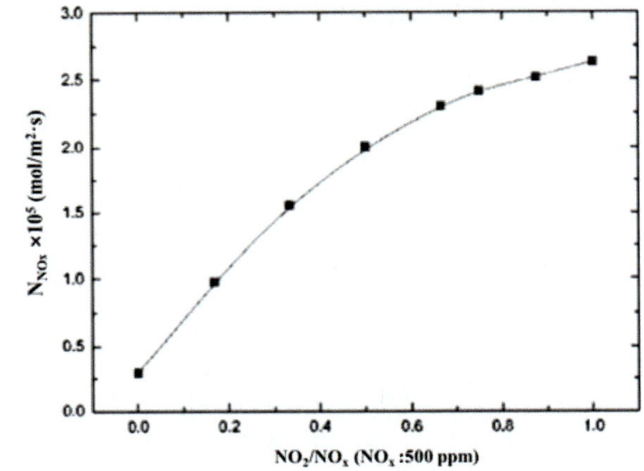

Fig. 4.70 Effect of NO_2/NO_x on the NO_x absorption rate. Gas–liquid interfacial area, 50.27 cm^2; T, 30 °C; inlet total concentration of NO + NO_2, 500 ppm [46]

of N_2O_4 increased, the latter of which was owing to the increased concentration of NO_2 in the gas-phase equilibrium mixture. Furthermore, the N_2O_3 concentration increased slightly. Overall, the NO_x absorption rate increases as the NO_2/NO_x ratio and the dissolvable composition of N_2O_4 and N_2O_3 in the gas phase increase.

Proposed pathways for NO_x absorption: NO_x absorption in a $CaCO_3$ slurry involves two absorption pathways (i.e., pure NO_2 absorption and simultaneous NO and NO_2 absorption); however, the absorption model of NO_x in a $CaCO_3$ slurry in the presence of SO_2 and O_2 is more complex. A proposed global pathway of NO_x absorption and nitrite/nitrate formation is shown in Fig. 4.71 and includes a variety of reversible, redox, and disproportionation reactions. Here, the main gaseous components include NO, NO_2, SO_2, and O_2. The solubility of NO is enhanced considerably by oxidation or the formation of N_2O_3. The absorption products of NO_2 or N_2O_4 are

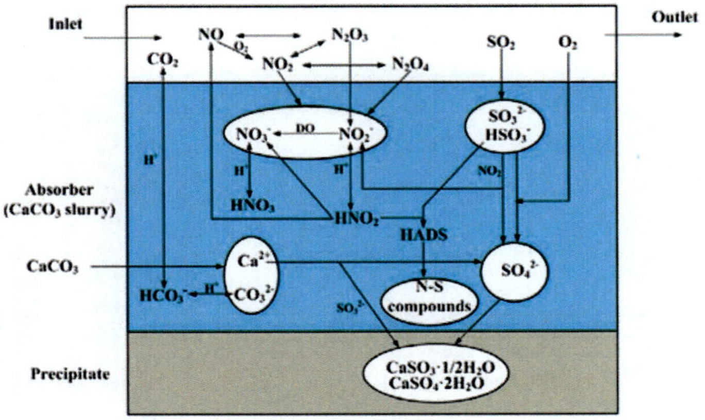

Fig. 4.71 Proposed pathways for NO_x absorption in $CaCO_3$ slurry in the presence of SO_2 and O_2 [45]

nitrite and nitrate, whereas only nitrite is formed from N_2O_3. Nitrite can combine with H^+ to produce nitrous acid, which is unstable in the liquid phase and partly decomposes, releasing NO. Aqueous NO_2^- is oxidized to NO_3^-. In the presence of SO_2, sulfite and bisulfate formation favors NO_2 absorption, with NO_2^- converting into NO_3^- and S(IV) converting into SO_4^{2-}. In addition, HNO_2 can react with HSO_3^- to form hydroxylamine disulfonate (HADS), a precursor to numerous sulfur–nitrogen compounds. As the slurry pH decreases due to the absorption of NO_x and SO_2, the dissolution rate of $CaCO_3$ increases, the carbonate transforms into bicarbonate, and the bicarbonate turns into CO_2, which escapes from the liquid phase. Calcium ions can also react with S(IV) and SO_4^{2-} to form precipitates, such as $CaSO_3 \cdot 1/2H_2O$ and $CaSO_4 \cdot 2H_2O$.

4.4.1.2 Enhancement of NO_x Reduction

a. *Preoxidation Combined with Chemical Absorption*

Newly developed combined preoxidation and chemical absorption systems oxidize NO in coal-fired flue gas without a strong oxidizer and then simultaneously absorb and remove NO_x and SO_2 during WFGD. These combined systems can reduce operating costs and avoid secondary pollution caused by adding a strong oxidizer, corrosion intensification, and oxidizer leaks. Preoxidation methods currently used in this combined technology include catalytic, plasma-discharge, and ozone oxidation.

Catalytic oxidation: Catalytic oxidation technology can be divided into conventional and photocatalytic oxidation technologies. The catalysts used for NO oxidation include transition metal oxides and various supported noble metals [47–49].

Plasma-discharge oxidation: Low-temperature plasma or nonequilibrium plasma has a high electron temperature and low gas temperature and can produce many active radials, making reactions impossible at normal temperatures possible. The NO_x oxidation mechanism by plasma is complex. High-energy electrons in plasma can transform oxygen and water in the air into O, OH, and other free radicals, which can oxidize NO in the flue gas [50, 51].

Ozone oxidation: Ozone oxidation takes advantage of the of ozone (O_3) to oxidize low-valent pollutants in flue gases to high-valent pollutants for removal. Compared with this plasma method, this method has a lower energy consumption and lower operating temperature. Moreover, O_3 has a strong oxidative removal effect on various volatile organic compounds (VOCs) to produce CO_2 and H_2O. Here, NO in the combustion flue gas is oxidized into higher-valence nitrogen oxides [52], mainly including NO_2 and N_2O_5, as shown in Fig. 4.72. These more-soluble nitrogen oxides can then be absorbed with SO_2 during WFGD.

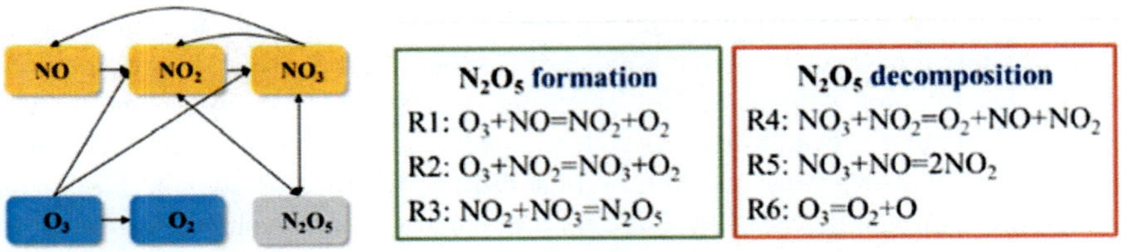

Fig. 4.72 Schematic of the main reaction pathway for NO oxidation by ozone (O_3) [42]

b. *Additive-Enhanced Absorption*

Liquid-phase oxidation technologies involve the addition of a strong oxidizer (i.e., $NaClO_2$ or $KMnO_4$) to the liquid phase to convert NO into more-soluble nitrogen oxides (NO_2, NO_2^-, and NO_3^-).

$NaClO_2$: Here, NO reacts with ClO_2^- to form NO_2; NO_3^- is produced from the reaction between ClO_2^- and NO_2 [53] as

$$2NO + ClO_2^- \rightarrow 2NO_2 + Cl^-, \tag{4.85}$$

$$4NO_2 + ClO_2^- + 2H_2O \rightarrow 4NO_3^- + Cl^- + 4H^+. \tag{4.86}$$

The overall reaction can thus be described as

$$4NO + 3ClO_2^- + 2H_2O \rightarrow 4NO_3^- + 3Cl^- + 4H^+. \tag{4.87}$$

$KMnO_4$: In a strongly alkaline aqueous solution [54],

$$NO + 2MnO_4^- + 2OH^- \rightarrow NO_2^- + MnO_4^{2-} + H_2O. \tag{4.88}$$

In a weakly alkaline aqueous solution,

$$NO_2^- + 2MnO_4^- + 2OH^- \rightarrow NO_3^- + 2MnO_4^{2-} + H_2O, \tag{4.89}$$

$$3NO_2^- + 2MnO_4^- + H_2O \rightarrow 3NO_3^- + 2MnO_2 + 2OH^-. \tag{4.90}$$

In an acidic aqueous solution,

$$5NO + 3MnO_4^- + 4H^+ \rightleftharpoons 5NO_3^- + 3Mn^{2+} + 2H_2O. \tag{4.91}$$

Many other chemical oxidizers, such as NaClO, ClO_2, Cl_2, $Na_2S_2O_8$, and H_2O_2, have been added into the desulfurization slurry to simultaneously remove SO_2 and NO_x. The overall reactions of NO oxidation using different oxidants are summarized in Table 4.15. As shown in Fig. 4.73, commonly used oxidants promote NO_x absorption as $NaClO_2 > KMnO_4 > H_2O_2 >$ NaClO.

Table 4.15 Reactions of NO oxidation with other oxidants in absorption solutions

Oxidizers	$NO \rightarrow NO_2$	$NO \rightarrow NO_3^-$
NaClO	$ClO^- + NO \rightarrow Cl^- + NO_2$	
ClO_2	$ClO_2 + NO \rightarrow ClO + NO_2$ $2ClO_2 + 5NO + H_2O \rightarrow 2HCl + 5NO_2$	$ClO_2 + 5NO + 4H_2O \rightarrow 5HNO_3 + 3HCl$
Cl_2	$Cl_2 + NO + H_2O \rightarrow 2Cl^- + NO_2 + 2H^+$	$2Cl_2 + 2NO + 3H_2O \rightarrow 4Cl^- + NO_3^- + 6H^+ + NO_2^-$
$NaClO/NaClO_2$		$ClO_2^- + 2NO + ClO^- + H_2O \rightarrow 2Cl^- + 2NO_3^- + 2H^+$
$Na_2S_2O_8$		$S_2O_8^{2-} \rightarrow 2SO_4^-$ $SO_4^- + NO + H_2O \rightarrow 2H^+ + SO_4^{2-} + NO_2^-$ $SO_4^- + NO_2^- + H_2O \rightarrow 2H^+ + SO_4^{2-} + NO_3^-$
H_2O_2	$NO + H_2O_2 \rightarrow NO_2 + H_2O$ $H_2O_2 + hv \rightarrow 2OH$ $NO + OH \rightarrow NO_2 + H$	$3H_2O_2 + 2NO \rightarrow 2H^+ + 2NO_3^- + 2H_2O$

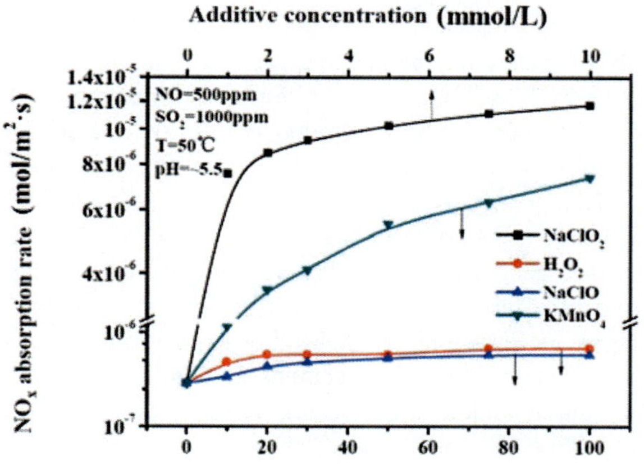

Fig. 4.73 Effect of different oxidants on NO_x absorption [55]

c. *Catalyst-Enhanced Absorption*

The results from an experimental study on the influence of metal ion catalysts (Fe^{2+}, Mn^{2+}, Co^{2+}, Mo^{6+}, Fe^{3+}, Al^{3+}, Ni^{2+}, and Zn^{2+}) on NO absorption are shown in Fig. 4.74 [55], under an NO_2 inlet concentration of 500 ppm, a temperature of 50 °C, a slurry pH of 5.5, and a metal ion catalyst concentration of 10 mmol/L. Overall, the NO absorption rate was improved using Fe^{2+} and Mn^{2+} catalysts, and absorption rate improvements were in the order of $Fe^{2+} > Mn^{2+} > Co^{2+} > Mo^{6+} > Fe^{3+} > Al^{3+} > Ni^{2+} > Zn^{2+}$. When the $CaCO_3$ slurry absorbs pure NO_2, NO gas is released, causing a negative NO absorption rate. As Fe^{2+} also significantly inhibited NO rerelease, the

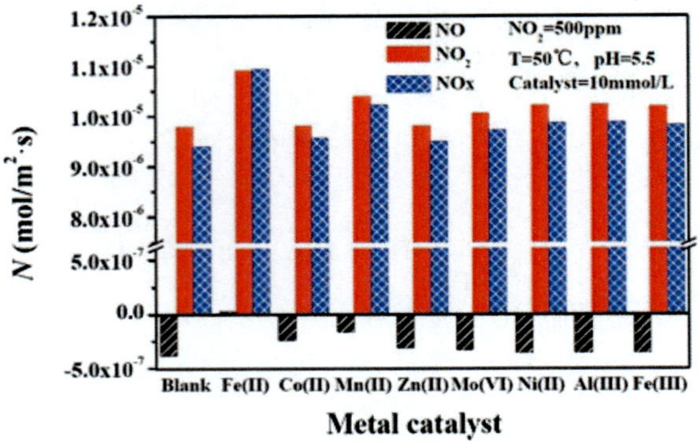

Fig. 4.74 Effects of different metal ions on NO_x absorption [55]

Fe^{2+} catalyst most effectively improves overall NO absorption by both promoting NO_x absorption and inhibiting NO rerelease.

4.4.2 Simultaneous Control of Hg

Toxic trace elements, such as Hg, As, Se, and Pb, released from coal combustion have high mobility and are known to be detrimental to environmental and human health; here, methods aiming to improve Hg removal during desulfurization are demonstrated to illustrate progress toward heavy metal pollutant control.

Heavy metals present in various forms in coal partially or completely vaporize when coal is burned in a furnace at temperatures around 1300–1500 °C. Hg present in flue gas as Hg^0 is highly volatile and cannot easily be adsorbed by fly ash or gypsum, although Hg may also be converted to Hg^{2+} or granular Hg^p (e.g., $HgCl_2$, HgO, $HgSO_4$, and HgS). Methods commonly used to remove Hg include preoxidation, absorption enhancement, and secondary conversion inhibition. The morphologic conversion of Hg during SCR is a complex heterogeneous oxidation reaction involving Hg, catalyst, and other flue gas components. SCR catalysts can effectively promote the oxidation of Hg in flue gas and transform insoluble Hg^0 into highly water-soluble Hg^{2+}, which can then be removed during WFGD [55]. In particular, vanadium-based SCR catalysts can convert Hg^0 to Hg^{2+} in flue gas and achieve high-efficiency deNO_x as was discussed in Sect. 2.4.1. In this system, HCl plays an important promoting role in the reaction, whereas the presence of NH_3 inhibits the morphogenetic transformation reaction of Hg [56]. Further Hg emission reduction can be achieved via an electrostatic precipitator (ESP) and/or desulfurization. As oxidized Hg^{2+} is soluble in water, it can easily be captured by the WFGD slurry and immobilized in the gypsum. In addition, O_2, NO_2, activated carbon, and Fe–Al compounds in fly ash can promote the conversion of remaining Hg^0 into Hg^p or Hg^{2+}. In this section, the movement patterns of Hg throughout the WFGD system and the effect of absorption process on Hg reduction are detailed.

Table 4.16 Parameters of CFPPs examined by Zheng et al. [57] to clarify Hg partitioning

Plant	Installed capacity (MW)	Boiler type	Coal type	Air pollution control devices
#1	1000	PC	Bit[a]	SCR + LLT[b]-ESP + WFGD + WESP
#2	660	PC	Bit[a]	SCR + LLT-ESP + WFGD + WESP
#3	660	PC	Bit[a]	SCR + CS[c]-ESP + WFGD and WESP
#4	660	PC	Bit[a]	SCR + LLT-ESP + WFGD + WESP
#5	50	CFB	Bit[a] + sludge	SNCR-SCR + EFF[d] + WFGD + WESP

[a]Bituminous coal
[b]Low-low temperature ESP
[c]Cold-side ESP
[d]Electric fabric filter

4.4.2.1 Site Description

The results discussed in this section were obtained from a comprehensive Hg partitioning analysis performed by Zheng et al. [57] in the solid and liquid streams in five typical ultra-low emission CFPPs. As summarized in Table 4.16, four of the facilities tested are pulverized coal (PC) boilers with installed capacities that range from 660 to 1000 MW, whereas the fifth is a lower-capacity CFB boiler that burns blended coal and waste sludge.

4.4.2.2 Partitioning of Hg in the WFGD System

The mean concentrations of Hg in the solid (gypsum, wastewater sludge, and limestone) and water (WFGD slurry water, wastewater, and process water) streams in the WFGD systems are shown in Table 4.17.

Table 4.17 Mean concentrations of Hg in WFGD [57]

Plant	Gypsum (mg/kg)	Wastewater sludge (mg/kg)	Limestone (mg/kg)	WFGD slurry water (μg/kg)	Wastewater (μg/kg)	Process water (μg/kg)
#1	0.048	9.232	0.009	2.17	154	< 1
#2	0.196	10.87	0.065	168	72.5	< 1
#3	1.153	7.233	0.003	2.25	2.25	< 1
#4	0.381	11.65	< 0.001	6.14	43.9	< 1
#5	0.084	2.93	0.050	42.9	57.5	9.2

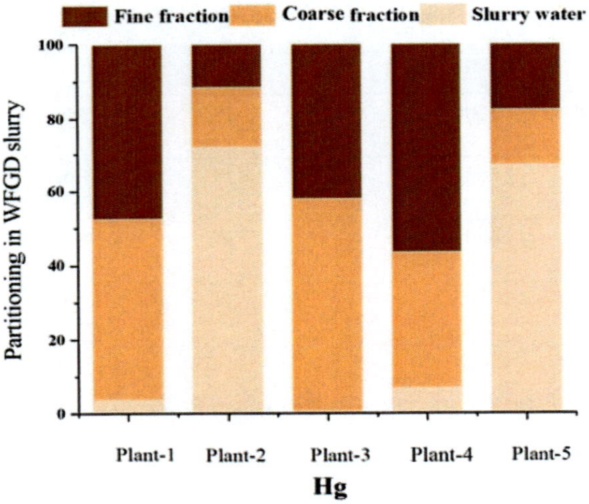

Fig. 4.75 Distribution of Hg in the desulfurized slurry [57]

The Hg distribution in the WFGD slurry (including fine fraction, coarse fraction, and slurry water) is presented in Fig. 4.75. In plants 2 and 5, 67% and 72% of the total Hg in the wet scrubber, respectively, was retained in the slurry water. In the other three plants, only 0.6–7% of the Hg remained in the liquid phase. Part of the Hg^{2+} in the flue gas that was absorbed by the slurry formed low-solubility sulfate and was then cured in the solid phase, whereas the rest was retained in the liquid phase, probably as $Hg(SO_3)_2{}^{2-}$, $HgCl_2$, and $Hg(OH)^+$. The different flue gas compositions and operating conditions in the WFGD may influence the Hg reaction pathways, Eqs. (4.92) to (4.96), thereby influencing the partitioning ratio between the solid and liquid phases.

$$SO_3{}^{2-} + Hg^{2+} \leftrightarrow HgSO_3, \tag{4.92}$$

$$SO_3{}^{2-} + HgSO_3 \leftrightarrow Hg(SO_3)_2^{2-}, \tag{4.93}$$

$$S(IV) + HgCl_2 \leftrightarrow Hg \cdot S(IV)\text{complexes}, \tag{4.94}$$

$$Hg^{2+} + OH^- \leftrightarrow Hg(OH)^+, \tag{4.95}$$

$$Hg(OH)^+ + OH^- \leftrightarrow Hg(OH)_2. \tag{4.96}$$

The distribution of Hg in the WFGD products (gypsum, wastewater, and sludge) is shown in Fig. 4.76. After WFGD, gypsum is produced by discharging part of the WFGD slurry from the wet scrubber through the gypsum cyclone station to vacuum belt dehydration. Wastewater, containing a certain amount of sludge, is discharged

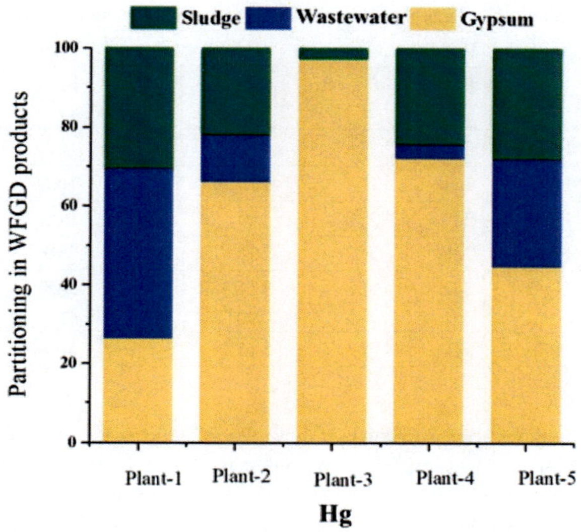

Fig. 4.76 Hg partitioning in WFGD products [57]

from the system via the wastewater cyclone station. The remaining part of the slurry returned to the scrubber. Gypsum is produced mainly from the coarse fractions in the WFGD slurry, whereas the waste sludge contains a large amount of fine fractions. The distributions of Hg in the produced gypsum, wastewater, and sludge varied by plant and were closely related to the size distribution in the WFGD slurry. Overall, approximately 2.6–35% of the Hg was retained in the waste sludge whereas 0.01–43% was retained in the wastewater. Special attention should be given to Hg during the wastewater treatment and waste sludge disposal processes.

4.4.2.3 Characteristics of Thermal Desorption of Hg in Solid Samples

The composition of Hg compounds retained in the solid products can be identified via temperature-programmed desorption (TPD), which is based on the different release curves of an elemental species during pyrolysis. The release temperature of Hg compounds increase as $HgI_2 < HgBr_2 < Hg_2Cl_2 = HgCl_2 < Hg(CN)_2 < HgCl_2O_8{\cdot}H_2O < Hg(SCN)_2 < HgS(red) < HgF_2 < Hg_2(NO_3)_2 \cdot 2H_2O < Hg(NO_3)_2{\cdot}H_2O < HgO$ (yellow, red) $< Hg_2SO_4 < HgSO_4$.

Hg compounds were identified by comparing the thermal desorption curves of the samples with the characteristic peak temperatures of reference Hg species. $HgBr_2$, $HgCl_2$, black HgS, red HgS, HgO, and $HgSO_4$ were identified in the fly ash and gypsum from plant 1, whereas $HgBr_2$ was not observed in sludge. Only $HgBr_2$, $HgCl_2$, and black HgS were present in WESP slag. Hence, different Hg compounds were found in various solid samples. Furthermore, the characteristic peak temperatures of the same compounds differed slightly between solid products, as detailed in Table 4.18. For example, the $HgBr_2$ characteristic peak ranged from 94 to 105 °C. The

Table 4.18 Peak temperature of Hg compounds in solid samples [58]

	Plant 1				Plant 5			
	Fly ash	Gypsum	Sludge	WESP slag	Fly ash	Gypsum	Sludge	WESP slag
$HgBr_2$	94	105	–	104	102	125	110	138
$HgCl_2$	161	163	159	135	163	190	214	178
HgS(b)	229	230	210	203	230	250	267	247
HgS(r)	288	287	262	–	290	310	325	316
HgO	342	358	324/383 (red)	–	345	355	383/427 (red)	376
$HgSO_4$	570	561	546	–	500	580	568	505

peak temperature of $HgCl_2$ in fly ash, gypsum, and sludge, was 161 °C, 163 °C, and 159 °C, respectively, whereas it was only 135 °C in WESP slag. Similarly, the peak temperature of black HgS in WESP slag was 203 °C, whereas it ranged between 210 and 230 °C in other solid products. The red HgS and HgO peaks were 262–288 °C and 324–358 °C, respectively. In addition, a small peak was observed after 383 °C owing to the release of red HgO. Among all Hg compounds present in the solid products, $HgSO_4$ showed the highest release temperature: desorption at 400 °C and peak temperature range of 546–570 °C.

Similarly, $HgBr_2$, $HgCl_2$, black HgS, red HgS, HgO, and $HgSO_4$ were present in the solid products sampled from plant 5, although the peak temperatures of the Hg species were slightly higher than those in plant 1. A small peak at 427 °C caused by red HgO was also observed in the sludge sampled from plant 5. The peak temperatures of $HgSO_4$ in fly ash and WESP slag were 500 °C and 505 °C, respectively, which were relatively lower than those in gypsum and sludge. The desorption temperature increased as $HgBr_2 < HgCl_2 <$ black HgS < red HgS < HgO < $HgSO_4$.

The mass distribution of Hg species in the solid products from plant 1 is presented in Fig. 4.77. In fly ash, the dominant Hg compounds were $HgCl_2$ and black HgS,

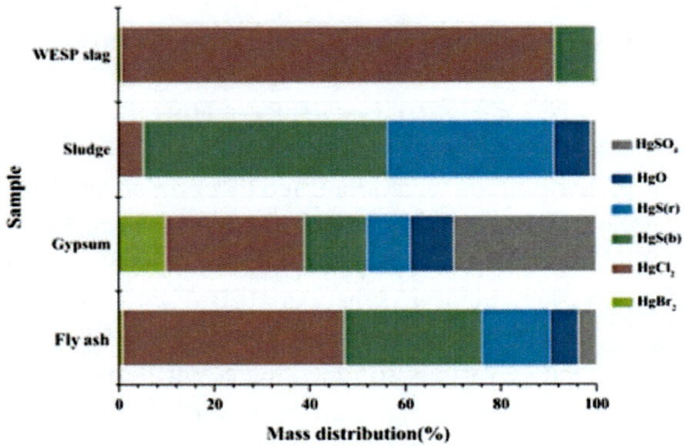

Fig. 4.77 Mass distribution of Hg compounds in solid samples from plant 1 [58]

accounting for 46.41 and 28.83% of the total Hg species, respectively. The high content of $HgCl_2$ was mainly attributed to the adsorption of Hg on the chlorinated active sites on the surface of the fly ash, whereas black HgS was formed by the interaction between Hg and unburned sulfur in the coal. Furthermore, $HgSO_4$ constituted 3.7% of the total Hg species in the fly ash likely owing to the formation of active acid sites caused by the condensation of SO_3 on the surface of fly ash in low-low temperature ESP (LL-ESP), which was conducive to Hg adsorption.

The most abundant Hg compounds in gypsum were $HgSO_4$ (29.74%), $HgCl_2$ (29.10%), and $HgBr_2$ (9.91%). The dissolution of Hg^{2+} in the flue gas in WFGD slurry and the bond formation with SO_4^{2-} contributed to the formation of $HgSO_4$. The presence of $HgCl_2$ and $HgBr_2$ was likely caused by the relatively high concentrations of chlorine and bromide ions in the WFGD slurry, respectively. Although it has been reported that Hg in gypsum is mainly in the form of black HgS [59], the high pH found in the WFGD slurry of plant 1 promoted the oxidation of sulfite and hydrosulfite to sulfate, thereby favoring interactions between sulfate and Hg.

The mass distribution of Hg in sludge was markedly different; black and red HgS accounted for 51.03 and 34.82% of the total Hg species, respectively, indicating that the HgS content of fine particles in the slurry of plant 1 was high. HgS is produced by the reaction between Hg^{2+} and H_2S in the liquid phase and is dependent on pH and redox potential in complex solutions. As the sludge was mainly composed of fine particles in the slurry, a high HgS content of sludge indicates that HgS was easily adsorbed onto the fine particles.

The primary Hg compound in WESP slag was $HgCl_2$ (90.68%), indicating that $HgCl_2$ can be easily enriched in fine particles. Owing to the relatively low stability and high absolute concentration of $HgCl_2$ in WESP slag, the discharge of these particles into the atmosphere may cause considerable harm to creatures and the environment.

Similarly, the mass distribution of Hg compounds in solid products sampled from plant 5 is shown in Fig. 4.78. As in plant 1, the composition of Hg compounds in fly ash mainly comprised $HgCl_2$ (41.63%) and black HgS (29.40%), although less concentration $HgSO_4$ (1.56%) was found than in plant 1. The dominant Hg species in gypsum was also HgS; black and red HgS accounted for 43.75 and 41.18% of the total

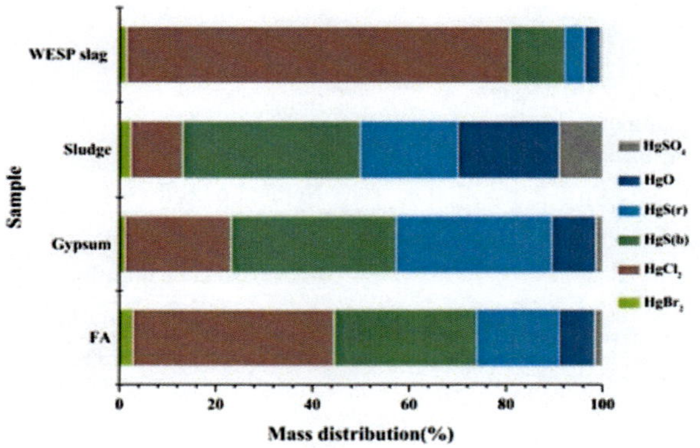

Fig. 4.78 Mass distribution of Hg compounds in solid samples from plant 5 [58]

HgS, respectively. Furthermore, $HgCl_2$ (28.22%) was present in gypsum. However, a much lower $HgSO_4$ fraction (1.74%) was present in plant 5 than in plant 1. Hg in sludge was mainly in the forms of black HgS (36.76%) and red HgS (20.11%); HgO and $HgSO_4$ were also present in higher fractions than in sludge (13.32% and 8.96%, respectively). The dominant form of Hg in WESP slag was $HgCl_2$ (79.17%), although minute quantities of red HgS (4.00%), HgO (3.29%), and $HgSO_4$ (0.46%) were also observed.

4.4.3 Applications

Here, the application of deep NO oxidation by ozone to a system comprising four 75 t/h CFB boilers is discussed, as detailed by Lin [60]. Before the system was applied, the operating SNCR system had an average NO_x emission concentration of approximately 150 mg/m^3 (ranging from 100 to 200 mg/m^3), which does not meet the ultra-low pollutant emission limit ($\leq$ 50 mg/m^3). The operating parameters of the boiler and desulfurization tower are shown in Table 4.19; a schematic of the ozone generation system, ozone mixing system, and wet scrubbing system is shown in Fig. 4.79. In the applied system, NO is oxidized by O_3 into high-valent NO_x (e.g., NO_2, NO_3, and N_2O_5), which can then be removed together with SO_2 by wet scrubbing because of their higher solubility. The total rate of flue gas was approximately 420,000 m^3/h.

In such multi-pollutant removal systems, ozone generators and air separators require large initial investment and operating costs. Minimizing the required O_3 dosage while satisfying NO_x emission standards could thus reduce emissions of primary pollutants in the flue gas, lower the investment and operating costs, and reduce the residual O_3 emissions. The amount of O_3 injection was adjusted by changing the ozone generator power. The resulting inlet and outlet NO_x concentration after applying the deep oxidation system is shown in Fig. 4.80. The applied system allowed for nearly all of the NO to be oxidized to NO_2 or even higher NO_x and an overall NO_x concentration below 50 mg/m^3. The outlet concentration of SO_2 satisfied the ultra-low emission limit. In addition, it can enhance the reduction of Hg.

Table 4.19 Design parameters of the desulfurization tower

Item	Parameter	Unit
Unit capacity	75	t/h
Number of boilers	4	
Inlet gas flow rate	420,000	m^3/h
Inlet NO_x concentration	100–200	mg/m^3, dry, 6% O_2
Outlet NO_x concentration	$\leq$ 50	mg/m^3, dry, 6% O_2
Flue gas temperature	130	°C

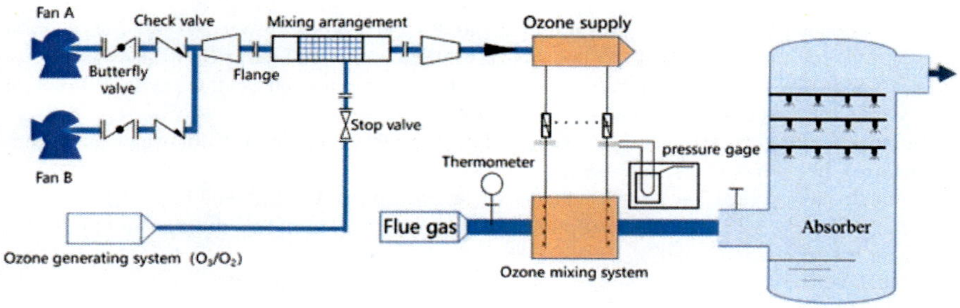

Fig. 4.79 Schematic of the NO deep oxidation system proposed by Lin [60]

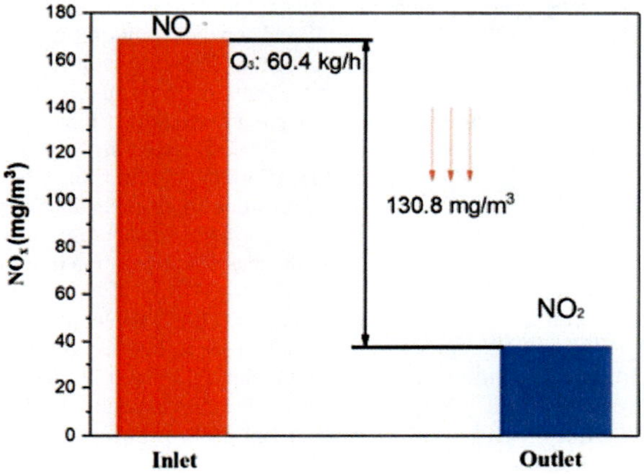

Fig. 4.80 Inlet and outlet NO_x concentration [60]

References

1. Toprac A, Rochelle G (1982) Limestone dissolution in stack gas desulfurization. Environ Prog Sustainable Energy 1(1):52–58
2. Pan D, Wu H, Yang L (2017) Investigation on the relationship between the fine particle emission and crystallization characteristics of gypsum during wet flue gas desulfurization process. J Environ Sci 55(005):303–310
3. Rezaei F et al (2015) SO_x/NO_x removal from flue gas streams by solid adsorbents: a review of current challenges and future directions. Energy Fuels 29(9):5467–5486
4. Zhou J, Wang H (2020) Study on efficient removal of SO_x and NO_x from marine exhaust gas by wet scrubbing method using urea peroxide solution. Chem Eng J 390:124567
5. Ding J, Zhong Q, Zhang S et al (2014) Simultaneous removal of NO_x and SO_2 from coal-fired flue gas by catalytic oxidation-removal process with H_2O_2. Chem Eng J 243:176–182
6. Zhao Yi et al (2010) Simultaneous removal of SO_2 and NO using M/$NaClO_2$ complex absorbent. Chem Eng J 160(1):42–47
7. Zhu J, Ye S, Bai J et al (2015) A concise algorithm for calculating absorption height in spray tower for wet limestone–gypsum flue gas desulfurization. Fuel Process Technol 129:15–23
8. Tian H et al (2021) Simultaneous magnesia regeneration and sulfur dioxide generation in magnesium-based flue gas desulfurization process. J Cleaner Prod 284:124720
9. Andreasen A, Mayer S (2007) Use of seawater scrubbing for SO_2 removal from marine engine exhaust gas. Energy Fuels 21(6):3274–3279

10. Gao X, Ding H, Du Z et al (2010) Gas–liquid absorption reaction between $(NH4)_2SO_3$ solution and SO_2 for ammonia-based wet flue gas desulfurization. Appl Energy 87(8):2647–2651
11. Kallinikos L et al (2010) Simulation of the operation of an industrial wet flue gas desulfurization system. Fuel Process Technol 91(12):1794–1802
12. Sander R (2015) Compilation of Henry's law constants (version 4.0) for water as solvent. Atmos Chem Phys 15(8):4399–4981
13. Whiteman W (1923) The two-film theory of gas absorption. Chem Metal Eng 29(1):147–157
14. Higbie R (1935) The rate of absorption of a pure gas into still liquid during short periods of exposure. Am Inst Chem Eng 31:365–389
15. Danckwerts P (1970) Gas-liquid reaction. McGraw-Hill Press, New York
16. Zhao Z, Zhang Y, Gao W et al (2021) Simulation of SO_2 absorption and performance enhancement of wet flue gas desulfurization system. Process Saf Environ Prot. https://doi.org/10.1016/j.psep.2021.04.032
17. Gao X, Guo R, Ding H et al (2009) Dissolution rate of limestone for wet flue gas desulfurization in the presence of sulfite. J Hazard Mater 168(2–3):1059–1064
18. Chen Y (2012) Experimental study on the promoting dissolution of limestone and strengthening SO_2 absorption by WFGD additives. Zhejiang University, China (in Chinese)
19. Guo R (2008) Study on the effects of limestone reactivity and flow field on wet flue gas desulfurization efficiency. Zhejiang University, China (in Chinese)
20. Guo R, Gao X, Wang J et al (2007) Reactivity of limestone in wet flue gas desulfurization system. J Combust Sci Technol 06:485–490 (in Chinese)
21. Yue L, Jin D, Xu Z (2003) Transition of shape and crystallization type of $CaCO_3$ composite synthesized by co-precipitation method. Acta Chim Sinica 61(10):1587–1591 (in Chinese)
22. Hong Y (2020) Study on wet flue gas desulfurization based on pH control technology. Zhejiang University, China (in Chinese)
23. Mori T, Matsuda S, Nakajima F et al (1981) Effect of Al^{3+} and F^- on desulfurization reaction in the limestone slurry scrubbing process. Ind Eng Chem Process Des Dev 20(1):144–147
24. Electric utility engineer's FGD manual. Volume 1: FGD process design. Final report, 4 Mar 1996, United States. https://digital.library.unt.edu/ark:/67531/metadc695907/
25. Liu C, Zhao Z, Gao W et al (2021) Process optimization of S (IV) oxidation in flue gas desulfurization scrubbers. Process Saf Environ Prot 149(3):610–618
26. Huo W (2009) Mechanism research on the absorption, oxidation and crystallization in the process of limestone-gypsum WFGD. Zhejiang University, China (in Chinese)
27. Gerbec M, Stergarsek A, Kocjancic R (1995) Simulation model of wet flue gas desulphurization plant. Comput Chem Eng 19(Supplement 1):283–286
28. Fick's LM, Law G-K (2000) Formula, and Heisenberg's equation of motion. Phys Rev Lett 85(12):2422–2425
29. Fuller E, Schettler P, Giddings J (1966) New method for prediction of binary gas-phase diffusion coefficients. Ind Eng Chem 58(5):18–27
30. Gao X, Huo W, Zhong Y et al (2008) Effects of magnesium and ferric ions on crystallization of calcium sulfate dihydrate under the simulated conditions of wet flue-gas desulfurization. Chem Res Chin Univ 24(6):688–693
31. Zhong Y, Gao X, Huo W et al (2008) A model for performance optimization of wet flue gas desulfurization systems of power plants. Fuel Process Technol 89(11):1025–1032
32. Stefan O et al (1993) A model for the absorption of sulphur dioxide into a limestone slurry. Chem Eng J 51(2):99–108
33. Tseng P, Rochelle G (1986) Calcium sulfite hemihydrate: crystal growth rate and crystal habit. Environ Prog 5(1):5–11
34. Pasiuk-Bronikowska W, Rudziński KJ (1991) Absorption of SO_2 into aqueous systems. Chem Eng Sci 46(9):2281–2291
35. Li C, Zhang J, Zhang Y et al (2015) Removal efficiency of high sulfur dioxide in WFGD based on dual-pH value control. Acta Sci Circum 35(12):4081–4087
36. Frandsen JB, Kiil S et al (2001) Optimization of a wet FGD pilot plant using fine limestone and organic acids. Chem Eng Sci 56(10):3275–3287

37. Xu C, Wu C, Duan S et al (2014) Research of composite additive on high efficient desulfurization. Energy Eng 4:53–58 (in Chinese)
38. Pan S et al (2017) Development of high-gravity technology for removing particulate and gaseous pollutant emissions: principles and applications. J Clean Prod 149:540–556
39. Lin Y (2006) Study on large scale limestone-gypsum spraying desulphurization technology and project application. Zhejiang University, China (in Chinese)
40. Chen X (2008) Experimental study on desulfurization and nitrification in spray scrubber. Zhejiang University, China (in Chinese)
41. Zhong Y (2008) Theoretical and experimental study of simultaneous removal of sulfur, nitrogen and mercury pollutant in WFGD system. Zhejiang University, China (in Chinese)
42. Lin F, Wang Z, Zhang Z et al (2020) Flue gas treatment with ozone oxidation: an overview on NO_x, organic pollutants, and mercury. Chem Eng J 382:123030
43. Dora J, Gostomczyk M, Jakubiak M et al (2009) Parametric studies of the effectiveness of NO oxidation process by ozone. Chem Process Eng 30(4):621–633
44. Wang Z, Zhou J, Zhu Y et al (2007) Simultaneous removal of NO_x, SO_2 and Hg in nitrogen flow in a narrow reactor by ozone injection: experimental results. Fuel Process Technol 88(8):817–823
45. Zheng C, Xu C, Zhang Y et al (2014) Nitrogen oxide absorption and nitrite/nitrate formation in limestone slurry for WFGD system. Appl Energy 129(Sept 15):187–194
46. Gao X, Du Z, Ding H et al (2010) Kinetics of NO_x absorption into $(NH_4)_2SO_3$ solution in an ammonia-based wet flue gas desulfurization process. Chem Commun 24(11):3996–4015
47. Karkront H, Rosenberg H (1984) Flue gas denitrification. Selective catalytic oxidation of nitric oxide to nitrous oxide. Ind Eng Chem Res 23(4):808–814
48. Corro G, Elizalde M, Velasco A (2002) Promotional effect of Sn on NO oxidation over Pt/Al_2O_3 as promising gas exhaust catalytic converter. Selective Pt and Sn dispersion determination. React Kinet Catal Lett 76(1):117–122
49. Ichiura H, Kitaoka T, Tanaka H (2003) Photocatalytic oxidation of NO_x using composite sheets containing TiO_2 and a metal compound. Chemosphere 51(9):855–860
50. Kim H, Han J, Kawaguchi I et al (2007) Simultaneous removal of NO_x and SO_2 by a nonthermal plasma hybrid reactor. Energy Fuels 21(1):141–144
51. Li K, Tang X, Yi H et al (2011) Mechanism of catalytic oxidation of NO over Mn–Co–Ce–O_x catalysts with the aid of nonthermal plasma at low temperature. Ind Eng Chem Res 50(19):11023–11028
52. Mok Y (2004) Oxidation of NO to NO_2 using the ozonization method for the improvement of selective catalytic reduction. J Chem Eng Jpn Off Publ Soc Chem Eng 37(11):1337–1344
53. Zheng C, Xu C, Gao X et al (2014) Simultaneous absorption of NO_x and SO_2 in oxidant-enhanced limestone slurry. Environ Prog Sustainable Energy 33(4):1171–1179
54. Sada E, Kumazawa H, Hayakawa N et al (1977) Absorption of NO in aqueous solutions of $KMnO_4$. Chem Eng Sci 32(10):1171–1175
55. Xu C (2014) Experimental study on the integrated and intensive absorption of NO_x/SO_2 from coal-fired flue gas. Zhejiang University, China (in Chinese)
56. He S, Zhou J, Zhu Y et al (2010) Effect of vanadium-based selective catalytic reduction catalysts on mercury speciation transformation. J Zhejiang Univ (Eng Sci) 44(09):1773–1780 (in Chinese)
57. Zheng C, Wang L, Zhang Y et al (2017) Partitioning of hazardous trace elements among air pollution control devices in ultra-low-emission coal-fired power plants. Energy Fuels 31(6):6334–6344
58. Zheng C, Xu L, Liu S et al (2018) Speciation and thermal stability of mercury in solid products from ultra-low emission air pollution control devices. Energy Fuels 32(12):12655–12664
59. Sui Z, Zhang Y, Li W et al (2015) Partitioning effect of mercury content and speciation in gypsum slurry as a function of time. J Therm Anal Calorim 119(3):1611–1618
60. Lin F (2018) Basic characteristics study on NO_x deep removal in the flue gas by active molecules (ozone)-catalytic method. Zhejiang University, China (in Chinese)

Chapter 5
Multi-pollutant Removal System and Technology Evaluations

5.1 Introduction

The negative influences of the production of air pollutants, such as NO_x, SO_2, PM, and heavy metals, during coal combustion have led to the implementation of strict emission limits. Pollutant control techniques have thus been developed to limit these pollutants, including selective catalytic reduction (SCR), flue gas desulfurization (i.e., scrubbers), electrostatic precipitator (ESPs), and bag filters. Single-pollutant removal systems are no longer the solution; combining pollutant removal techniques in a multi-pollutant removal system can help meet strict emission requirements while maintaining high efficiency, high reliability, and low cost.

Researchers investigating simultaneous multi-pollutant removal techniques aim to remove two or more pollutants in a single reactor or system; significant research efforts have been dedicated to developing these multi-pollutant control systems. The national air pollutant control policy for developed countries has imposed increasingly strict limits on pollution emissions, leading to challenges meeting these requirements, particularly in CFPPs. Researchers from domestic and international scientific research institutes [1–9] (Zhejiang University, Tsinghua University, Huazhong University of Science and Technology, Southeast University, North China Electric Power University, Chinese Academy of Sciences, University of Cincinnati, Georgia Institute of Technology, Kyoto University, US EPA, DOE, EPRI, German KIT, European Commission JRC, etc.) and enterprises have aimed to develop and optimize technologies to drastically reduce flue gas pollutants (including NO_x, SO_2, and PM and SO_3, Hg, VOCs) and allow CFPPs to reach emissions near that of natural gas combustion under low-cost, stable, and reliable operation. In China, strict air pollutant emission standards, variations in coal quality (e.g., sulfur content, ash content, and chemical components), and load fluctuations contribute to design challenges for CFPP flue gas pollutant control technologies that require high efficiency, adaptability to varying flue gas conditions, reliability, and economical operation.

Multi-pollutant removal systems can thus be used to meet ULE requirements for varying fuel, capacity characteristics, and furnace types. A typical ULE system

X. Gao et al., *Multi-Pollutant Control for Flue Gases*, Advanced Topics in Science and Technology in China 63, https://doi.org/10.1007/978-981-16-1518-4_5

comprises several typical subsystems, including SCR, dry ESP, wet flue gas desulfurization (WFGD), and wet ESP (WESP). Other technologies and designs can also play significant roles, such as selective noncatalytic reduction (SNCR), semidry FGD, bag-filters, electrostatic-fabric integrated precipitators (which is also named electrostatic fabric filters (EFFs)), the integration of WFGD and ESP, temperature control equipment, and global optimization systems.

As detailed in previous chapters, synergistically removing one or more pollutants during the established removal process of a different pollutant can significantly improve the overall emission control performance. For example, SCR has been widely used to drastically reduce NO_x emissions but can also be used to reduce Hg and VOC emissions. Similarly, WFGD systems developed to remove SO_2, Hg^{2+}, and PM from flue gas. ESP and wet ESP can not only remove particulate matters but also other pollutants, such as SO_3, Hg, etc.

In this chapter, the techniques used for the simultaneous control of multiple pollutants to meet ultra-low emission limit are introduced, along with their mechanisms, influencing factors, and optimization methods. Four typical ULE technology routes are introduced considering load variations and the simultaneous removal of various pollutants. Furthermore, four typical applications of multi-pollutant removal techniques in CFPPs are discussed using ULE performance evaluations to guide the design, operation, and analysis of the system. Emission mitigation effects are also discussed.

5.2 Multi-pollutant Removal System

5.2.1 Ultra-Low Emission (ULE) Technologies

Different removal methods are used for different types of pollutants and must be targeted to various coal, capacity characteristics, and furnace types. A typical ULE system for CFPPs is equipped with several typical subsystems, such as SCR, a dry ESP, WFGD, and a WESP, as shown in Fig. 5.1. During operation, each system targets the removal of or one or more pollutants and synergistically targets other pollutants, as shown in Fig. 5.2 [10]. In this section, the synergistic equipment mechanisms used to target NO_x, SO_2, PM, SO_3, and Hg and the factors influencing the removal efficiency are detailed discussed based on recent research to provide a basis for further optimization of ULE systems.

NO_x removal: Denitrification is mainly performed using SCR, which chemically reduces NO_x into molecular nitrogen and water vapor at a theoretical removal efficiency of up to 90%. As introduced in Chap. 2, the factors that generally affect the denitrification efficiency include the flue gas temperature, catalyst characteristics, flue gas composition, flow uniformity, and mixing of the reductant/flue gas.

In addition, part of the insoluble NO can be oxidized to water-soluble higher-valence nitrogen oxides, e.g., NO_2 and N_2O_5, via catalytic oxidation, corona

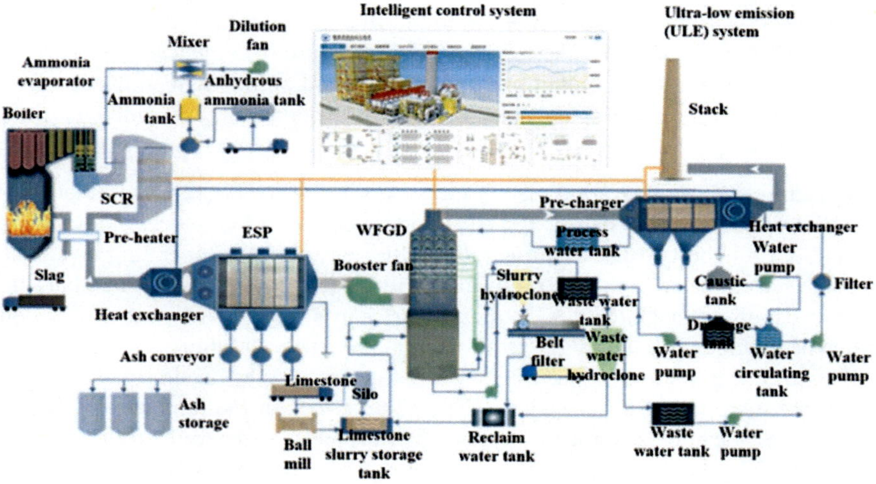

Fig. 5.1 Schematic of a typical ULE system

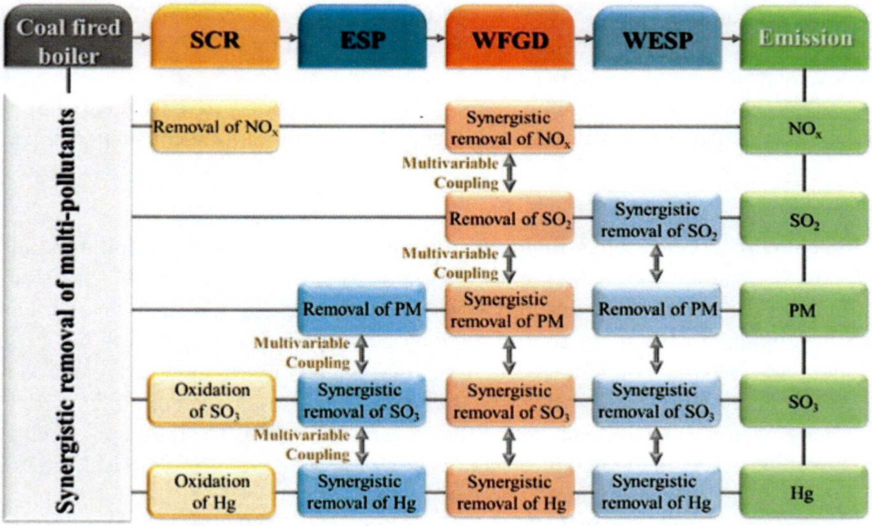

Fig. 5.2 Synergistic removal of five main pollutants in a typical ULE system by subsystem [10]

discharge oxidation, or ozone oxidation. After passing through the SCR, preoxidizing, and dedusting devices, these higher-valence NO_x are absorbed and removed along with SO_2 during desulfurization by the slurry or solution in the scrubber.

Owing to the presence of reaction intermediate sulfites during limestone-based WFGD, NO_2 is absorbed and converted to NO_2^- and NO_3^-, whereas SO_3^{2-} is oxidized to SO_4^{2-} and SO_3^{2-} is generated owing to the absorption of SO_2 by the limestone slurry. As the NO_2/NO ratio in the flue gas increases, the removal efficiency of NO first increases and then decreases. The use of some additives, such as magnesium sulfate, ferric sulfate, sodium sulfate, and ammonium sulfate, as well as some of their complexes, can enhance NO_2 absorption by the insoluble calcium sulfite slurry.

SO_2 removal: Desulfurization mainly occurs during WFGD, where the limestone-gypsum wet scrubber system is the most widely deployed FGD technology owing to its high reliability, high desulfurization efficiency, and low operating cost. The WFGD system has a high SO_2 removal efficiency (up to 99%) even when burning coal with high sulfur contents. As introduced in Chap. 4, key factors of desulfurization efficiency include the liquid/gas ratio, flue gas SO_2 concentration, residence time, mixing of flue gas/slurry, and combinations of different spray levels.

Furthermore, a WESP provides a synergistic effect on the removal of SO_2. In a WESP, SO_2 can be absorbed by spray water droplets, as determined by effective mass transfer between the gas and liquid phases and enhanced by corona discharge. In the discharge zone, some of the high-energy free electrons around the corona electrode may break the molecular chemical bonds and thus generate strong oxidizing free radicals, such as O_3, OH, or O, which can oxidize SO_2. Owing to the low temperature and high humidity, SO_2 can be oxidized to gaseous sulfuric acid and quickly condensed to form acid mist, which is removed by the WESP system. The operating voltage and the SO_2 concentration at the entrance of the device are the key factors affecting the efficiency of the WESP for the cooperative removal of SO_2. When the flue gas temperature, humidity and other conditions are constant, a higher inlet concentration of SO_2 and a higher operating voltage provide higher removal efficiency.

PM removal: Dry ESPs, bag-filters, electrostatic-fabric integrated precipitators, and WESPs are among the technologies are typically used to control PM emissions in exhaust gas streams. Increasingly strict PM emission requirements have pushed the development of particle control technology and significant progress has been made. Particle collection efficiency can be improved by optimizing ESP electrodes and filter materials, as well as by particle coagulation, particle conditioning, particle precharging, power supply, etc. High-efficiency particle control technologies, such as electrostatic-fabric integrated precipitators, low–low temperature ESPs (LLT-ESPs), and WESPs, are becoming more popular.

WFGD also helps remove PM owing to the severe collision between sprayed slurry droplets and capturing PM in the gas flow. The PM removal efficiency in WFGD is related to a variety of factors, including the effective height of the spray layer, the residence time, the density of the spray slurries, the turbulence intensity, and the temperature difference between droplets and particles. In general, larger fly ash particles can easily be captured by the slurry droplets, whereas the Brownian diffusion mechanism is crucial in the process of trapping particles $< 10\ \mu m$ in diameter in the slurry droplets, and thermophoretic forces dominate the trapping process of submicron particles.

SO_3 removal: As SO_2 is often oxidized to SO_3 during combustion or SCR process, SO_3 removal should also be considered. When the temperature is lower than the acid dew point in an ESP, some of the SO_3 condenses into sulfuric acid mist or is absorbed by fly ash. The formation of sulfuric acid on the surface of the particles strengthens the adhesion between them, increasing agglomeration and facilitating removal via ESP. The SO_3 removal efficiency by ESPs is related to factors such as the amount and surface area of fly ash, as well as the temperature of the flue gas. SO_3 can also be used to reduce particle resistivity in electrostatic dust removal devices.

In the scrubber, SO_3 existing as gaseous sulfuric acid quickly cools to the dew point by the slurry spray and may then condense through homogeneous nucleation to form a sulfuric acid mist. Some of the sulfuric acid molecules are neutralized by the limestone slurry, however, owing to the high humidity and high flow rate, the neutralization and dissolution rate of the formed sulfuric acid mist is lower than the formation rate. The submicron-sized sulfuric acid mist cannot be removed effectively by the spray slurry or solution, which limits the removal efficiency of SO_3 via WFGD. The primary factors affecting the removal efficiency of SO_3 include the flue gas temperature, inlet SO_3 concentration, liquid/gas ratio, residence time, and tower internal components, such as sieve trays or a turbulence flow grid.

In a WESP, the removal efficiency of sulfuric acid mist needs to be improved, owing to its small diameter ($<0.1\ \mu m$). Key factors affecting the removal efficiency in a WESP include the operating power supply voltage, specific collection area (SCA), inlet temperature, humidity, and SO_3 concentration, etc.

Hg removal: Mercury present in the exhaust gas exists as particulate mercury (Hg^p), elemental mercury (Hg^0), and oxidized mercury (Hg^{2+}). There are two main approaches for Hg removal: oxidation of Hg^0 to Hg^{2+}, and solid adsorption of Hg^p that is then removed downstream WFGD and in ESP devices.

Although HCl in the flue gas is not a strong oxidizing agent, catalysts used in SCR can promote reaction of HCl and oxygen to oxide Hg^0 to Hg^{2+}. The largest factor affecting Hg removal is thus the chlorine content; the higher chlorine content in bituminous coal results in a higher content of Hg^{2+}. Key factors influencing Hg oxidation and control SCR include the flue gas HCl concentration, temperature, and load.

Hg^0 is not removed during WFGD whilst gaseous Hg^{2+} is water-soluble and thus can be removed. However, some Hg^{2+} is reduced to Hg^0 and thus is not removed. The overall removal efficiency of Hg in the WFGD system is thus affected by many factors, including the reaction temperature, pH, and Cl^- and SO_2 concentration, etc.

The particulate-bound Hg^p is captured with a high removal efficiency in the upstream ESP or WESP. In addition, some Hg^0 is converted from gaseous $Hg^0(g)$ to adsorbed Hg^0 (ads) and interacts with CaO, MgO, Al_2O_3, TiO_2, and Fe in the fly ash to form stable adherent materials, which can then be removed by the dust collector.

5.2.2 Technical Routes

5.2.2.1 Conditions

As the properties of coal vary by geographic location, so do the burning conditions. Low-quality coal reserves, including high-sulfur, high-ash, or high-alkali coal are widely distributed; in Southwestern China, for example, the proportion of high-sulfur coal (3–8%) exceeds 30%, as shown in Fig. 5.3. Burning high-sulfur coal requires high desulfurization efficiency. Furthermore, the high SO_3 concentration in the exhaust gas increases the equipment load, deactivation of SCR catalysts, and air preheater blockages, all of which negatively affect the operation of the power generation system. Power stations thus must be designed considering the coal content.

The total load on a power unit varies with the electricity demand, which varies daily, monthly, and seasonally, and thus should also be considered when selecting a technological route. The typical monthly load of a 600 MW power plant is shown in Fig. 5.4.

5.2.2.2 Technological Processing

When designing the technological route to meet the ULE standards adopted in China in 2014, the types and quality fluctuations of the coal (e.g., sulfur content, ash content, and chemical components), the furnace type, and the local landscape must be considered. The interactions between subprocesses must also be considered to achieve high efficiency, high load adaptability, low operating energy consumption, low economic cost, and the synergies between all facilities. First, when selecting a technical route, the coal types used and their quality fluctuations must be considered to ensure that

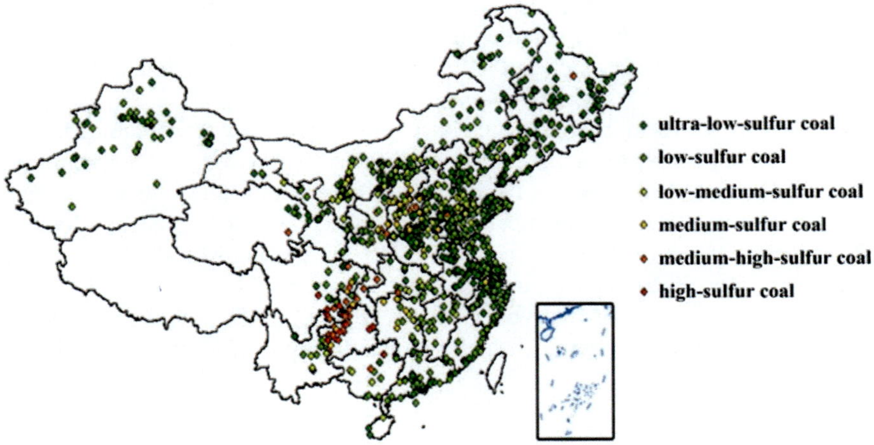

Fig. 5.3 Sulfur content of coal distribution of China

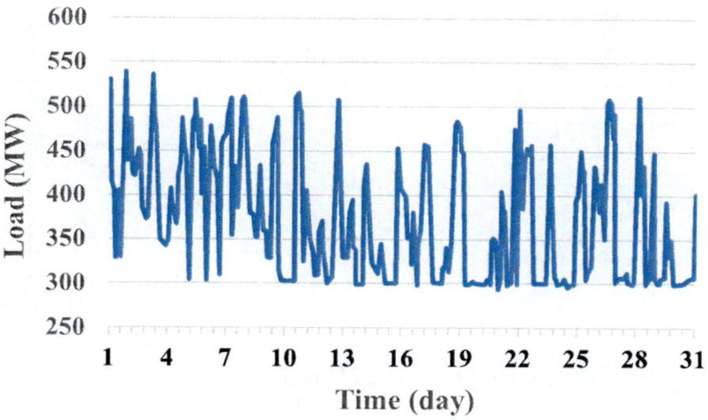

Fig. 5.4 Typical monthly load of a 660 MW power unit in China

ULE can be realized under different kinds of burning conditions. In a facility burning coal having a relatively stable quality with low ash content, for example, a combination of LLT-ESP and an scrubber system is preferred as the technical route for ultra-low PM emissions. Conversely, when the quality of coal varies, a bag-filter or electrostatic-fabric integrated precipitator is more suitable; furthermore, a WESP may be installed according to the PM concentration at the outlet of the dust collector and any synergistic dust removal effects.

The influence of the furnace type on the fly ash composition and emitted NO_x and SO_2 concentrations should also be taken into account. When burning inferior fuel (e.g., high-ash content, large diameter when compared with the furnace) in a circulating fluidized bed (CFB) boiler, the high ash content, larger particle size, and higher temperature of the exhaust gas necessitate an electrostatic-fabric integrated precipitator or bag-filter for dust removal; when using coal with a high calorific value in a CFB boiler, LLT-ESP is a good choice. SNCR-SCR coupling technology can be adopted when limited space is available. For the controlling of NO_x, SNCR technology is preferred; When it comes to the controlling of SO_2, the desulfurization in furnace technology is a good approach. The application of semidry FGD or WFGD can also help meet ULE requirements for SO_2 emissions.

Local conditions are also significant; an ESP, a WESP, and dual-tower or pH dual-loop desulfurization processes all require space. The influence of atmospheric conditions of CFPPs located in high altitudes must also be considered when selecting appropriate pollutant removal devices.

Overall, as each facility affects each other in these integrated ULE systems, the synergy between each subsystem should be investigated during the design, construction, and operation to ensure overall process optimization. Not only must the emission standards be met, but the future implementation of stricter emission standards and economic and technological developments should be considered when selecting a suitable ULE technology route.

ULE standards address some main (NO_x, SO_2, and PM) and other air pollutants (e.g., SO_3 and Hg). As many technical options exist for controlling main air pollutants, as shown in Fig. 5.5, synergies between different facilities should be considered

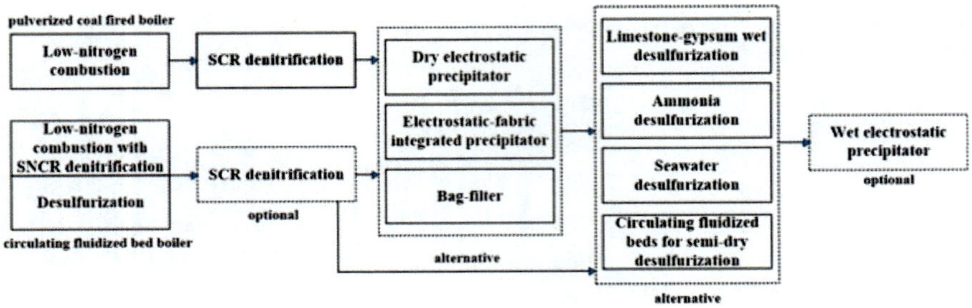

Fig. 5.5 ULE technique routes for CFPPs [11]

when selecting appropriate technical routes according to the specific conditions of different CFPPs.

a. *Ultra-low NO_x Emission*

Low-nitrogen combustion technology, wherein the NO_x concentration at the outlet of the furnace is reduced, is preferred for controlling NO_x emissions. The conditions of a pulverized coal-fired boiler can be optimized by modifying the burner and boiler parameters to ensure outlet NO_x emissions of < 550 mg/m^3, whereas the NO_x concentration at the outlet of a CFB boiler should be < 200 mg/m^3. The technical routes of different furnace types with low NO_x emissions are provided in Table 5.1 [11]. According to the NO_x concentration, flue gas volume, temperature, flow rate, and dust concentration of the flue gas at the boiler outlet, the appropriate denitrification processes and parameters can then be selected. At the downstream of the boiler, SCR, SNCR, and SNCR-SCR are the most widely applied. When adopting a CFB, it is better to install SNCR and, if necessary, another SCR device to meet the ULE requirements. If the required denitrification efficiency of the flue gas (i.e., owing to a higher NO_x concentration) is below 50%, SNCR or SCR (with few layers) can be considered; if a higher denitrification efficiency is required, SCR with more layers or SNCR coupled with SCR may be required.

Table 5.1 Technical routes of ultra-low NO_x emission [11]

Type of boiler	Inlet concentration (mg/m^3)	Denitrification efficiency (%)	Typical number of denitrification catalyst layers[a]
Pulverized coal fired boiler	< 200	80	2 + 1
	200–350	80–86	3 + 1
	350–550	86–91	
CFB boiler	< 200	60–80	SNCR(+SCR)

In the form $n + 1$, n represents the number of catalyst layers and 1 represents the installation space needed for the preparation of the catalyst

b. *Ultra-low SO_2 Emission*

Limestone–gypsum WFGD is often the best option for desulfurization; ammonia- and magnesium-based WFGD also perform well in some cases. Still, the desulfurization process must be adopted for different inlet SO_2 concentrations. When considering desulfurization routes, the flue gas volume and composition, liquid/gas ratio, desired desulfurization efficiency, and required outlet SO_2 concentration should be considered. Some process selections to achieve SO_2 emissions of $\leq$ 30 mg/m^3 based on constant inlet SO_2 concentrations are provided in Table 5.2, although the maturity and economics of the selected system should also be considered. The removal efficiency can also be enhanced by optimizing the flow field and considering the synergistic effects of any additives.

c. *Ultra-low PM Emission*

High-efficiency dust removal devices used before WFGD typically include a dry ESP, an electrostatic-fabric integrated precipitator, and bag-filter technology. A dust removal efficiency of $\geq$ 99.85% can be reached by combining a high-efficiency power supply, advanced dust removal methods, and an LLT-ESP. When adopting an electrostatic-fabric integrated precipitator, the PM removal efficiency can be as high as 99.99%. In addition, a WESP should be used after WFGD. A limestone–gypsum wet desulfurization composite tower equipped with an efficient demister or a wet dust removal device (such as a WESP) can achieve a synergistic removal efficiency of $\geq$ 70%.

In practice, the characteristics of various technologies combined with their maturity, applicability, costs, and secondary pollution must be considered when selecting a suitable ultra-low PM emission technology route. Some technical routes are summarized by application in Table 5.3 [11], where more stars correspond to a higher recommendation for usage, considering the efficiency, costs, secondary pollution, and overall system applicability.

d. *Typical ULE Technical Routes for CFPPs*

The discussion thus far in this section has only considered emissions of a single pollutant type at a time. However, in a ULE system, multiple pollutants must be

Table 5.2 Technical routes of ultra-low SO_2 emission [11]

Inlet concentration (mg/m^3)	Technologies	Desulfurization efficiency (%)
$\leq$ 1000	High-efficiency spray scrubber	97
$\leq$ 2000	Scrubber with tray	98.5
$\leq$ 3000	Scrubber with double tray, turbulence flow grid	99
$\leq$ 6000	Single-tower dual-loop scrubber, double tray, turbulent pipe grid	99.5
$\leq$ 10,000	Double tower technology	99.7

Table 5.3 Technical routes of ultra-low PM emissions [11]

Type of boiler	Capacity (MW)	Inlet dust concentration (mg/m³)	Before WFGD system			After WFGD system	
			ESP (efficiency ≥ 99.85%)	Electrostatic-fabric integrated precipitator (efficiency ≥99.9%)	Bag-filter (efficiency ≥ 99.9%)	WESP (efficiency ≥ 70%)	WFGD (synergistic efficiency ≥ 70%)
Pulverized coal-fired boiler (tangential firing, wall firing)	≤ 200	≥ 30,000	★	★★★	★★★	★★★	★
		20,000–30,000	★★	★★	★★	★★	★★
		≤ 20,000	★★★	★	★	★	★★★
	200–600	≥ 30,000	★	★★★	★★	★★★	★
		20,000–30,000	★★	★★	★	★★	★★
		≤ 20,000	★★★	★	★	★	★★★
	≥ 600	≥ 30,000	★	★★★	★	★★★	★
		20,000–30,000	★★	★★	★	★★	★★
		≤ 20,000	★★★	★	★	★	★★★
Pulverized coal-fired boiler (W-flame combustion)		≥ 30,000	★	★★★	★★	★★★	★
		20,000–30,000	★★	★★★	★	★★	★★
		≤ 20,000	★★★	★★	★	★	★★★
Circulating fluidized bed boiler			★	★★★	★★	★★★	★

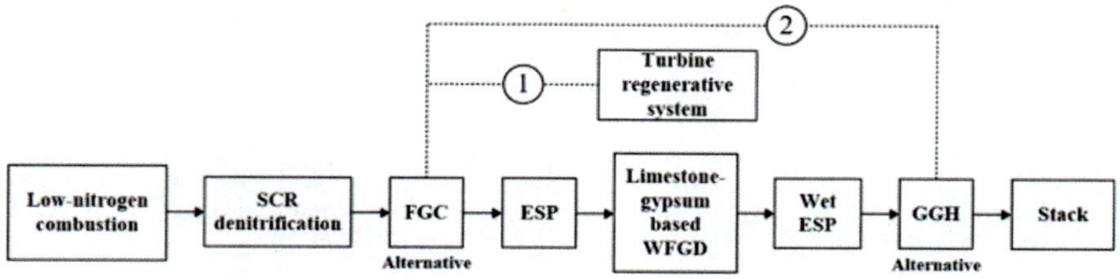

Fig. 5.6 Typical ULE technical route 1

removed; this requires an overall design. To provide some examples, four typical technical routes are introduced in the following subsections.

Typical technical route 1: Considering a unit subjected to large fluctuations in coal quality and unit load and required to have high adaptability for future modification to ULE standards, this technical route employs a low-nitrogen combustion technology, SCR for denitrification, limestone–gypsum WFGD for desulfurization, and an ESP and WESP to remove PM, as summarized in Fig. 5.6. In addition, a tubular GGH and flue gas cooler (FGC) could be installed as optional devices. In summary, this route fully considers the synergy of the existing flue gas pollutant removal of coal-fired power units and is especially suitable for power units that are greatly affected by the large fluctuations of coal quality and unit load with a requirement of high adaptability for ultra-low emissions.

These technologies used for NO_x, SO_2, and PM control when using boilers with different coal quality, capacity, and original flue gas pollutant concentration were selected using the information provided in Tables 5.1, 5.2 and 5.3 as references.

A dry ESP, an electrostatic-fabric integrated precipitator, or a bag-filter could be adopted upstream of the WFGD system. In general, the ESP should be designed to have an outlet PM concentration of ≤ 30 mg/m^3; units limited by certain engineering conditions (e.g., space constraints), should be designed to have an outlet PM concentration of ≤ 50 mg/m^3. The heat from the FGC can be recovered to the GGH or to the turbine regenerative system in systems without a GGH (see (2) and (1) in Fig. 5.6, respectively). If a bag-filter or electrostatic-fabric integrated precipitator is adopted, the FGC should instead be installed downstream of the ESP. Installing the GGH downstream of the WESP efficiently mitigates plume opacity. Employing a state-of-the-art FGC reduces the inlet temperatures of the ESP and desulfurization system, thereby increasing the PM removal efficiency and synergistic SO_3 removal efficiency in the ESP as well as the desulfurization efficiency and systematic dedusting efficiency during WFGD and reducing water consumption.

In this technical route, outlet NO_x, SO_2, and PM concentrations of ≤ 50, ≤ 35, and ≤ 5 mg/m^3 can be achieved, respectively. With better coal quality and more stable operating conditions, they can be further reduced to ≤ 30, ≤ 20, and ≤ 3 mg/m^3, respectively.

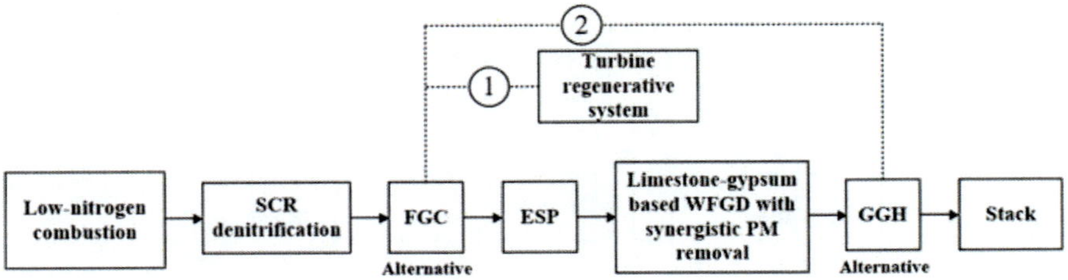

Fig. 5.7 Typical ULE technical route 2

Typical technical route 2: Considering a unit subjected to stable coal quality and small load fluctuations, this technical route, summarized in Fig. 5.7, employs a low-nitrogen combustion technology, SCR for denitrification, limestone–gypsum WFGD for desulfurization and synergistic PM removal, and an ESP for PM removal. A tubular GGH and FGC can be installed as optional devices.

When choosing ultra-low emission technology for NO_x, SO_2, and particulate matters, taking Tables 5.1, 5.2 and 5.3 as references. A dry ESP, an electrostatic-fabric integrated precipitator, or a bag-filter could be adopted upstream of the WFGD system. In general, the ESP should be designed to have an outlet PM concentration of $\leq$ 30 mg/m^3. The synergistic dedusting performance of the limestone–gypsum WFGD system then allows for the further reduction of PM emission to $\leq$ 10 mg/m^3. The heat from the FGC can be recovered to the GGH or to the turbine regenerative system in systems without a GGH (see (2) and (1) in Fig. 5.7, respectively). If a bag-filter or electrostatic-fabric integrated precipitator is adopted, the FGC should be installed downstream of the ESP. As in route 1, employing a state-of-the-art FGC reduces the inlet temperatures of the ESP and desulfurization system, thereby increasing the PM removal efficiency and synergistic SO_3 removal efficiency in the ESP as well as the desulfurization efficiency and systematic dedusting efficiency during WFGD and reducing water consumption.

In this technical route, outlet NO_x, SO_2, and PM concentrations of $\leq$ 50, $\leq$ 35, $\leq$ 10 mg/m^3 can be achieved, respectively; with better coal quality and more stable operating conditions, they can be further reduced to $\leq$ 30, $\leq$ 20, and $\leq$ 5 mg/m^3, respectively.

Typical technical route 3: Considering a unit with good adaptability to coal quality and flue gas working conditions and subjected to coal with high ash content and poor dust-collecting capabilities, this technical route, summarized in Fig. 5.8, employs a low-nitrogen combustion technology, SCR for denitrification, limestone–gypsum WFGD for desulfurization and synergistic PM removal, and a high-efficiency electrostatic-fabric integrated precipitator for PM removal. A tubular GGH and FGC can be installed as optional devices.

When choosing ultra-low emission technology for NO_x, SO_2, and particulate matters, taking Tables 5.1, 5.2 and 5.3 as references. The electrostatic-fabric integrated precipitator should be designed to have an outlet PM concentration of $\leq$ 10 mg/m^3, and the limestone–gypsum WFGD system should be designed to ensure

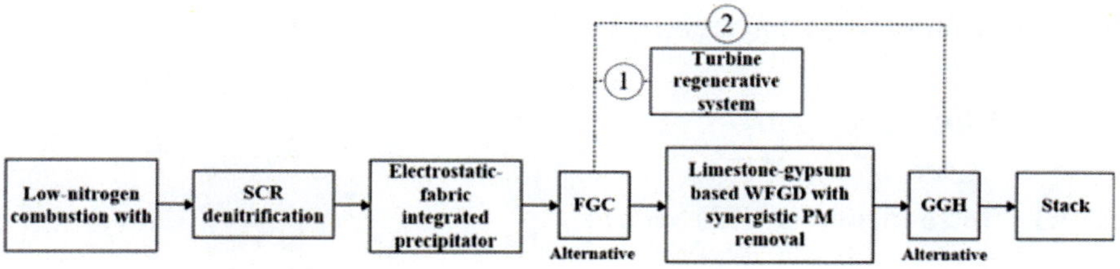

Fig. 5.8 Typical ULE technical route 3

that the concentration of PM, such as fly ash, acid mist, or gypsum particles does not increase. The heat from the FGC can be recovered to the GGH or to the turbine regenerative system in systems without a GGH (see (2) and (1) in Fig. 5.8, respectively). As in route 1, Employing a state-of-the-art FGC reduces the inlet temperatures of the ESP and desulfurization system, thereby increasing the PM removal efficiency and synergistic SO_3 removal efficiency in the ESP as well as the desulfurization efficiency and systematic dedusting efficiency during WFGD and reducing water consumption. In this technical route, outlet NO_x, SO_2, and PM concentrations of $\leq$ 50, $\leq$ 35, and $\leq$ 10 mg/m^3 can be achieved, respectively.

Typical technical route 4: To minimize water consumption in small and medium-scale facilities in water-restricted areas, this technical route, summarized in Fig. 5.9, employs a CFB boiler with in-furnace desulfurization, SNCR-SCR for denitrification, a semidry desulfurization absorption tower, and a bag-filter for PM removal.

When choosing ultra-low emission technology for NO_x, SO_2, and particulate matters, taking Tables 5.1, 5.2 and 5.3 as references. Limestone Injection into the Furnace and Activation of Calcium (LIFAC) is an optional technology and the inlet SO_2 concentration should be no more than 1500 mg/m^3. Usually, a bag-filter is selected for the removal of particulate matter, while the emission requirement of particulate matter and SO_2 becomes much stricter, the installation of a wet ESP downstream from the bag-filter could be a solution for further pollutant removal. In this technical route, outlet NO_x, SO_2, and PM concentrations of $\leq$ 50, $\leq$ 35, and $\leq$ 10 mg/m^3 can be achieved, respectively.

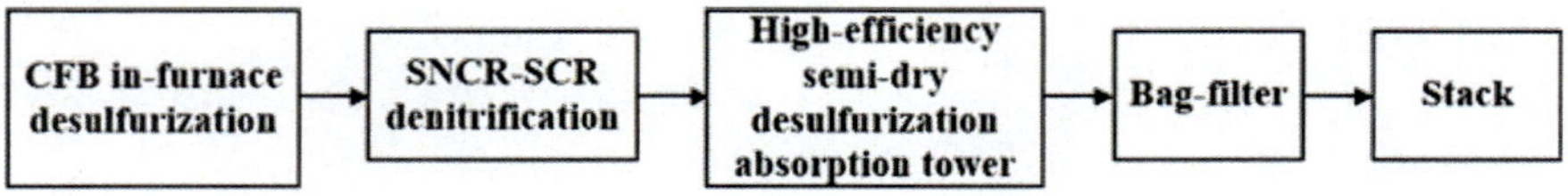

Fig. 5.9 Typical ULE technical route 4

5.3 Engineering Practice of ULE System

5.3.1 1000 MW Coal-Fired Power Unit

5.3.1.1 Introduction of ULE System

A ULE system was implemented on two 1000 MW ultra-supercritical coal-fired units of the Jiahua power plant (units #7 and #8) in the Zhejiang province in 2013 to reduce SO_2, NO_x, and PM emissions to meet the emission requirements of a natural gas turbine unit (i.e., ≤ 35, ≤ 50, and ≤ 5 mg/m^3, respectively) while also reducing SO_3 and Hg emissions. The temperature of the flue gas entering the stack was raised by a GGH to ≥ 72 °C to avoid slurry carryover and white plume. Unit #8 was the first demonstration unit applying ULE technology in China. Parameters of the feed coal and flue gas emitted from the boiler to the SCR system are summarized in Tables 5.4 and 5.5, respectively. Photographs of the unit and the layout of the implemented ULE system are shown in Fig. 5.10.

As discussed throughout this chapter, ULE technology is based on the efficient removal of multiple pollutants and involves the synergistic removal of SO_2, NO_x, and PM in each subsystem. Here, a low-nitrogen burner with SCR, a temperature-regulating high-efficiency ESP, an enhanced WFGD system, and a WESP are integrated to reduce the emissions of NO_x, SO_2, $PM_{2.5}$, SO_3, and Hg; a summary of the key subsystems used is provided in Table 5.6. By optimizing the low-nitrogen burner of the denitrification system, the designed outlet NO_x concentration is ≤ 300 mg/m^3 for a denitrification efficiency of $> 85\%$. Using a LLT-ESP and high-frequency power supply increased the designed PM efficiency to over 99.89%. The designed desulfurization efficiency increased from 95 to 98.5% by using an enhanced SO_2 removal scrubber. A WESP and a tubular two-stage GGH were added after desulfurization. In the #8 unit, a new catalyst replaced the conventional catalyst for the synergistic

Table 5.4 Parameters of designed and actual coal fed to the boiler

Item	Symbol	Unit	Designed coal	Actual coal
Total moisture	M_t	%	14 ± 4	13.5
Air-dried moisture	M_{ad}	%	7.30	4.5
Volatile matter	V_{daf}	%	36.50 ± 5	35.93
Arb ash	A_{ar}	%	11.00 ± 5	14.51
Arb carbon	C_{ar}	%	60.33	58.77
Arb hydrogen	H_{ar}	%	3.62	3.61
Arb oxygen	O_{ar}	%	9.94	8.20
Arb nitrogen	N_{ar}	%	0.70	0.78
Total sulfur	$S_{t,ar}$	%	0.41 ± 0.3	0.63
Net calorific value as received basis	$Q_{ar.net}$	MJ/kg	22.76 ± 2.3	22.24

Table 5.5 Flue gas parameters at the inlet to the SCR system

Item	Unit	BMCR	100% THA	75% THA	50% THA	35% BMCR
Actual volume flow	m^3/h	3,062,214	2,924,161	2,196,977	1,692,613	1,567,719
Mass flow rate	kg/h	4,098,643	3,912,527	2,939,555	2,264,717	2,097,608
Temperature	°C	365	360	335	322	308
Economizer outlet pressure	kPa	− 1.47	/	/	/	/
Carbon dioxide content	%	14.88	14.88	14.06	12.59	11.20
Oxygen content	%	2.52	2.52	3.50	5.27	6.97
Sulfur dioxide content	%	0.04	0.04	0.04	0.03	0.03
Water content	%	9.49	9.49	9.09	8.35	7.63

BMCR: boiler maximum continue rate, THA: turbine heat acceptance

Fig. 5.10 ULE system of unit #8 at the Jiahua facility

oxidation of Hg, allowing the designed Hg oxidation efficiency to reach ≥ 70%. The resulting design targets of NO_x, SO_2, and PM emissions are shown in Table 5.7.

5.3.1.2 Performance of the 1000 MW ULE System

The China National Environmental and Zhejiang Ecological and Environmental Monitoring Centers conducted on-site monitoring of the implemented ULE system, designed for maximum NO_x, SO_2, PM, and Hg outlet concentrations of 30 mg/m^3, 10 mg/m^3, 1.5 mg/m^3, and 1.5 $\mu g/m^3$, respectively. In all cases, the observed emissions were near or within their target limits, as summarized in Table 5.8, at observed loads of 1000 and 750 MW.

Further, continuous online monitoring of the NO_x, SO_2, and PM concentrations of unit #8 by the Continuous Emission Monitoring System (CEMS) demonstrated

Table 5.6 Main pollutant removal technologies applied to unit #8 of the Jiahua facility

System	Item	Content
Denitrification	Low-emission combustion	• Outlet concentration < 300 mg/m^3
	SCR system	• Three layers of catalyst (two-layer with wide-temperature catalyst, one layer with high Hg^0 oxidation efficiency) • High uniformity velocity distribution and enhanced ammonia mixture system
Desulfurization	WFGD system	• Two layers of the tray for uniform flow distribution and higher SO_2 absorption rate • Four spray layers with high spray density
Dust removal	Dry ESP	• LLT-ESP with four electric fields (designed flue gas temperature is between 85 and 95 °C) • High-frequency (20 kHz) power supply • Flow field optimization component
	WESP	• WESP with one field and precharger • High-frequency (20 kHz) power supply • Flow field optimization component

Table 5.7 ULE design targets considering boiler outlet and optimized system design

Pollutant	Design condition of boiler outlet	Target (mg/m^3)
NO_x	Concentration of 300 mg/m^3	≤ 50
SO_2	0.8% Sulphur; concentration of 1745 mg/m^3	≤ 35
PM	18% ash; concentration of 23.4 g/m^3	≤ 5

stable emissions well within their ULE design targets, as shown in Fig. 5.11. From April through June of 2015, the average concentrations of emitted NO_x, SO_2, and PM were 29.47, 12.63, and 1.52 mg/m^3, respectively, for total emissions of 193.98, 104.5, and 12.04 t, respectively; this represents a reduction in NO_x, SO_2, and PM emissions by 41.06%, 69.9%, and 84.8%, respectively, compared with the emissions in the same working condition when the ultra-low emission limit is not implemented.

5.3.2 600 MW Coal-Fired Power Unit

5.3.2.1 Introduction of ULE System

A ULE system was applied to a 600 MW coal-fired unit at the Yueqing power plant (unit #1) in the Zhejiang province in 2014 to act as a demonstration project of emissions reduction to ULE limits for the National Energy Administration. The feed coal to this facility is summarized in Table 5.9. This ULE system adopted key ULE technologies, including low-nitrogen combustion, SCR, a LLT-ESP, optimized

Table 5.8 Observed inlet and outlet parameters of power units #7 and #8 at the Jiahua facility

Unit	Load (MW)	Parameters of coal feedstock			Outlet pollutant emission concentrations			
		S_{ar} (%)	A_{ar} (%)	Hg_{ar} (ng/g)	NO_x (mg/m^3)	SO_2 (mg/m^3)	PM (mg/m^3)	Hg (μg/m^3)
#8	1000	0.35	11.94	22.1	20.58	9.02	1.25	1.29
	1000	0.35	9.09	27.8	23.13	7.87	0.425	1.31
	750	0.35	9.07	75.6	27.67	8.76	3.58	0.723
	750	0.42	11.39	86.6	16.01	8.87	1.45	0.653
#7	1000	0.71	15.16	203	15.59	16.83	Undetected	1.78
	1000	0.68	15.21	192	31.75	13.38	0.875	1.65
	750	0.75	16.13	244	13.15	11.81	1.25	2.18
	750	0.72	15.75	278	12.35	9.92	1.78	1.28

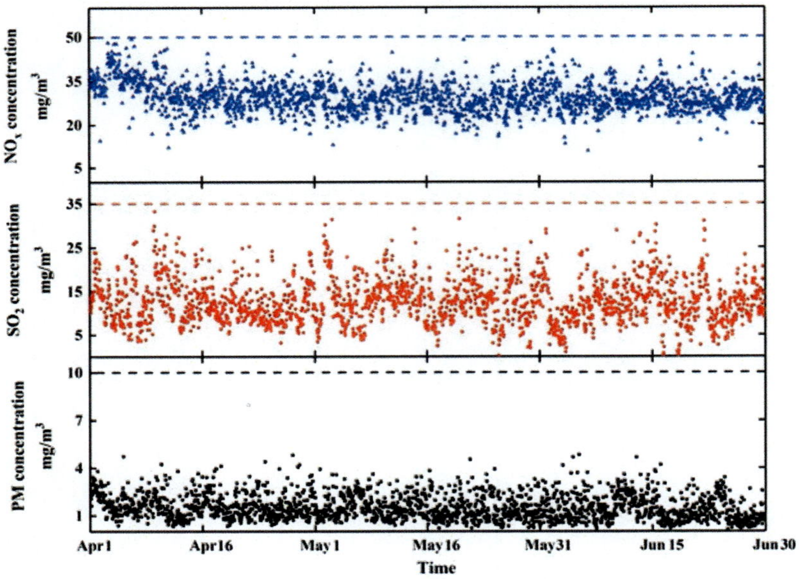

Fig. 5.11 Observed hourly average concentrations of major pollutants (NO_x, SO_2, and PM) of unit #8 at the Jiahua facility from April 1 to June 30 in 2015

Table 5.9 Parameters of designed and actual coal fed to the boiler

Item	Symbol	Unit	Designed coal	Actual coal
Total moisture	M_t	%	14.0 ± 4	14.0
Air-dried moisture	M_{ad}	%	5.90	5.0
Volatile matter	V_{daf}	%	33.19 ± 5	38.0
Arb ash	A_{ar}	%	7.04 ± 5	15.0
Arb carbon	C_{ar}	%	63.25	58.0
Arb hydrogen	H_{ar}	%	3.40	3.50
Arb oxygen	O_{ar}	%	11.18	7.90
Arb nitrogen	N_{ar}	%	0.64	0.80
Total sulfur	$S_{t,ar}$	%	0.7 ± 0.3	0.63
Net calorific value as received basis	$Q_{ar.net}$	MJ/kg	23.39 ± 2.3	22.00

limestone–gypsum WFGD, and a WESP, as detailed in Table 5.10. Finally, the design targets of NO_x, SO_2, and PM emissions are shown in Table 5.11, and the photographs of the power unit and the ULE system layout are shown in Fig. 5.12.

Table 5.10 Main pollutant removal technology applied to unit #1 of the Yueqing facility

System	Item	Content
Denitrification	Low emission combustion	• Outlet concentration < 330 mg/m^3
	SCR system	• Two layers of wide-temperature catalyst • High uniformity velocity distribution
Desulfurization	WFGD system	• One layer of the tray for uniform flow distribution and higher SO_2 absorption rate • Five spray layer with high spray density • pH dual-loop technology with an auxiliary tank
Dust removal	Dry ESP	• Low-temperature ESP with four electric fields (designed flue gas temperature is between 89 and 95 °C) • High-frequency (20 kHz) power supply • Flow field optimization component
	Wet ESP	• WESP with two electric fields and precharger • High-frequency (20 kHz) power supply • Flow field optimization component

Table 5.11 ULE design targets considering boiler outlet and optimized system design

Pollutant	Design condition of boiler outlet	Target (mg/m^3)
NO_x	Concentration of 330 mg/m^3	≤ 50
SO_2	0.85% sulfur; concentration of 1860 mg/m^3	≤ 35
PM	18% ash; concentration of 20.2 g/m^3	≤ 5

Fig. 5.12 ULE system of unit #1 at the Yueqing facility

5.3.2.2 Performance of ULE System

The Zhejiang Ecological and Environmental Monitoring Center tested the concentration of pollutants emitted from the implemented ULE system under loads of 100%

and 75%; this ULE was designed for maximum NO_x, SO_2, and PM outlet concentrations of 40, 20, and 2 mg/m^3, respectively. Table 5.12 suggests that the observed emissions were within their designed limits.

Continuous online monitoring of the NO_x, SO_2, and PM emissions for over two months also demonstrated emissions within these targets, as shown in Fig. 5.13. During this period, the average concentrations of emitted NO_x, SO_2, and PM were 27.55, 8.74, and 2.71 mg/m^3, respectively, for a total emission of 136.87, 43.42, and 13.46 t, respectively; this represents a reduction in NO_x, SO_2, and PM emissions by 44.9%, 75.03%, and 72.9%, respectively, compared with the emissions in the same working condition when the ultra-low emission limit is not implemented.

Table 5.12 Emission concentrations of unit #1 at the Yueqing facility

Pollutants	NO_x (mg/m^3)		SO_2 (mg/m^3)		PM (mg/m^3)	
Monitoring period	First	Second	First	Second	First	Second
100% load	39.0	39.3	10.3	13.1	1.39	1.32
75% load	32.3	34.0	17.0	19.6	1.57	1.47

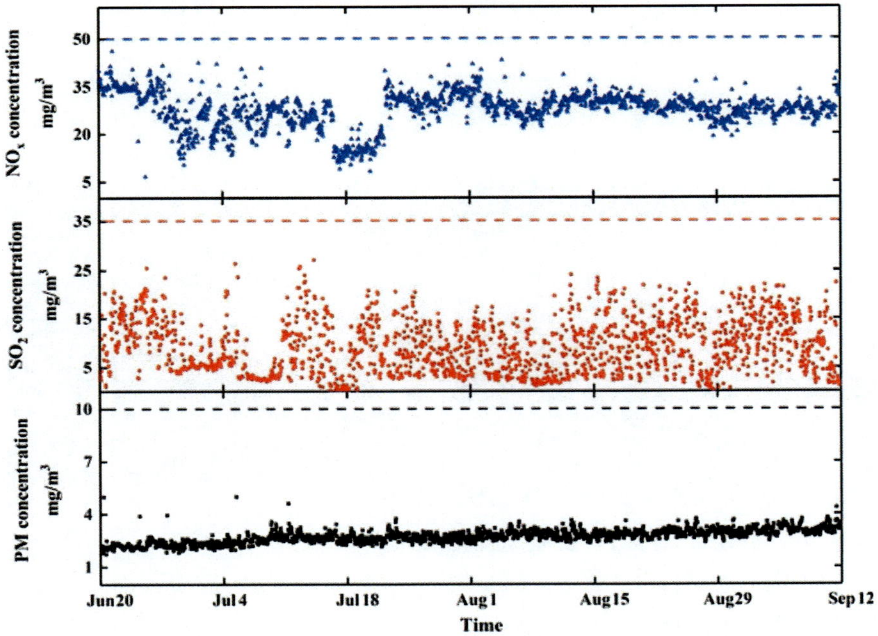

Fig. 5.13 Hourly average concentration of major pollutants (NO_x, SO_2, and PM) of unit #1 at the Yueqing facility from June 20 to September 12 in 2019

5.3.3 300 MW Thermal Power Unit

5.3.3.1 Introduction of ULE System

Beginning in 2008, two 300 MW subcritical coal-fired combined heat and power units at the Wusha thermal power station in Guangdong province were operating in compliance with prior regulations via dust removal, desulfurization, and denitrification systems. In 2014, each of these subsystems and the overall system was upgrated to meet the ULE emission standards for NO_x, SO_2, and PM emissions, which are stricter than the previous limits established for thermal plants (GB13223-2011). The feed coal to this facility is summarized in Table 5.13.

To meet the ULE standards, several key technologies were adopted, including low-nitrogen combustion combined with enhanced SCR, enhanced WFGD, and a novel WESP, as detailed in Table 5.14. Specifically, the SCR system was modified to include three layers of catalysts, thereby improving the designed denitrification efficiency to 87.5% with a corresponding NO_x concentration of 30 mg/m^3. In the enhanced WFGD system, a gas–liquid mass transfer enhanced component and pH dual-looping technology was added, thereby improving the designed desulfurization efficiency to over 98.5% for a corresponding SO_2 concentration at the scrubber outlet of < 30 mg/m^3. The downstream WESP can further reduce the SO_2 concentration to $\leq$ 15 mg/m^3, thereby improving the adaptability of the system to feedstock sulfur concentration. Furthermore, including a WESP reduces the designed PM emission concentration to $\leq$ 3 mg/m^3, ensures that the gypsum carryover phenomenon at the stack outlet can be avoided, and effectively removes $PM_{2.5}$, SO_3, and Hg. Overall, these modifications allowed for a system designed to meet the ULE targets, representing the emission requirements of a natural gas turbine unit (i.e., $NO_x \leq 50$ mg/m^3, $SO_2 \leq 35$ mg/m^3, PM ≤ 5 mg/m^3), as summarized in Table 5.15. Photographs of the unit and the general layout of the implemented ULE system are shown in Fig. 5.14.

Table 5.13 Parameters of designed and actual coal fed to the boiler

Item	Symbol	Unit	Designed coal	Actual coal
Air-dried moisture	M_{ad}	%	11	2.85
Volatile matter	V_{daf}	%	35	28
Arb ash	A_{ar}	%	8.25	25.09
Arb carbon	C_{ar}	%	58.93	53.41
Arb hydrogen	H_{ar}	%	3.71	3.06
Arb oxygen	O_{ar}	%	10.20	6.64
Arb nitrogen	N_{ar}	%	0.83	0.72
Total sulfur	$S_{t,ar}$	%	0.58	0.63
Net calorific value as received basis	$Q_{ar.net}$	MJ/kg	23	20.35

Table 5.14 Main pollutant removal technologies applied to the Wusha thermal power unit

System	Item	Content
Denitrification	Low emission combustion	• Outlet concentration < 240 mg/m^3
	SCR system	• Three layers (the first layer of the catalyst was regenerated) • High uniformity velocity distribution • Enhanced ammonia mixture system
Desulfurization	WFGD system	• One layer of the tray with uniform velocity distribution • Four spray layers with high spray density • pH dual-loop technology with an auxiliary tank
Dust removal	Dry ESP	• LLT-ESP with four electric fields • High-frequency (20 kHz) power supply • Flow field optimization component
	Wet ESP	• WESP with two electric fields • Optimized electrodes • High-frequency (20 kHz) power supply • Flow field optimization component

Table 5.15 ULE design targets considering boiler outlet and optimized system design

Pollutant	Design condition of boiler outlet	Target (mg/m^3)
NO_x	Concentration of 450 mg/m^3	≤ 50
SO_2	0.8% sulfur content; concentration of 1948 mg/m^3	≤ 35
PM	Concentration of 22.5 g/m^3	≤ 5

Fig. 5.14 ULE system at the Wusha thermal power unit

5.3.3.2 Performance of ULE System

Third-party testing of this 300 MW coal-burning ULE system demonstrated that emissions were within the designed limits ($NO_x \leq 50$ mg/m^3, $SO_2 \leq 35$ mg/m^3, and PM ≤ 2.5 mg/m^3), as detailed in Table 5.16.

Continuous online monitoring of the NO_x, SO_2, and PM concentrations for over two months also demonstrated emissions within these targets, as shown in Fig. 5.15. During this period, the average concentrations of emitted NO_x, SO_2, and PM were 30.59, 12.52, and 2.15 mg/m^3, respectively, for a total emission of 79.29, 32.45, and 5.57 t, respectively; this represents a reduction in NO_x, SO_2, and PM emissions by 38.82%, 64.23%, and 78.5%, respectively, compared with the emissions in the same working condition when the ultra-low emission limit is not implemented.

Table 5.16 Designed and observed emissions at the Wusha thermal power unit

Item	Designed value (mg/m^3)	Observed value (mg/m^3)
NO_x	50	17.15
SO_2	35	17.49
PM	5	2.09

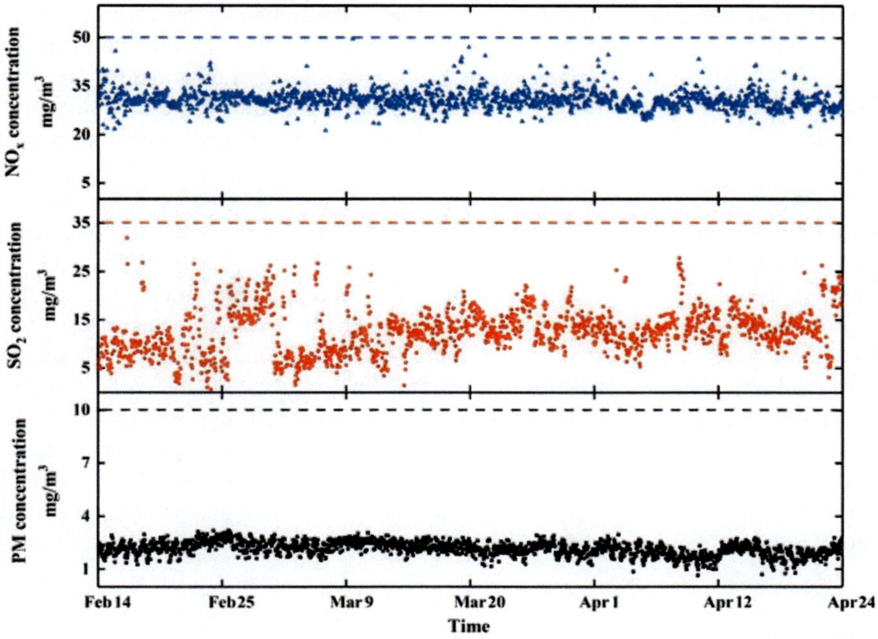

Fig. 5.15 Hourly average concentration of major pollutants (NO_x, SO_2, and PM) at the Wusha thermal power unit from Feb 14th to April 24th in 2017

Table 5.17 Parameters of the designed coal for the boiler

Items	Symbol	Unit	Designed coal
Total moisture	M_t	%	16.9
Air-dried moisture	M_{ad}	%	4.63
Volatile matter	V_{daf}	%	27.18
Arb ash	A_{ar}	%	9.94
Total sulfur	$S_{t,ar}$	%	0.4
Net calorific value as received basis	$Q_{ar.net}$	MJ/kg	22.724

5.3.4 50 MW Coal and Sludge Co-combustion Unit

5.3.4.1 Introduction of ULE System

The Jies thermal power plant in Jiaxing in Zhejiang province comprises four 220-t/h high-temperature, high-pressure CFB coal-fired boilers, two 220-t/h high-temperature, high-pressure CFB sludge incinerators, and one 130-t/h high-temperature, high-pressure biomass incinerator and can simultaneously produce heat, electricity, and compressed air. Here, sludge is sourced from nearby water treatment plants and local industrial facilities (e.g., textile); the sulfur content fluctuates greatly (0.5–4%). The sludge is dried before fed to the boiler to maintain moisture content of 40–50%. The mixed combustion proportion of sludge was generally $\leq$ 50%. The feedstock to this unit #4 of this facility is summarized in Table 5.17. In 2014, this facility was redesigned to meet the ULE emission standards for thermal power boilers.

To meet ULE standards, several key technologies were adopted in unit #4, including in-furnace calcium-injection desulfurization, SNCR-SCR integrated denitrification, an electrostatic-fabric integrated precipitator, limestone–gypsum, WFGD, and a WESP, as detailed in Table 5.18. Specifically, adopting an integrated SNCR-SCR denitrification system allowed for a designed NO_x emission concentration of $\leq$ 50 mg/m^3. An enhanced limestone–gypsum WFGD and simultaneous dust removal system allowed for a designed SO_2 removal efficiency of $\geq$ 99% with a corresponding SO_2 concentration at the scrubber outlet of $\leq$ 35 mg/m^3. Using a WESP reduced the designed emitted PM concentration to $\leq$ 5 mg/m^3 and the slurry droplet concentration to $\leq$ 30 mg/m^3. Overall, these modifications allowed for a system designed to meet the ULE targets, as summarized in Table 5.19. Photographs of the unit and partial layout of implemented ULE system are shown in Fig. 5.16.

5.3.4.2 Performance of ULE System

The Zhejiang Ecological and Environmental Monitoring Center tested the concentration of pollutants emitted from the implemented ULE system on unit #4 at the

Table 5.18 Main pollutant removal technologies applied to unit #4 of the Jies facility

System	Item	Content
Denitrification	Low emission combustion	• Outlet concentration < 400 mg/m^3
	SNCR-SCR integrated technology	• One layer of denitrification catalyst
Desulfurization	In-furnace calcium-injection desulfurization WFGD technology	• One tray layer with uniform velocity distribution • Four spray layers with high spray density • pH dual-loop technology with an auxiliary tank
Dust removal	Electrostatic-fabric integrated precipitator	• ESP with one electric field • Microporous fabric filter • Flow field optimization
	Wet ESP	• WESP with one field and precharger • High-frequency (20 kHz) power supply • Flow field optimization component

Table 5.19 ULE design targets considering boiler outlet and optimized system design

Pollutant	Design condition of boiler outlet	Target (mg/m^3)
NO_x	Concentration of 400 mg/m^3	≤ 50
SO_2	Concentration of 3000 mg/m^3	≤ 35
PM	Concentration of 27.27 g/m^3	≤ 5

Fig. 5.16 ULE system at the Jies thermal power unit

Jies facilities under full load; as summarized in Table 5.20, the outlet concentrations of PM, SO_2, and NO_x were within the ULE limits at 2.47, 1.0, and 27 mg/m^3, respectively.

Online monitoring of the NO_x, SO_2, and PM concentrations for three months also demonstrated emissions within the ULE targets, as shown in Fig. 5.17. During this period, the average concentrations of emitted NO_x, SO_2, and PM were 29.11,

Table 5.20 Emission concentrations and removal efficiency of pollutants

Item	Emission concentration of pollutants		
	NO_x	SO_2	PM
Concentration (mg/m^3)	27	< 1	2.47
Removal efficiency (%)	84.83	99.9	> 99.99

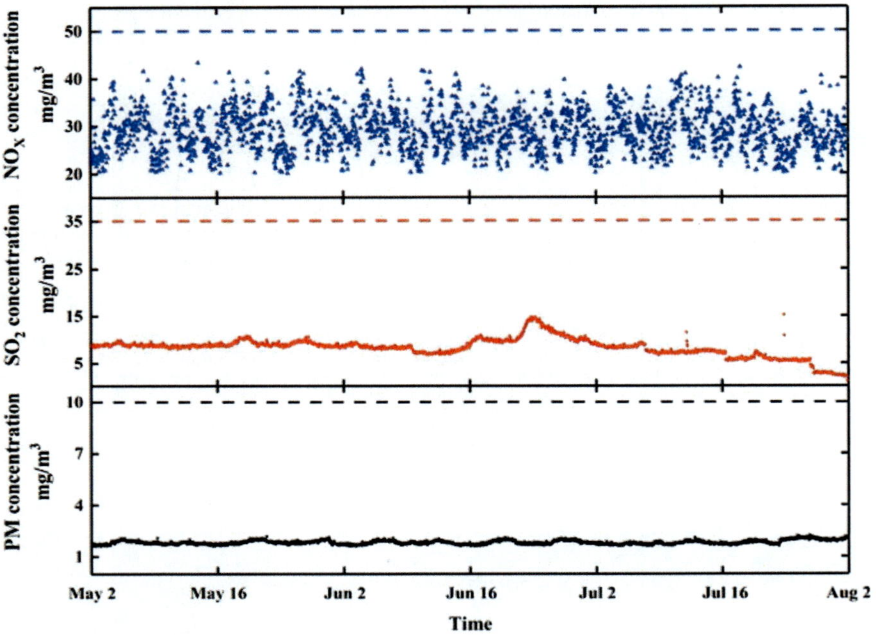

Fig. 5.17 Hourly average concentration of major pollutants (NO_x, SO_2, and PM) of unit #4 at the Jies thermal power unit from May 2 to August 2 in 2015

8.21, and 1.81 mg/m^3, respectively, for total emissions of 18.86, 5.32, and 1.17 t, respectively; this represents a reduction in NO_x, SO_2, and PM emissions by 41.78%, 74.4%, and 81.9%, respectively, compared with the emissions in the same working condition when the ultra-low emission limit is not implemented.

5.4 Technology Evaluations

In general, technologies are evaluated in three areas: technological or engineering readiness, economic impact, and environmental impact. As ULE systems have been demonstrated in prior sections to maintain concentrations of NO_x, SO_2, and PM emissions within the ULE limits (50, 35, and 5 mg/m^3, respectively) on full-scale systems, they can be regarded as mature and stable. In this section, therefore, the economic (i.e., investment and operating costs) and environmental (i.e., emission reduction, industry trends) impacts of ULE systems are discussed.

5.4.1 Economic Evaluation

5.4.1.1 Cost Analysis Model

Installing and operating pollutant control systems place an economic burden on the coal-fired power sector and thus reduce corporate profits. Investment costs associated with constructing and installing pollutant control systems are high and often involve long-term bank loans. Operating pollutant control systems consumes electricity and other resources (e.g., water and catalysts). Therefore, an accurate estimation model and an effective strategy are required by the power industry to quantify the influence of the operating cost of such systems on power generation cost and reduce these costs, respectively. Furthermore, accurate cost estimations can support those responsible for forming and implementing environmental protection regulations and emission standards [12].

Cost analysis models for pollutant control systems are designed to calculate the added cost of a certain power unit when pollutant control systems are installed to meet emission standards. Here, operating costs indicate ongoing expenses involved during operation of the pollutant control system and include variable and fixed costs [13–15], where variable costs represent costs of resources and energy consumed, such as electricity and absorbent, and fixed costs include depreciation [16], maintenance, employee wages and welfare [17], finance, and other costs that are not explicitly related to the operation of pollutant control systems, such as taxes, management costs, and insurance. Among these terms, depreciation cost estimates the decrease in value of the pollutant control system during its service time and depends on the initial investment and service life, maintenance costs refer to the costs incurred during an estimated amount of daily and breakdown maintenance, and finance costs represent the interest exchange for the investment loan from a bank. The total operating cost is thus the sum of the variable and fixed costs [18]. The most common indicator of economic analysis in CFPPs is the cost per unit of electricity generated [19]. Thus, the added cost of the pollutant control system can be calculated, can be then compared with the power generation cost, and can serve as a reference for power plant managers as they compare the cost of pollution control systems with the environmental subsidies provided by the government. In China, the environmental subsidies for NO_x, SO_2, and PM control are 10, 15, and 2 CNY/MWh in China, respectively [20]; in some regions, an additional 10 CNY/MWh is added for meeting the ULE standards ahead of schedule. For a direct comparison of the cost and environmental subsidy, the following cost model can be developed.

$$\mathrm{COST} = \mathrm{COST}_f + \mathrm{COST}_v, \tag{5.1}$$

$$\mathrm{COST}_f = \sum_i \sum_j \mathrm{COST}_{f,i,j}, \tag{5.2}$$

$$\mathrm{COST}_v = \sum_i \sum_j \mathrm{COST}_{v,i,k}, \tag{5.3}$$

where COST represents the operating cost of a pollutant control system per power generation in a power unit, COST_f represents fixed costs, $\mathrm{COST}_{f,i,j}$ represents the jth sub-cost in the fixed cost of the ith pollutant control system, COST_v represents the variable cost, and $\mathrm{COST}_{v,i,k}$ represents the kth subcost of the variable cost in the ith pollutant control system. The sub-costs of each fixed cost represent the depreciation, maintenance, employee wage and welfare, and finance and other costs can be calculated as

$$\mathrm{COST}_{f,i,1} = \frac{\mathrm{TCI}_i \times y}{n_i} \times \frac{1}{Qh}, \tag{5.4}$$

$$\mathrm{COST}_{f,i,2} = \frac{\mathrm{COST}_{f,i,break} + \mathrm{COST}_{f,i,routine}}{Qh}, \tag{5.5}$$

$$\mathrm{COST}_{f,i,3} = \frac{N_i W}{Qh}, \tag{5.6}$$

$$\mathrm{COST}_{f,i,4} = \left(\frac{TCI_i \times x_i \times (1+r)^{n_i}}{n_i}\right)\frac{1}{Qh}, \tag{5.7}$$

respectively, where TCI_i represents the total investment cost, n_i represents the service time of the pollutant control system, y represents the rate of the fixed asset, Q represents the unit capacity, h represents the annual service hours, $\mathrm{COST}_{f,i,break}$ and $\mathrm{COST}_{f,i,routine}$ represent the breakdown and routine maintenance costs, respectively, N_i represents the number of workers, W represents the annual wage and welfare per worker, x_i represents the loan proportion of the total investment, r represents the interest rate, and $\mathrm{COST}_{f,i,routine}$ includes other costs, such as the management cost. Variable costs are determined by the pollutant control system.

For example, for SO_2 control by limestone-based WFGD, the sub-costs can be calculated as

$$\mathrm{COST}_{v,1,\mathrm{CaCO_3}} = \frac{\left(\left(c_{1,in} - c_{1,out}\right) \times V \times \frac{M_{\mathrm{CaCO_3}}}{M_{\mathrm{SO_2}}} \times \frac{\delta_1}{\lambda}\right) \times P_{\mathrm{Ca}}}{q}, \tag{5.8}$$

$$\mathrm{COST}_{v,1,electricity} = \frac{e_1\left(\eta_1, c_{1,in}, c_{1,out}, V, q\right) \times q \times 1 \times P_E}{q}, \tag{5.9}$$

$$\mathrm{COST}_{v,1,water} = \left(\left(\left(c_{1,in} - c_{1,out}\right) \times V \times \frac{M_{\mathrm{CaCO_3}}}{M_{\mathrm{SO_2}}} \times \frac{\delta_1}{\lambda}\right) \times \frac{2M_{\mathrm{H_2O}}}{M_{\mathrm{CaCO_3}}} \times \alpha\right.$$

$$+\left(w_{evaporation} + w_{waste} + w_{liquiddrop}\right) \times V\Big) \times P_W \times \frac{1}{q}, \quad (5.10)$$

where $COST_{v,1,CaCO_3}$ represents the absorbent cost, based on the material balance and the calcium–sulfur ratio during the operating process, $COST_{v,1,electricity}$ represents the electricity cost incurred by operating pumps, booster fans, and other devices and is based on a survey of the electric equipment used, and $COST_{v,1,water}$ represents the added water consumption costs for, e.g., crystal and slurry water in the gypsum slurry, evaporated water, desulfurization wastewater, and the entrained liquid drops in the flue gas. Here, $c_{1,in}$ and $c_{1,out}$ represent the inlet and outlet concentrations of SO_2, respectively, V represents the flue gas flow in the standard state, M represents the molar mass, δ_1 represents the calcium–sulfur ratio, λ represents the limestone purity, q represents the unit load, P represents the price, e represents the station service power consumption rate, η represents the SO_2 removal efficiency, and α represents the quantity ratio of the slurry water.

For NO_x control by an SCR system, the sub-costs can be calculated as

$$COST_{v,2,NH_3} = \frac{\left(\left(c_{1,in} - c_{1,out}\right) \times \frac{2}{3} \times V \times \frac{M_{NH_3}}{M_{NO}} \times \frac{1}{\delta_2} \times \frac{ae \times M_{NH_3}}{V_m}\right) \times P_{NH_3}}{q}, \quad (5.11)$$

$$COST_{v,2,catalyst} = \frac{CV \times P_C}{3Qh}, \quad (5.12)$$

$$COST_{v,2,steam} = \frac{\beta \times CV \times P_S}{q}, \quad (5.13)$$

$$COST_{v,2,electricity} = \frac{e_2\left(\eta_2, c_{2,in}, c_{2,out}, V, q\right) \times q \times 1 \times P_E}{q}, \quad (5.14)$$

where $COST_{v,2,NH_3}$ represents the reductant cost, based on the material balance and the NH_3–NO_x ratio during the operating process, $COST_{v,2,catalyst}$ represents the catalyst cost, $COST_{v,2,steam}$ represents the cost of steam used for cleaning the deposited dust on the catalyst surface, and $COST_{v,2,electricity}$ represents the electricity cost for the operation of pumps, booster fan, and other devices and is based on a survey of the electric equipment. Here, $c_{2,in}$ and $c_{2,out}$ represent the inlet and outlet concentrations of NO_x, respectively, V represents the flue gas flow in the standard state, δ_2 represents the NH_3–NO_x ratio, ae represents the ammonia escape rate, CV represents the amount of catalyst, and β represents the amount of steam per unit catalyst.

For PM control using the ESP and WESP systems, the sub-costs can be calculated as follows:

$$COST_{v,3\&4,electricity} = \frac{\left(e_3\left(\eta_3, c_{3,out}, V, q\right) + e_4\left(c_{4,out}, V, q\right)\right) \times q \times 1 \times P_E}{q}, \quad (5.15)$$

$$\mathrm{COST}_{v,3\&4,w} = \frac{w \times Pw}{q}, \tag{5.16}$$

$$\mathrm{COST}_{v,3\&4,Na} = \frac{Na \times P_{Na}}{q}, \tag{5.17}$$

where $\mathrm{COST}_{v,3\&4,electricity}$ represents the electricity cost for the operation of electric fields, pumps, booster fan, and other devices and is based on a survey of the electric equipment, $\mathrm{COST}_{v,3\&4,w}$ represents the water cost, and $\mathrm{COST}_{v,3\&4,Na}$ represents the cost of NaOH used for neutralizing the acidic component in the flue gas. Here, $c_{3,out}$ and $c_{4,out}$ represent the PM outlet concentrations of the ESP and WESP system, respectively, w represents the water consumption, and Na is the amount of NaOH.

5.4.1.2 Cost Analysis of the Pollutant Removal System

a. *General Investment and Operating Costs*

The general investment and operating costs of an installed ULE system for some typical coal types, typical load conditions, and different ULE technical routes are illustrated in Table 5.21. Both the investment and operating costs per unit of energy decrease significantly with increase in the plant capacity. Investments costs for 100- and 1000 MW CFPPs are 758–913 and 260–303 CNY/kW, respectively, and the corresponding operating costs are 0.0508–0.0632 to 0.0244–0.0294 CNY/kWh, respectively [21].

b. *Investment of Pollutant Removal System*

The variations in investment costs per unit of energy of a limestone-based WFGD system, SCR system, and PM control system with different unit capacities are shown in Fig. 5.18a–c, respectively. In all cases, the investment costs decrease significantly with increase in the capacity, owing to the scale effect. In all cases, meeting the ULE limits increased investment costs. Achieving the stricter emission limits during WFGD requires higher SO_2 removal efficiency desulfurization equipment, such as an scrubber with a tray or a dual-loop absorber, consequently increasing the investment costs. Other equipment or equipment modifications may also be required to meet

Table 5.21 General investment and operating costs of typical ULE systems

Capacity (MW)	Investment (CNY/kW)	Operating cost (CNY/kWh)
100	758–913	0.0508–0.0632
200	526–620	0.0387–0.0473
300	425–509	0.0332–0.0408
600	332–386	0.0276–0.0335
1000	260–303	0.0244–0.0294

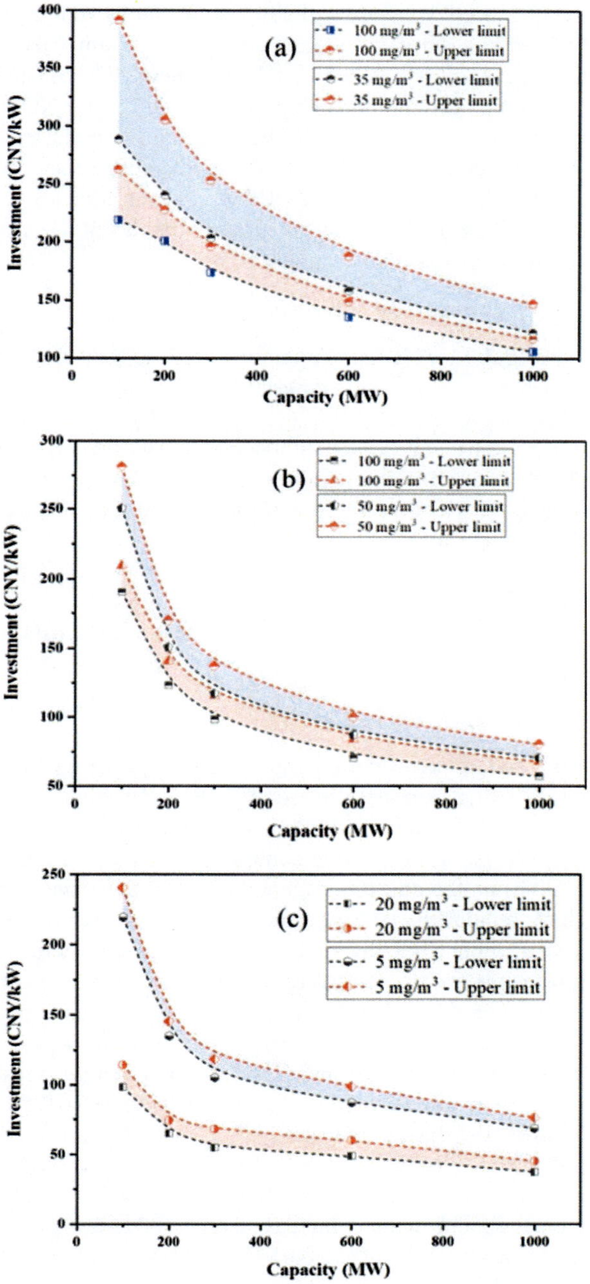

Fig. 5.18 Investment costs per unit of energy at different emission limits of **a** WFGD, **b** SCR, and **c** PM control systems

the increased demand for the limestone slurry and pressure drop, such as additional booster fans, slurry circulation pumps, or gypsum ball mill pumps. In the SCR system, the ULE limit is reached by adding a new layer of catalyst. Less added investment costs are required for meeting the stricter limits for NO_x emissions in the SCR system than in the limestone-based WFGD system. To reach the ULE limits in the PM control system, a five-electric field ESP (removal efficiency of $\geq$ 99.8%) is combined with a WESP to control the outlet PM concentration to $\leq$ 5 mg/m^3. The added costs of installing the WESP or integrating the desulfurization and dust removal systems cause the increased investment costs.

c. *Operating Cost of the Pollutant Removal System*

For a 1000 MW ULE unit operating at 58–62% load when the SO_2, NO_x, and PM inlet concentrations are in the range 1900–2200 mg/m^3, 200–250 mg/m^3, and 23–25 g/m^3, respectively, the operating costs of the WFGD, SCR, and PM control systems are 0.0153–0.0172, 0.0056–0.0078, and 0.0047–0.0056 CNY/kWh, respectively. A more detailed variation in operating costs by unit capacity, control system, and emission limit is provided in Table 5.22; in general, larger units have lower operating costs than smaller units.

5.4.1.3 Key Factors Influencing Operating Costs

a. *SO_2 Control System*

Inlet SO_2 concentration and annual operating hours: Considering a 350 MW CFPP using limestone-based WFGD and a high-efficiency spray scrubber with a designed SO_2 inlet and outlet concentrations of 2000 and $\leq$ 35 mg/m^3, respectively, the effects of inlet sulfur content (0.5–1.2%) and annual operating hours (4000–6000 h) on the unit power generation and SO_2 control costs are shown in Fig. 5.19. Here, annual operating hours indicate the effective operating time of the unit in a year, which is converted into operating time with a full load. Overall, the unit power generation cost increases with increase in the sulfur content but decreases with increase in annual operating hours, whereas unit SO_2 control cost decreases with increase in the sulfur content or annual operating hours. For example, at 4000 annual operating hours and as the sulfur content increases from 0.5 to 1.2%, the unit power generation cost increases from 0.0156 to 0.0178 CNY/kWh while the unit SO_2 control cost decreases from 7726 to 3607 CNY/t. At 5000 annual operating hours and as the sulfur content increases from 0.5 to 1.2%, the unit power generation cost increases from 0.0140 to 0.0162 CNY/kWh. At a sulfur content of 1% and as the annual operating hours increase from 4000 to 6000 h, the unit power generation cost decreases from 0.0173 to 0.0147 CNY/kWh and the unit SO_2 control cost decreases from 4199 to 2346 CNY/t.

Absorbent price: The sensitivity of the unit operating and SO_2 control costs to the limestone price is detailed in Table 5.23, assuming a benchmark limestone price of

Table 5.22 Operating cost of the pollutant control system

(a) Operating cost of the limestone-based WFGD system

Capacity (MW)	Emission limit	
	100 mg/m^3	35 mg/m^3
100	0.0186–0.0230	0.0265–0.0321
200	0.0154–0.0184	0.0220–0.0257
300	0.0134–0.0159	0.0193–0.0224
600	0.0116–0.0133	0.0168–0.0190
1000	0.0105–0.0118	0.0153–0.0172

(b) Operating cost of the SCR system

Capacity (MW)	Emission limit	
	100 mg/m^3	50 mg/m^3
100	0.0119–0.0158	0.0154–0.0205
200	0.0077–0.0105	0.0100–0.0135
300	0.0063–0.0088	0.0084–0.0116
600	0.0049–0.0069	0.0066–0.0091
1000	0.0042–0.0058	0.0056–0.0078

(c) Operating cost of the PM control system

Capacity (MW)	Emission limit	
	20 mg/m^3	5 mg/m^3
100	0.0061–0.0075	0.0089–0.0106
200	0.0042–0.0053	0.0067–0.0081
300	0.0034–0.0044	0.0055–0.0068
600	0.0028–0.0036	0.0042–0.0054
1000	0.0024–0.0031	0.0035–0.0044

103 CNY/t. In this system, the SO_2 inlet and outlet concentrations are 1000 and 35 mg/m^3, respectively, annual operating hours = 6500 h, and the costs of electricity and water are 0.3734 CNY/kWh and 1 CNY/t, respectively. The sensitivity of the unit operating and SO_2 control costs to the limestone price are both 0.074.

Electricity price: The sensitivity of the unit operating and SO_2 control costs to the electricity price is detailed in Table 5.24, assuming a benchmark electricity price of 0.3734 CNY/kWh. The system specifications are the same as detailed above. The cost of the absorbent is 103 CNY/t. The sensitivity of the unit operating and SO_2 control costs to the electricity price are both 0.456.

b. *NO_x Control System*

NO_x inlet concentration and annual operating hours: Considering a 1000 MW CFPP using SCR with a urea reductant with a designed NO_x outlet concentration limit of 50 mg/m^3, the effects of inlet NO_x concentration (200–600 mg/m^3) and

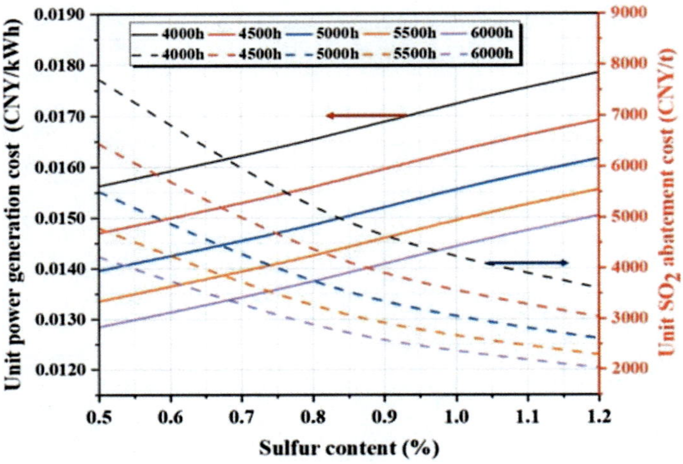

Fig. 5.19 Operating costs of the SO_2 control system of a 350 MW unit affected by the SO_2 concentration and annual operating hours

Table 5.23 Operating costs of the SO_2 control system of 350 MW unit affected by absorbent price

Sensitivity analysis	Benchmark data	– 20%	– 10%	10%	20%
lsp (CNY/t)	103	82.4	92.7	113.3	123.6
GC (CNY/kWh)	0.0120	0.0118	0.0119	0.0121	0.0122
PC (CNY/t)	3989	3930	3959	4018	4048
ΔGC/GC (%)	0	– 1.48%	– 0.74%	0.74%	1.48%
ΔPC/PC (%)	0	– 1.48%	– 0.74%	0.74%	1.48%
$S_{GC\text{-}lsp}$	0	0.074	0.074	0.074	0.074
$G_{PC\text{-}lsp}$	0	0.074	0.074	0.074	0.074

lsp: limestone price, GC: operating cost, PC: unit SO_2 control cost, ΔGC/GC: changing rate of operating cost, ΔPC/PC: changing rate of unit SO_2 control cost, $S_{GC\text{-}lsp}$: sensitivity of limestone price to the operating cost, $G_{PC\text{-}lsp}$: sensitivity of limestone price to the unit SO_2 control cost

Table 5.24 Operating costs of the SO_2 control system of 350 MW unit affected by electricity price

Sensitivity analysis	Benchmark data	– 20%	– 10%	10%	20%
pwp (CNY/t)	0.3734	0.2987	0.3361	0.4107	0.4481
GC (CNY/kWh)	0.0120	0.0109	0.0115	0.0126	0.0131
PC (CNY/t)	3989	3625	3807	4171	4353
ΔGC/GC (%)	0	– 9.12%	– 4.56%	4.56%	9.12%
ΔPC/PC (%)	0	– 9.12%	– 4.56%	4.56%	9.12%
$S_{GC\text{-}pwp}$	0	0.456	0.456	0.456	0.456
$G_{PC\text{-}pwp}$	0	0.456	0.456	0.456	0.456

pwp: electricity price, $S_{GC\text{-}pwp}$: sensitivity of electricity price to the operating cost, $G_{PC\text{-}pwp}$: sensitivity of electricity price to the unit SO_2 control cost

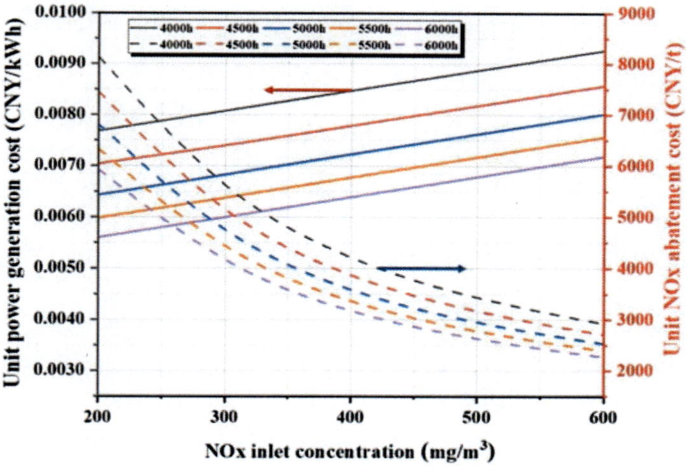

Fig. 5.20 Operating costs of the NO_x control system of a 1000 MW unit affected by the NO_x inlet concentration and annual operating hours

annual operating hours (4000–6000 h) on the unit power generation and NO_x control costs are illustrated in Fig. 5.20. The annual operating hours ranged from 4000 to 6000 h. In all cases, the unit power generation cost is lower than 0.01 CNY/kWh (i.e., the environmental electricity subsidy price). Overall, the unit power generation cost increases with increase in the NO_x inlet concentration but decreases with increase in annual operating hours, whereas the unit NO_x control cost decreases with increase in the NO_x inlet concentration or annual operating hours. For example, at 4000 annual operating h and as the NO_x inlet concentration increases from 200 to 600 mg/m^3, the unit power generation cost increases from 0.0077 to 0.0093 CNY/kWh and the unit NO_x control cost decreases from 8123 to 2923 CNY/t. At 6000 annual operating h and as the NO_x inlet concentration increases from 200 to 600 mg/m^3, the unit power generation cost increases from 0.0056 to 0.0072 CNY/kWh and the unit NO_x control cost decreases from 5920 to 2266 CNY/t. At the NO_x inlet concentration of 400 mg/m^3 and as the annual operating hours decrease from 6000 to 4000 h, the unit power generation cost decreases from 0.0085 to 0.0064 CNY/kWh and the unit NO_x control cost increases by 32.6%.

Catalyst price: The sensitivity ((− 20%)–(+ 20%)) of the unit operating and NO_x control costs to the catalyst price is detailed in Table 5.25, assuming a benchmark catalyst price of 33,400 CNY/m^3. In this system, annual operating hours = 7042 h, the NO_x inlet and outlet concentrations are 270 and 50 mg/m^3, respectively, the electricity price is 0.3679 CNY/kWh, and the costs of the reductant (urea) and steam are 1585 and 180 CNY/t, respectively. The sensitivity of the unit operating and NO_x control costs to the catalyst price are both 0.19.

Electricity price: The sensitivity of the unit operating and NO_x control costs to the electricity price is detailed in Table 5.26, assuming a benchmark electricity price of

Table 5.25 Operating costs of the NO_x control system of the 1000 MW unit affected by catalyst price

Sensitivity analysis	Benchmark data	– 20%	– 10%	10%	20%
ctp (CNY/t)	33,400	26,720	30,060	36,740	40,080
GC (CNY/kWh)	0.0066	0.0063	0.0064	0.0067	0.0068
PC (CNY/t)	6121	5889	6005	6238	6354
ΔGC/GC (%)	0	– 3.8%	– 1.9%	1.9%	3.8%
ΔPC/PC (%)	0	– 3.8%	– 1.9%	1.9%	3.8%
$S_{GC\text{-}ctp}$	0	0.19	0.19	0.19	0.19
$G_{PC\text{-}ctp}$	0	0.19	0.19	0.19	0.19

ctp: catalyst price, SGC ctp: sensitivity of catalyst price to the operating cost, GPC ctp: sensitivity of catalyst price to the unit NO_x control cost

Table 5.26 Operating costs of the NO_x control system of the 1000 MW unit affected by electricity price

Sensitivity analysis	Benchmark data	– 20%	– 10%	10%	20%
pwp (CNY/t)	0.3679	0.2943	0.3311	0.4047	0.4415
GC (CNY/kWh)	0.0066	0.0064	0.0065	0.0067	0.0068
PC (CNY/t)	6121	5927	6024	6218	6315
ΔGC/GC (%)	0	– 3.16%	– 1.58%	1.58%	3.16%
ΔPC/PC (%)	0	– 3.16%	– 1.58%	1.58%	3.16%
$S_{GC\text{-}pwp}$	0	0.158	0.158	0.158	0.158
$G_{PC\text{-}pwp}$	0	0.158	0.158	0.158	0.158

$G_{PC\text{-}pwp}$: sensitivity of electricity price to the unit NO_x control cost

0.3679 CNY/kWh. The system specifications are the same as described above. The sensitivity of the unit operating and NO_x control costs to the electricity price are both 0.158.

Reductant price: The sensitivity of the unit operating and unit NO_x control costs to the reductant (urea) price is detailed in Table 5.27, assuming a benchmark reductant price of 1585 CNY/t. The system specifications are the same as described above, and the catalyst price is 33400 CNY/m^3. The sensitivity of the unit operating and NO_x control costs to the reductant price are both 0.172.

c. *PM Control System*

Unit capacity and annual operating hours: Considering a ULE system employing an ESP and WESP with an assumed electricity price of 0.3734 CNY/kWh and designed PM inlet and outlet concentrations of 15,000 and 5 mg/m^3, respectively, the effects of the unit capacity (100–1000 MW) and annual operating hours (4000–6500 h) on the unit power generation costs as detailed in Fig. 5.21. Both the unit

Table 5.27 Operating costs of the NO_x control system of the 1000 MW unit affected by reductant price

Sensitivity analysis	Benchmark data	– 20%	– 10%	10%	20%
rp (CNY/t)	1585	1268	1427	1744	1902
GC (CNY/kWh)	0.0066	0.0063	0.0065	0.0067	0.0068
PC (CNY/t)	6121	5910	6016	6227	6332
ΔGC/GC (%)	0	– 3.44%	– 1.72%	1.72%	3.44%
ΔPC/PC (%)	0	– 3.44%	– 1.72%	1.72%	3.44%
$S_{GC\text{-}rp}$	0	0.172	0.172	0.172	0.172
$G_{PC\text{-}rp}$	0	0.172	0.172	0.172	0.172

rp: reductant price, $S_{GC\text{-}rp}$: sensitivity of reductant price to the operating cost, $G_{PC\text{-}rp}$: sensitivity of reductant price to the unit NO_x control cost

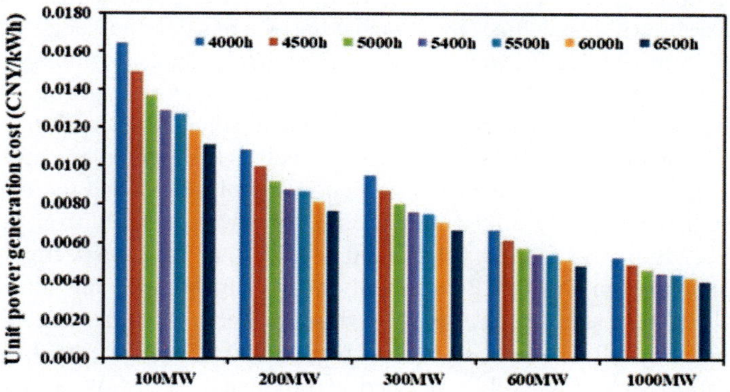

Fig. 5.21 Operating costs of the PM control system affected by the annual operating hours and unit capacity [12]

capacity and number of annual operating hours have a pronounced influence on the operating cost; as the number of annual operating hours and plant capacity increase, the unit operating costs decrease.

Electricity price: The sensitivity of the unit operating and PM control costs to the electricity price is detailed in Table 5.28, assuming a benchmark electricity price of 0.3734 CNY/kWh. In this system, an ESP and a WESP are operated, the designed PM inlet and outlet concentrations are 15,000 and 5 mg/m^3, respectively, annual operating hours = 5500 h, and the costs of NaOH (for WESP) and water are 2000 and 1 CNY/t, respectively. The sensitivity of the unit operating and the PM control costs to the electricity price are both 0.345.

Table 5.28 Operating costs of the PM control system of the 1000 MW unit affected by electricity price

Sensitivity analysis	Benchmark data	− 20%	− 10%	10%	20%
pwp (CNY/t)	0.3734	0.2987	0.3361	0.4107	0.4481
GC (CNY/kWh)	0.0043	0.0040	0.0042	0.0045	0.0046
ΔPC/PC (%)	0	− 6.90%	− 3.45%	3.45%	6.90%
$S_{GC\text{-}pwp}$	0	0.345	0.345	0.345	0.345

5.4.2 *Environmental Evaluation*

5.4.2.1 Promotion of ULE Technology

ULE regulations lead to a systematic reduction in air pollutant emissions for the power generation industry, thereby contributing to considerable social benefits in terms of environmental improvement and technological progress in emission control. Technological improvements in air pollutants control devices have allowed emissions have to meet the ULE limits in recent years. The denitrification, desulfurization, and dedusting devices employed in CFPP plants, as well as their synergetic effects when combined, have helped reduce air pollution in many places around the world. As a result, the China's unit NO_x, SO_2, and PM emissions were reduced from 1.98, 1.86, and 0.34 g/kWh in 2013 to 0.20, 0.19, and 0.04 g/kWh in 2019, respectively. A map of the installed capacity of CFPP units applying ULE technology in China is shown in Fig. 5.22. In some areas, such as Beijing–Tianjin–Hebei, Yangtze River Delta, and Pearl River Delta, all CFPPs have been equipped with ULE systems. In 2019, the installed capacity of ULE CFPPs reached approximately 890 GW, accounting for 86% of the total installed CFPP capacity. By progressively conducting technological improvements and applying highly efficient pollutant removal technologies, these enterprises can make groundbreaking progress toward reducing global emissions and building the world's largest ULE coal-fired power generation system.

The 2015–2019 installed capacity of CFPPs and ULE CFPPs is illustrated in Fig. 5.23. During this time, the capacity of CFPPs meeting the ULE limits increased drastically from 160 to 890 GW. The ULE technology implementation rate, i.e., the capacity of CFPPs meeting the ULE limits divided by that of the total capacity of CFPPs, increased from 18% in 2015 to 86% in 2019.

5.4.2.2 Emission Reduction Benefits

a. *Effect on Total Pollutant Emissions*

The 2013–2019 pollutant emissions from the thermal power industry in China are illustrated in Fig. 5.24 [22]. During this time, the annual thermal power NO_x, SO_2, and PM emissions decreased by 7.41, 6.91, and 1.24 million tons, respectively, for

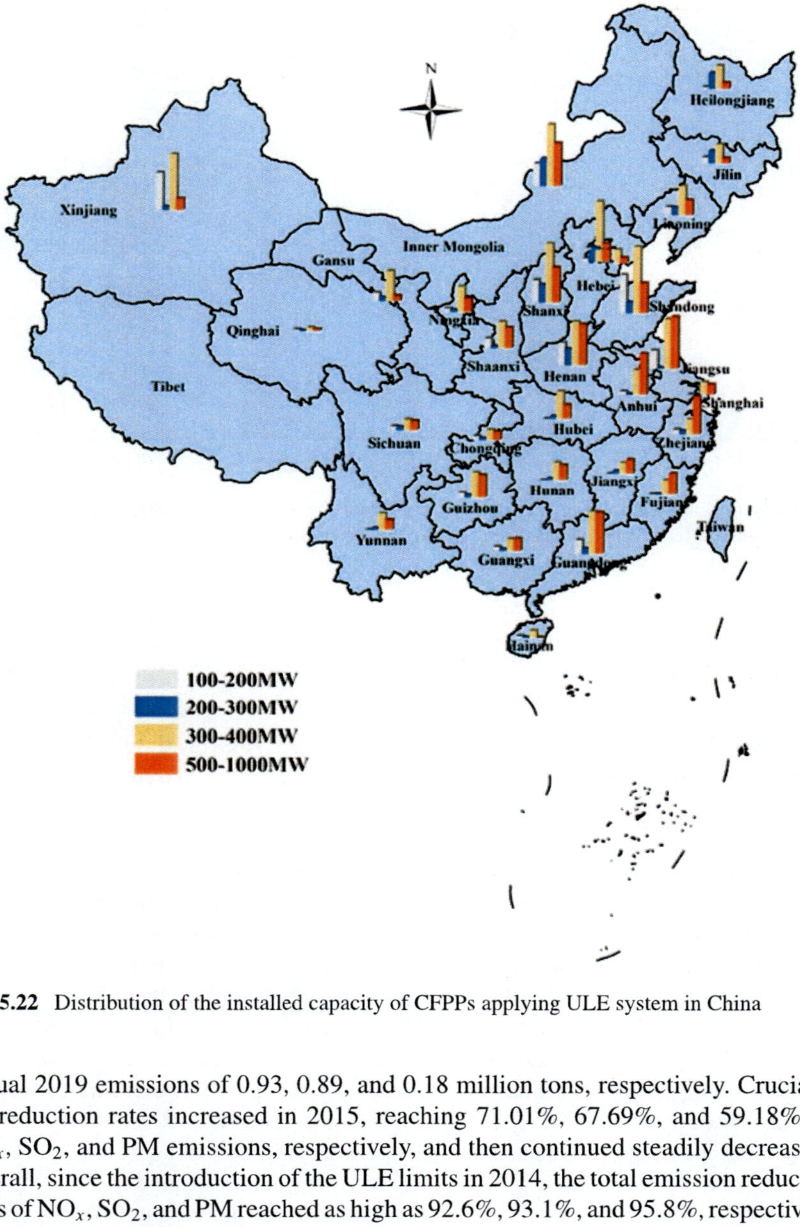

Fig. 5.22 Distribution of the installed capacity of CFPPs applying ULE system in China

annual 2019 emissions of 0.93, 0.89, and 0.18 million tons, respectively. Crucially, the reduction rates increased in 2015, reaching 71.01%, 67.69%, and 59.18% for NO_x, SO_2, and PM emissions, respectively, and then continued steadily decreasing. Overall, since the introduction of the ULE limits in 2014, the total emission reduction rates of NO_x, SO_2, and PM reached as high as 92.6%, 93.1%, and 95.8%, respectively, over the whole ULE period.

The share of NO_x, SO_2, and PM emissions from the power sector remained low in 2019 (approximately 8%, 19%, and 2%). Furthermore, the power sector emissions for these three conventional pollutants declined markedly from 2013 to 2019, with NO_x emissions contributing most to nationwide emission reductions (from 37% in

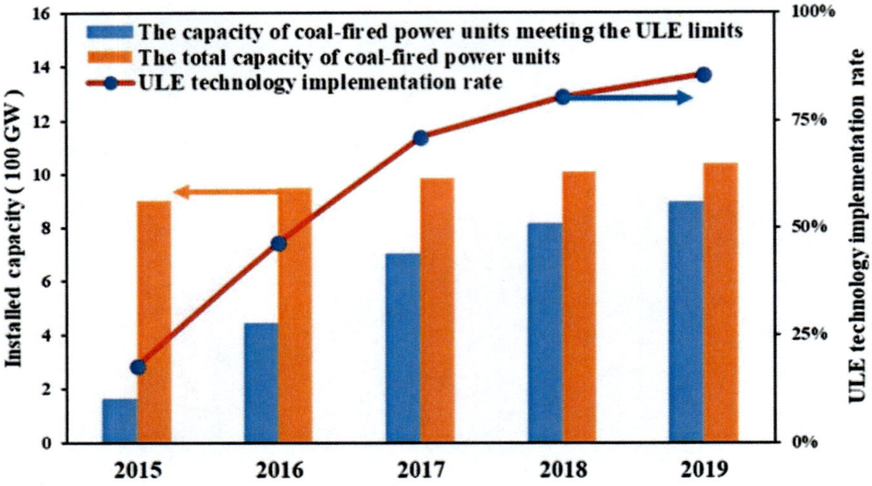

Fig. 5.23 Installed capacity of ULE technology for CFPPs from 2015 to 2019. Data from China Environmental Status Bulletin and China Ecological Environment Status Bulletin

2013 to 8% in 2019, a drop of 29%), followed by SO_2 emissions (from 38% in 2013 to 19% in 2019, a drop of 19%), and PM emissions (from 11% in 2013 to 2% in 2019, a drop of 9%). Overall, the share of overall emissions of NO_x, SO_2, and PM from the power industry reduced from 25% in 2013 to 10% in 2019, reflecting the significant improvements made by the thermal power industry in meeting the ULE standards and reducing emissions.

b. *Effect on Pollutant Emission Concentration*

China's CEMS network allows performing real-time measurements of the emission concentrations of a variety of air pollutants at many participating power plants. CEMS data suggest that systematic reductions in stack concentrations at Chinese thermal power plants have occurred since the introduction of the ULE limits in 2014. The stack concentrations of 4622 power plant stacks monitored by the CEMS network declined clearly and continuously from 2014 to 2017, with mean annual reductions of NO_x, SO_2, and PM concentrations of 28.29%, 33.34%, and 38.06%, respectively, as shown in Fig. 5.25 [23]. The percentages of total capacity decreasing the annual average concentrations of NO_x, SO_2, and PM below their respective ULE limits (the samples on the left of the red dashed lines in the histograms of Fig. 5.25), increased from 10.47%, 15.63%, and 15.79% in 2014 to 70.64%, 74.54%, and 87.50% in 2017, respectively. Since 2014, the mean NO_x, SO_2, and PM concentrations from Chinese CFPPs hit 52.00, 35.30, and 5.70 mg/m^3 at the end of 2017 [23].

c. *Effect on the Pollutant Emission Performance*

The NO_x, SO_2, and PM emissions per kWh of thermal power plant capacity in the recent decade (2010–2019) are shown in Fig. 5.26. Considerable progress has been

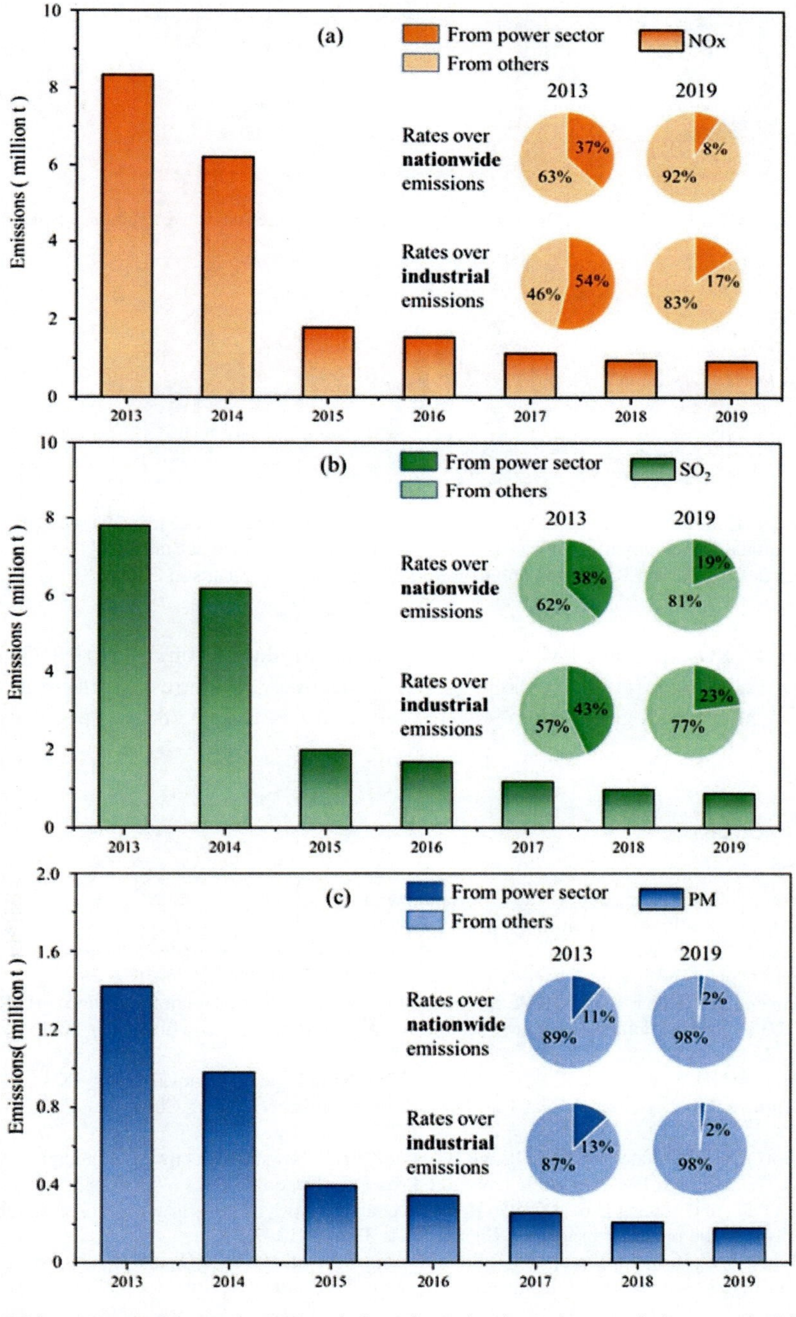

Fig. 5.24 **a** NO_x, **b** SO_2, and **c** PM emissions from the thermal power industry and emission rates from the power sector over nationwide emissions and industrial emissions. Data from Annual Development Report of China Power Industry, China Environmental Status Bulletin, and China Ecological Environment Status Bulletin

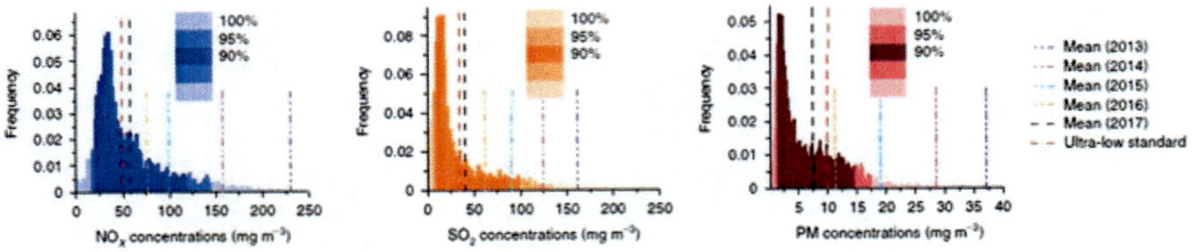

Fig. 5.25 Concentrations of NO_x, SO_2, and PM for different thermal power plant stacks [23]

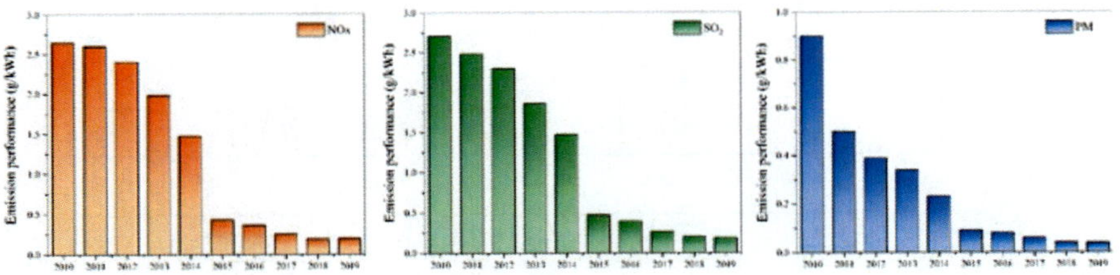

Fig. 5.26 Emission performance of NO_x, SO_2, and PM in the recent decade (2010–2019). Data from Annual Development Report of China Power Industry and Current Status and Challenges of Atmospheric Pollution Prevention and Control of Thermal Power Plants in China

made; the NO_x, SO_2, and PM emission performance reduced from 2.65 to 0.195, from 2.71 to 0.187, and from 0.9 to 0.038 g/kWh, respectively, for corresponding reduction rates, which are all greater than 90% (92.64%, 93.1%, and 95.78%, respectively).

References

1. Fang P, Cen C, Tang Z et al (2011) Simultaneous removal of SO_2 and NO_x by wet scrubbing using urea solution. Chem Eng J 168(1):52–59
2. Han D, Wu Q, Wang S et al (2021) Distribution and emissions of trace elements in coal-fired power plants after ultra-low emission retrofitting. Sci Total Environ 754:142–285
3. Hu B, Liu Y, Ren F et al (2016) Experimental study on simultaneous control of fine particle and SO_3 by low-low temperature electrostatic precipitator. Proc Chin Soc Electr Eng 36(16):4319–4325
4. Hutson ND, Krzyzynska R, Srivastava RK (2008) Simultaneous removal of SO_2, NO_x, and Hg from coal flue gas using a $NaClO_2$-enhanced wet scrubber. Ind Eng Chem Res 47(16):5825–5831
5. Jaworek A, Marchewicz A, Sobczyk AT et al (2018) Two-stage electrostatic precipitators for the reduction of PM2.5 particle emission. Prog Energy Combust Sci 67:206–233
6. Pan D, Gu C, Zhang D et al (2019) Removal characteristics of sulfuric acid aerosols in the wet electrostatic precipitator system. Energy Fuels 33(8):7813–7818
7. Wang Z, Zhou J, Zhu Y et al (2007) Simultaneous removal of NO_x, SO_2 and Hg in nitrogen flow in a narrow reactor by ozone injection: experimental results. Fuel Process Technol 88(8):817–823
8. Wu H, Pan D, Zhang R et al (2017) Reducing fine particle emissions by heterogeneous vapor condensation after wet desulfurization process. J Chem Technol Biotechnol 92(9):2342–2350

9. Zheng C, Wang Y, Liu Y et al (2019) Formation, transformation, measurement, and control of SO_3 in coal-fired power plants. Fuel 241:327–346
10. Huang Y, Zheng C, Li Q et al (2020) Numerical simulation of the simultaneous removal of particulate matter in a wet flue gas desulfurization system. Environ Sci Pollut Res 27(2):1598–1607
11. Ministry of Ecology and Environment of the People's Republic of China (2017) Guideline on available technologies of pollution prevention and control for thermal power plant (in Chinese)
12. Yang H, Zhang Y, Zheng C et al (2018) Cost estimate of the multi-pollutant abatement in coal-fired power sector in China. Energy 161:523–535
13. Mussatti DC, Srivastava R, Hemmer PM et al (2002) EPA air pollution control cost manual. Air Quality Strategies and Standards Division of the Office of Air Quality Planning and Standards, US Environmental Protection Agency, Research Triangle Park, NC 27711
14. Dong L, Dong HJ, Fujita T et al (2015) Cost-effectiveness analysis of China's sulfur dioxide control strategy at the regional level: regional disparity, inequity and future challenges. J Clean Prod 90:345–359
15. Xu YA (2011) Improvements in the operation of SO_2 scrubbers in China's coal power plants. Environ Sci Technol 45(2):380–385
16. Chen CJ, Wang LFS (2001) Cost-benefit analysis of electrocyclone and cyclone. Resour Conserv Recycl 31(4):285–292
17. Zhao Y, Zhang J, Wang Z (2020) Cost analysis of environmental protection price of coal-fired plants in China. Environ Sci Pollut Res 27(15):18729–18742
18. Liu XY, Lin BQ, Zhang YJ (2016) Sulfur dioxide emission reduction of power plants in China: current policies and implications. J Clean Prod 113:133–143
19. Jiang P, Khishgee S, Alimujiang A et al (2020) Cost-effective approaches for reducing carbon and air pollution emissions in the power industry in China. J Environ Manag 264:110452
20. National Development and Reform Commission of China, Ministry of Environmental Protection of China (2014) Measures on environmental electricity price for coal-fired power units and supervision of operation of environmental facilities (in Chinese)
21. Jin K (2016) Study on cost-benefit of environmental protection equipment in coal-fired power plants. Zhejiang University (in Chinese)
22. China Electricity Council (2013–2019) Annual development report of China Power Industry (in Chinese)
23. Tang L, Qu JB, Mi ZF et al (2019) Substantial emission reductions from Chinese power plants after the introduction of ultra-low emissions standards. Nat Energy 4(11):929–938